Thermal Conductivity

Theory, Properties, and Applications

PHYSICS OF SOLIDS AND LIQUIDS

Editorial Board: Jozef T. Devreese • *University of Antwerp, Belgium*
Roger P. Evrary • *University of Liège, Belgium*
Stig Lundqvist • *Chalmers University of Technology, Sweden*
Gerald D. Mahan • *University of Tennessee, USA*
Norman H. March • *University of Oxford, England*

Current Volumes in the Series

CRYSTALLINE SEMICONDUCTING MATERIALS AND DEVICES
Edited by Paul N. Butcher, Norman H. March, and Mario P. Tosi

ELECTRON CORRELATION IN MOLECULES AND CONDENSED PHASES
N. H. March

ELECTRONIC EXCITATIONS AT METAL SURFACES
Ansgar Liebsch

EXCITATION ENERGY TRANSFER PROCESSES IN CONDENSED MATTER: Theory and Applications
Jai Singh

FRACTALS
Jens Feder

INTERACTION OF ATOMS AND MOLECULES WITH SOLID SURFACES
Edited by V. Bortolani, N. H. March, and M. P. Tosi

IMPURITY SCATTERING IN METALLIC ALLOYS
Joginder Singh Galsin

LOCAL DENSITY THEORY OF POLARIZABILITY
Gerald D. Mahan and K. R. Subbaswamy

MANY-PARTICLE PHYSICS, Third Edition
Gerald D. Mahan

ORDER AND CHAOS IN NONLINEAR PHYSICAL SYSTEMS
Edited by Stig Lundqvist, Norman H. March, and Mario P. Tosi

PHYSICS OF LOW-DIMENSIONAL SEMICONDUCTOR STRUCTURES
Edited by Paul Butcher, Norman H. March, and Mario P. Tosi

QUANTUM TRANSPORT IN SEMICONDUCTORS
Edited by David K. Ferry and Carlo Jacoboni

THERMAL CONDUCTIVITY: Theory, Properties, and Applications
Edited by Terry M. Tritt

A Continuation Order Plan is available for this series. A continuation order will bring delivery of each new volume immediately upon publication. Volumes are billed only upon actual shipment. For further information please contact the publisher.

Thermal Conductivity

Theory, Properties, and Applications

Edited by

Terry M. Tritt

Clemson University
Clemson, South Carolina

Kluwer Academic / Plenum Publishers
New York, Boston, Dordrecht, London, Moscow

Library of Congress Cataloging-in-Publication Data

Thermal conductivity: theory, properties, and applications/[edited by] Terry Tritt.
  p. cm. — (Physics of solids and liquids)
  Includes bibliographical references and index.
  ISBN 0-306-48327-0
  1. Thermal conductivity. I. Tritt, Terry M. II. Series.

QC176.8.T39T48 2004
536'.2012—dc22

2003064141

ISBN 0-306-48327-0

©2004, Kluwer Academic/Plenum Publishers, New York
233 Spring Street, New York, New York 10013

http://www.wkap.nl

10 9 8 7 6 5 4 3 2 1

A C.I.P. record for this book is available from the Library of Congress

All rights reserved

No part of this book may be reproduced, stored in a retrieval system, or transmitted in any form or by any means, electronic, mechanical, photocopying, microfilming, recording, or otherwise, without written permission from the Publisher, with the exception of any material supplied specifically for the purpose of being entered and executed on a computer system, for exclusive use by the purchaser of the work.

Permissions for books published in Europe: *permissions@wkap.nl*
Permissions for books published in the United States of America: *permissions@wkap.com*

Printed in the United States of America

PREFACE

It has been almost thirty years since a book was published that was entirely dedicated to the theory, description, characterization and measurement of the thermal conductivity of solids. For example, the excellent texts by authors such as Berman[1], Tye[2] and Carlslaw & Jaeger[3] remain as the standards in the field of thermal conductivity. Tremendous efforts were expended in the late 1950's and 1960's in relation to the measurement and characterization of the thermal conductivity of solid-state materials. These efforts were made by a generation of scientists, who for the most part are no longer active, and this expertise would be lost to us unless we are aware of the great strides they made during their time.

I became interested in the field of thermal conductivity in the mid 1990's in relation to my own work on the next generation thermoelectric materials, of which the measurement and understanding of thermal conductivity is an integral part.[4] In my search for information, I found that most of the books on the subject of thermal conductivity were out of print. Much of the theory previously formulated by researchers such as Klemens[5] and Slack[6] contain considerable theoretical insight into understanding thermal conductivity in solids. However, the discovery of new materials over the last several years which possess more complicated crystal structures and thus more complicated phonon scattering mechanisms have brought new challenges to the theory and experimental understanding of these new materials. These include: cage structure materials such as skutterudites and clathrates, metallic glasses, quasicrystals as well as many of new nano-materials which exist today. In addition the development of new materials in the form of thin film and superlattice structures are of great theoretical and technological interest. Subsequently, new measurement techniques (such as the 3-ω technique) and analytical models to characterize the thermal conductivity in these novel structures were developed. Thus, with the development of many new and novel solid materials and new measurement techniques, it appeared to be time to produce a more current and readily available reference book on the subject of thermal conductivity. Hopefully, this book, *Thermal Conductivity-2004: Theory, Properties and Applications*, will serve not only as a testament to those researchers of past generations whose great care in experimental design and thought still stands today but it will also describe many of the new developments over the last several years. In addition, this book will serve as an extensive resource to the next generation researchers in the field of thermal conductivity.

First and foremost, I express great thanks to the authors who contributed to this book for their hard work and dedication in producing such an excellent collection of chapters. They were very responsive to the many deadlines and requirements and they were a great group of people to work with. I want to personally acknowledge my many conversations with Glen Slack, Julian Goldsmid, George Nolas and Ctirad Uher, as well as many other colleagues in the field, as I grasped for the knowledge necessary to personally advance in this field of research. I also want to acknowledge the support and encouragement of my own institution, Clemson University (especially my Chairman: Peter Barnes), during this editorial and manuscript preparation process. I am truly indebted to all my graduate students, for their contributions to these volumes, their hard work, for the patience and understanding they exemplified during the editorial and writing process. A special thanks goes to the publishers and editors at Kluwar Press for their encouragement and patience in all stages of development of these manuscripts for publication. I am indebted to my assistant at Clemson University, Lori McGowan, whose attention to detail and hard work (copying, reading, filing, corresponding with authors, etc.) helped make this book possible. I especially wish to acknowledge my wonderful wife, Penny, and my great kids for their patience and understanding for the many hours that I spent on this work.

Terry M. Tritt
September 18, 2003
Clemson University
Clemson, SC

REFERENCES

1. *Thermal Conduction in Solids*, R. BERMAN, Clarendon Press, Oxford, 1976.
2. *Thermal Conductivity Vol. I and Vol. II*, edited by R. P. TYE, Academic Press, New York, 1969.
3. H. S. Carslaw and J. C. JAEGER, *"Conduction of Heat in Solids"* (Oxford University Press, Oxford, 1959).
4. *"Recent Ternds in Thermoelectric Materials Research"* TERRY M. TRITT, editor, (Academic Press, Volumes **69–71**, *Semiconductors and Semimetals*, 2002).
5. See for example P. G. KLEMENS, Chapter 1, p1 of *Thermal Conductivity Vol. I*, edited by R. P. TYE, Academic Press, New York, 1969.
6. *Solid State Physics*, G. A. SLACK, **34**, 1, Academic Press, New York, 1979.

LIST OF SYMBOLS

Ch. 1.1 Yang

κ	– total thermal conductivity
α	– excitation
Q	– heat flow rate/ heat flux vector
T	– absolute temperature
c	– heat capacity
n	– concentration of particles
∇T	– temperature gradient
v	– velocity
$\partial E/\partial t$	– Time rate of change of energy
τ	– relaxation time
C	– total heat capacity
l	– particle mean free path
k	– electron wave vector
E	– electric field
f^0_k	– distribution function
E_F	– Fermi energy constant
k_B	– Boltzmann's constant
v_k	– electron velocity
e	– electron charge
E_{eff}	– effective field
J	– elecrical current density
K_n	– general integral
σ	– electrical conductivity
L_o	– standard Lorentz number
θ_D	– Debye temperature for phonons
k_f	– electron wave number at Fermi surface
q_D	– phonon Debye wave number
G'	– constant representing strength of electron-phonon interaction
$\hbar$	– Planck constant
m*	– electron effective mass
n_c	– number unit cells per unit volume
v_F	– electron velocity at Fermi surface
ρ	– resistivity
W_e	– electronic thermal resistivity

LIST OF SYMBOLS

W_i	– constant
ρ_o	– residual electrical resistivity
E_G	– energy gap
f^0_e	– equilibrium distribution for electron
f^0_h	– equilibrium distribution for holes
E_F^e	– distances of Fermi level below bottom of conduction band
E_F^h	– distances of Fermi level above top of valence band
K_n	– integral for electrons
K'_n	– integral for holes
α	– constant
h	– holes
μ	– mobility
$\omega_{\vec{q}}$	– phonon frequencey
$\frac{\partial \omega_{\vec{q}}}{\partial q}$	– group velocity
q	– wave vector
$N_{\vec{q}}$	– phonon distribution function
$\vec{v}_g$	– phonon group velocity
τ_q	– phonon scattering relaxation time
θ	– angle between $\vec{v}_g$ and ∇T
κ_L	– lattice thermal conductivity
ω_D	– Debye frequency (resonant frequency ?)
$\vec{\lambda}$	– constant vector-determine s anisotropy of phonon distribution
β	– parameter defined in the text (units of time)
B	– constant
γ	– Grüneisen constant
M	– average atomic mass
V	– volume per atom
m_i	– mass of atom
f_i	– fraction of atoms with mass mi
$\bar{m}$	– average mass of all atoms
d	– sample size for single crystal or grain size for a polycrystalline sample
N_D	– number dislocation lines per unit area
r	– core radius
B_D	– Burgers vector of dislocation
C	– constant proportional to concentration of resonant defects
ϵ	– electron-phonon interaction constant or deformation potential
d	– mass density
G	– proportionality constant
Δ	– chemical shift related to splitting of electronic states
r_0	– mean radius of localized state
L_e	– mean free path of electron
λ_p	– wavelength of phonon

LIST OF SYMBOLS

Ch. 1.2 Uher

κ	– total thermal conductivity
κ_e	– thermal conductivity of electrons
κ_p	– thermal conductivity of phonons
σ	– electrical conductivity
ρ	– resistivity
L	– constant that relates thermal conductivity of pure metals to their electrical conductivity
J_e	– electrical current density
n	– number of electrons per unit volume
v	– velocity
t	– time
ϵ	– electrical field
τ	– relaxation time/ collision time
l_e	– mean free path between electrons
m	– electron mass
J_Q	– thermal current density
c_v	– electronic specific heat per unit volume
c_M	– molar specific heat
ζ	– chemical potential/ Fermi level/ Fermi energy (E_F)
E	– particular energy state
D	– density
U	– internal energy of electron
k_B	– Boltzmann constant
T_F	– Fermi temperature
N_A	– Avagardo's number
Z	– valence of electron
R	– universal gas constant
γ	– coefficient of linear specific heat of metals#Sommerfield constant
$\hbar$	– Planck's constant
k	– Bloch wave vector of an electron
E(k)	– Bloch energy of an electron
Ψ_k	– wavefunction
U_k	– internal energy
H'	– perturbation Hamiltonian
$\Phi(E)$	– varying function of E ?
S	– Seebeck coefficient#thermoelectric power
Π	– Peltier coefficient
m*	– effective mass
A(E)	– vector quantity
ρ_{imp}	– impurity resistivity
$W_{kk'}$	– collision probability
V	– volume
k_{TF}	– Thomas-Fermi screening parameter
k_F	– Fermi wave-vector (momentum)
a_0	– Bohr radius
n_i	– density of impurities
q	– wave vector
ω_q	– frequency of vibration

ω_D	–	Debye frequency
V_0	–	sample volume
k	–	incoming Bloch state
k'	–	outgoing Bloch state
G	–	reciprocal lattice vector
r	–	point of electron
R_i	–	point of lattice ion
H_{e-ion}	–	potential energy of an electron in the field of all ions of lattice–interaction Hamiltonian
$\hat{e}_{q\alpha}$	–	unit polarization vectors
c	–	speed of sound
θ_D	–	Debye temperature
Θ_s	–	temperature related to longitudinal sound velocity
d	–	dimensionality of metallic system
L_{exp}	–	Wiedemann-Franz ratio
W^{exp}	–	total measured electronic thermal resistivity
ρ^{exp}	–	electrical resistivity
Δ_U	–	umklapp electron-electron value of the effectiveness parameter Δ
W_{e-p}	–	ideal thermal resistivity of metals
M	–	mass of an ion
C_j^2	–	coupling constant
M_a	–	average atomic mass
n_c	–	number of atoms in a primitive cell
γ	–	high temperature Grüneisen constant
θ_0	–	low temperature heat capacity
N_D	–	dislocation density
B	–	Burgers vector of dislocation
$T_{L.T.}$	–	low temperature domain
$T_{H.T.}$	–	high temperature regime (T = ∞)
k	–	electron wave vector
q	–	phonon small wave vector
C	–	constant
D	–	constant

Ch. 1.3 Murashove & White

κ	–	thermal conductivity
C	–	heat capacity per unit volume
v	–	average phonon group velocity
κ	–	heat transfer in gases
m	–	polarization branch
ω_D	–	Debye cut-off frequency
λ	–	mean free path of phonons
τ	–	relaxation time
k	–	wave vectors
n	–	number of atoms
ν_D	–	Debye cut-off frequency
ω	–	frequency
G	–	reciprocal lattice vector

LIST OF SYMBOLS

N	–	phonon concentration
t	–	time
d	–	dimension
A'	–	temperature-independent parameter
V_0	–	effective volume of defects
M	–	mass of regular elementary unit of substance
ΔM	–	difference in mass btw defect and regular unit
θ_D	–	Debye characteristic temperature
??	–	static binding energy
r_0	–	nearest-neighbor distance
ω_0	–	characteristic resonant frequency
l	–	interatomic spacing
D	–	coefficient depticting strength of host-guest coupling
θ_E	–	Einstein temperature
θ_C	–	cut-off frequency
η	–	refractive index
σ	–	Stefan-Boltzmann constant
k_r	–	Rosseland mean absorption coefficient

Ch. 1.4 Nolas & Goldsmid

τ	–	relaxation time
E	–	energy
r	–	constant
τ_0	–	constant
i	–	electric current
-e	–	electronic charge
u	–	velocity of charge carriers in x direction
f(E)	–	Fermi distribution function
g(E)	–	carrier distribution function
ζ	–	Fermi energy
m*	–	effective mass of carriers
σ	–	electrical conductivity
E	–	electric field
α	–	Seebeck coefficient
κ_e	–	electronic thermal conductivity
g	–	density of states
ξ	–	$E/K_B T$
F_n	–	Fermi-Dirac integral
K_S	–	integrals
η	–	reduced Fermi energy
Γ	–	gamma function
μ	–	mobility of charge characters
L	–	Lorenz number
r	–	scattering parameter
i	–	electric current density
κ_L	–	lattice contribution to total thermal conductivity
l_t	–	mean free path of phonon
c_v	–	specific heat per unit volume

v	–	speed of sound
θ_D	–	Debye temperature
ω	–	angular frequency of phonons
β	–	constant
A	–	constant
B	–	constant
C	–	constant
χ_i	–	concentration of unit cells
M_i	–	mass of unit cells
$\bar{M}$	–	average mass per unit cell
M	–	average atomic mass
N	–	number of cells per unit volume
ω_D	–	Debye frequency
L	–	grain size
λ_D	–	lattice conductivity
V	–	average atomic volume
γ	–	Grüneisen parameter
α_T	–	thermal expansion coefficient
χ	–	compressibility
ρ	–	density
R	–	gas constant
T_m	–	melting temperature
ϵ_m	–	melting energy
N_A	–	Avagadro's number
A_m	–	mean atomic weight

Ch. 1.5 Nolas

Z	–	thermoelectric figure of merit
α	–	Seebeck coefficient
σ	–	electrical conductivity
κ	–	thermal conductivity
κ_E	–	electronic contribution to thermal conductivity
T	–	temperature
y_{max}	–	maximum filling fraction
y	–	filling fraction
G	–	guest ion
κ_L	–	lattice thermal conductivity
U_{iso}	–	mean square displacement average over all directions
$\square$	–	vacancy
L	–	fitting parameter#represents inelastic scattering length

Ch. 1.6 Mahan

J_Q	–	heat flow
σ_B	–	boundary conductance
R_B	–	boundary resistance
K'	–	thermal conductivity at zero electric field

LIST OF SYMBOLS

K	– thermal conductivity
S_B	– boundary impedence
K_B	– boundary impedence
λ_q	– phonon mean free path
R_i	– site location
v_i	– site velocity
ω_λ	– frequency
q	– wave vector
λ	– polarization
N	– number unit cells
ξ	– polarization vector
$l_{\lambda p}$	– mean free path of phonon
$\hbar$	– Planck's constant
T	– temperature
T_e	– temperature for electrons
T_p	– temperature for phonons
$K_{e,p}$	– bulk thermal conductivity of electrons and phonons
P	– constant
ρJ^2	– Joule heating
λ	– length scale
r	– point in time
t	– time
$v_{z\lambda}(q)$	– velocity of phonon
$\tau_\lambda(q)$	– lifetime
$\eta_B(w_\lambda(q))$	– Boson occupation function

Ch. 1.7 Yang and Chen

Z	– figure of merit
κ	– total thermal conductivity
κ_E	– thermal conductivity of electrons
κ_L	– electrical conductivity
S	– Seebeck coefficient

Ch. 2.1 Tritt & Weston

κ_L	– lattice contribution to total thermal conductivity
κ_E	– electronic contribution to total thermal conductivity
κ_{TOT}	– the total thermal conductivity,
P_{SAM}	– the power flowing through the sample,
L_S	– the sample length between thermocouples,
ΔT	– the temperature difference measured and
A	– the cross sectional area of the sample through which the power flows.
P_{LOSS}	– the power that is lost (radiation, conduction etc.)
P_{IN}	– is the power input
α	– thermopower
ΔT	– temperature gradient
σ_{S-B}	– the Stephan-Boltzman constant $\sigma_{S-B} = 5.7 \times 10^{-8}$ W/m^2-K^4)

XIV   LIST OF SYMBOLS

ϵ – emissivity
T_{SAM} – the temperature of the sample
T_{SURR}– the temperature of the surroundings
σ – the electrical conductivity
ρ – the electrical conductivity
L_0 – the Lorentz number
C – heat capacity
I – electric current
r_1, r_2 – sample radii for the radial flow method
D – thermal diffusivity
C_p – heat capacity
K – thermal conductance
ΔT_{PP} – temperature gradient peak to peak
dQ/dT– the time rate of change of heat in the heater
T_1 – the heater temperature
C – the heater heat capacity
R – the heater resistance
K – the sample thermal conductance
T_0 – bath temperature
τ – relaxation time
V_A – adiabatic voltage
V_{IR} – resistive voltage
V_{TE} – thermoelectric voltage
Q_P – Peltier Heat
R_{C1} – Contact Resistance
ZT – figure of merit

Ch. 2.2 Borca-Tasciuc and Chen

ω – angular modulation frequency
I_0 – amplitude
P – power
R_h – resistance
$T_{2\omega}$ – amplitude of AC temperature rise
φ – phase shift
C_π – temperature coefficient of resistance
R_0 – heater resistance under no heating conditions
V – voltage
$V_{1\omega}$ – amplitude of the voltage applied across the heater
b – half-width of the heater
p/w – power amplitude dissipated per unit length of the heater
d_F – film thickness
k_F – cross-plane thermal conductivity of the film
T_s – temperature rise at the film-substrate interface
α_s – thermal diffusivity of the substrate
η – constant ~ 1
f_{linear} – linear function of $\ln\omega$
K_{Fxy} – ratio between the in-plane and cross-plane thermal conductivity
k_y – cross-plane thermal conductivity

LIST OF SYMBOLS

k_x — in-plane thermal conductivity
d_h — heater thickness
$(\rho c)_h$ — heat capacitance
R_{th} — thermal boundary resistance
ΔT — average complex temperature rise
L — distance from the heater to the heat sink
T_h — heater temperature rise
T_s — temperature of the sink
$R_{Membraam}$ — thermal resistance of the membrane
R_{Heater} — total thermal resistance
$T(x,t)$ — temperature response of a thin film membrane
α — thermal diffusivity along the membrane
2A — magnitude of the heat flux along x
T_{ac} — AC temperature
θ — complex amplitude of the ac temperature rise
q — amplitude of the heat flux generated by the heater
m — $(i2\omega/\alpha)^{1/2}$ as defined in the text
σ — Stefan-Boltzmann constant
T_0 — ambient and heat sink temperature
R — bridge electrical resistance
ϵ — emissivity
τ — time constant of the bridge
Σ — cross-section
k — thermal conductivity
r — reflectivity

Ch. 3.1 Sun & White

κ — thermal conductivity
T_c — superconducting transition temperature
ZT — figure of merit
S — Seebeck coefficient
σ — electrical conductivity
Z — average coordination number
⟨r⟩ — mean coordination number

Ch. 3.2 Pope and Tritt

κ_L — lattice contribution to total thermal conductivity
κ_E — electronic contribution to total thermal conductivity
κ_{TOT} — the total thermal conductivity,
α — thermopower
ΔT — temperature gradient
σ — the electrical conductivity
ρ — the electrical conductivity
L_0 — the Lorentz number [2.45×10^{-8} (V/K)^2]
ZT — figure of merit

Ch. 3.3 Savage & Rao

σ – electrical conductivity
σ_{tot} – total electrical conductivity
σ_e – electron conductivity
σ_h – hole conductivity
n – electron concentration
μ – carrier mobility
μ_h – hole mobility
ρ – electrical resistivity
ρ_L – electrical resistivity due to scattering by phonons
ρ_R – residual electrical resistivity due to scattering by impurities and defects
R – electrical resistance
ρ_E – electrical resistivity for empty single wall carbon nanotube bundles
ρ_F – electrical resistivity for filled single wall carbon nanotube bundles
k – Boltzmann constant
e – electron charge
E_F – Fermi energy
E_g – energy gap
S – thermopower
S_d – diffusion thermopower
S_g – phonon-drag thermopower
$S_{tot,met}$ – total thermopower in metals
S_{sem} – thermopower in semiconductors
T – temperature
p – hole concentration
κ – thermal conductivity
C – heat capacity
ℓ – phonon mean free path
j_Q – rate of heat flux
U – energy
C_p – heat capacity at constant pressure
C_v – heat capacity at constant volume
C_{ph} – heat capacity contribution due to phonons
C_{el} – heat capacity contribution due to electons
H – magnetic field
ZT – figure of merit
LA – longitundinal acoustic

CONTENTS

Section 1. – Overview of Thermal Conductivity in Solid Materials

Chapter 1.1 – Theory of Thermal Conductivity (Jihui Yang)
- Introduction   1
- Simple Kinetic Theory   2
- Electronic Thermal Conduction   3
    - Lattice Thermal Conductivity   9
- Summary   17
- References   17

Chapter 1.2 – Thermal Conductivity of Metals (Ctirad Uher)
- Introduction   21
    - Carriers of Heat in Metals   22
    - The Drude Model   24
- Specific Heat of Metals   29
- The Boltzmann Equation   32
    - Transport Coefficients   35
    - Electrical Conductivity   40
    - Electrical Thermal Conductivity   44
- Scattering Processes   46
    - Impurity Scattering   46
    - Electron-Phonon Scattering   50
    - Electron-Electron Scattering   61
        - Effect of e-e Processes on Electrical Resistivity   64
        - Effect of e-e Processes on Thermal Resistivity   69
- Lattice Thermal Conductivity   73
    - Phonon Thermal Resistivity Limited By Electrons   73
    - Other Processes Limiting Phonon Thermal Conductivity in Metals   77
- Thermal Conductivity of Real Metals   79
    - Pure Metals   79
    - Alloys   86
- Conclusion   87
- References   88

Chapter 1.3 – Thermal Conductivity of Insulators and Glasses
(Vladimir Murashov and Mary Anne White)
- Introduction   93
- Phononic Thermal Conductivity in Simple, Crystalline Insulators   94
  - Acoustic Phonons Carry Heat   94
  - Temperature-Dependence of κ 96
  - Impurities   97
- More Complex Insulators: The Role of Optic Modes   97
  - Molecular and Other Complex Systems   97
  - Optic-Acoustic Coupling   99
- Thermal Conductivity of Glasses   100
  - Comparison with Crystals   100
  - More Detailed Models   100
  - The Exception: Recent Amorphous Ice Results   101
- Minimum Thermal Conductivity   101
- Radiation   102
- References   102

Chapter 1.4 – Thermal Conductivity of Semiconductors
(G. S. Nolas and H. J. Goldsmid)
- Introduction   105
- Electronic Thermal Conductivity in Semiconductors   106
  - Transport Coefficients for a Single Band   106
  - Nondegenerate and Degenerate Approximations   109
  - Bipolar Conduction   110
  - Separation of Electronic and Lattice Thermal Conductivities   112
- Phonon Scattering in Impure and Imperfect Crystals   114
  - Pure Crystals   114
  - Scattering of Phonons by Impurities   115
  - Boundary Scattering   117
- Prediction of the Lattice Thermal Conductivity   118
- References   120

Chapter 1.5 – Semiconductors and Thermoelectric Materials
(G. S. Nolas, J. Yang, and H. J. Goldsmid)
- Introduction   123
- Established Materials   124
  - Bismuth Telluride and Its Alloys   124
  - Bismuth and Bismuth-Antimony Alloys   126
  - IV-VI Compounds   127
  - Silicon, Germanium, and Si-Ge Alloys   128
- Skutterudites   129
  - Binary (Unfilled) Skutterudites   130
  - Effect of Doping on the Co Site   132
  - Filled Skutterudites   133
- Clathrates   137
- Half-Heusler Compounds   141
  - Effect of Annealing   142
  - Isoelectronic Alloying on the M and Ni Sites   142
  - Effect of Grain Size Reduction   144

CONTENTS

- Novel Chalcogenides and Oxides — 145
  - Tl_9GeTe_6 — 146
  - Tl_2GeTe_5 and Tl_2SnTe_5 — 146
  - $CsBi_4Te_6$ — 147
  - $NaCo_2O_4$ — 147
- Summary — 149
- References — 149

Chapter 1.6 – Thermal Conductivity of Superlattices (G. D. Mahan)
- Introduction — 153
- Parallel to Layers — 154
- Perpendicular to Layers — 154
  - Thermal Boundary Resistance — 154
  - Multilayer Interference — 156
  - What is Temperature? — 157
  - Superlattices with Thick Layers — 159
- "Non-Kapitzic" Heat Flow — 161
  - Analytic Theory — 162
- Summary — 163
- References — 164

Chapter 1.7 – Experimental Studies on Thermal Conductivity of Thin Film and Superlattices (Bao Yang and Gang Chen)
- Introduction — 167
- Thermal Conductivity of Metallic Thin Films — 169
- Thermal Conductivity of Dielectric Films — 171
  - Amorphous SiO_2 Thin Films — 171
  - Thin Film Coatings — 173
  - Diamond Films — 174
  - Multilayer Interference — 174
- Thermal Conductivity of Semiconductor and Semimetal Thin Films — 174
  - Silicon Thin Films — 175
  - Semimetal Thin Films — 177
- Semiconductor Superlattices — 178
- Conclusions — 182
- Acknowledgments — 182
- References — 182

Section 2 – Measurement Techniques

Chapter 2.1 – Measurement Techniques and Considerations for Determining Thermal Conductivity of Bulk Materials (Terry M. Tritt and David Weston)
- Introduction — 187
- Steady State Method (Absolute Method) — 188
  - Overview of Heat Loss and Thermal Contact Issues — 189
  - Heat Loss Terms — 191
- The Comparative Technique — 193
- The Radial Flow Method — 195
- Laser-Flash Diffusivity — 197

- The Pulse-power Method ("Maldonado" Technique) — 199
- Parallel Thermal Conductance Technique — 200
- Z-Meters or Harman Technique — 201
- Summary — 202
- References — 202

Chapter 2.2 – Experimental Techniques for Thin–Film Thermal Conductivity Characterization (T. Borca-Tasciuc and G. Chen)
- Introduction — 205
- Electrical Heating and Sensing — 208
  - Cross-Plane Thermal Conductivity Measurements of Thin Films — 208
    - The 3ω Method — 208
    - Steady-State Method — 213
  - In-Plane Thermal Conductivity Measurements — 214
    - Membrane Method — 216
    - Bridge Method — 222
    - In-Plane Thermal Conductivity Measurement without Substrate Removal — 225
- Optical Heating Methods — 225
  - Time Domain Pump-and-Probe Methods — 226
  - Frequency–Domain Photothermal and Photoacoustic Methods — 230
    - Photothermal Reflectance Method — 230
    - Photothermal Emission Method — 230
    - Photothermal Displacement Method — 231
    - Photothermal Defelection Method (Mirage Method) — 231
    - Photoacoustic Method — 231
- Optical-Electrical Hybrid Methods — 232
- Summary — 233
- Acknowledgments — 234
- References — 234

Section 3 – Thermal Properties and Applications of Emerging Materials

Chapter 3.1 - Ceramics and Glasses (Rong Sun and Mary Anne White)
- Introduction — 239
- Ceramics — 239
  - Traditional Materials with High Thermal Conductivity — 240
    - Aluminum Nitride (AlN) — 240
    - Silicon Nitride (Si_3N_4) — 243
    - Alumina (Al_2O_3) — 244
  - Novel Materials with Various Applications — 244
    - Ceramic Composites — 244
    - Diamond Film on Aluminum Nitride — 244
    - Silicon Carbide Fiber-reinforced Ceramic Matrix Composite (SiC-CMC) — 244
    - Carbon Fiber-incorporated Alumina Ceramics — 245
    - Ceramic Fibers — 245
    - Glass-ceramic Superconductor — 245
  - Other Ceramics — 246

• Rare-earth Based Ceramics	246
• Magnesium Silicon Nitride (MgSiN$_2$)	247
• Thermoelectric Ceramics	247
• Glasses	248
• Introduction	248
• Chalcogenide Glasses	248
• Other Glasses	249
• Conclusions	250
• References	250

Chapter 3.2 – Thermal Conductivity of Quasicrystalline Materials (A. L. Pope and Terry M. Tritt)

• Introduction	255
• Contributions to Thermal Conductivity	257
• Low-Temperature Thermal Conduction in Quasicrystals	257
• Poor Thermal Conduction in Quasicrystals	258
• Glasslike Plateau in Quasicrystalline Materials	258
• Summary	259
• References	259

Chapter 3.3 – Thermal Properties of Nanomaterials and Nanocomposites (T. Savage and A. M. Rao)

• Nanomaterials	262
• Carbon Nanotubes	262
• Electrical Conductivity, σ	262
• Thermoelectric Power (TEP)	265
• Thermal Conductivity, κ	271
• Heat Capacity, C	274
• Nanowires	276
• Electrical Conductivity	276
• Thermoelectric Power	277
• Thermal Conductivity and Heat Capacity	278
• Nanoparticles	278
• Nanocomposites	279
• Electrical Conductivity	279
• Thermal Conductivity	280
• Applications	280
• References	282

Index   285

Chapter 1.1

THEORY OF THERMAL CONDUCTIVITY

Jihui Yang

Materials and Processes Laboratory, GM R&D Center, Warren, MI, USA

1. INTRODUCTION

Heat energy can be transmitted through solids via electrical carriers (electrons or holes), lattice waves (phonons), electromagnetic waves, spin waves, or other excitations. In metals electrical carriers carry the majority of the heat, while in insulators lattice waves are the dominant heat transporter. Normally, the total thermal conductivity κ can be written as a sum of all the components representing various excitations:

$$\kappa = \sum_{\alpha} \kappa_{\alpha}, \qquad (1)$$

where α denotes an excitation. The thermal conductivities of solids vary dramatically both in magnitude and temperature dependence from one material to another. This is caused by differences in sample sizes for single crystals or grain sizes for polycrystalline samples, lattice defects or imperfections, dislocations, anharmonicity of the lattice forces, carrier concentrations, interactions between the carriers and the lattice waves, interactions between magnetic ions and the lattice waves, etc. The great variety of processes makes the thermal conductivity an interesting area of study both experimentally and theoretically.

Historically thermal conductivity measurement was used as a powerful tool for investigating lattice defects or imperfections in solids. In addition to opportunities for investigating exciting, intriguing physical phenomena, thermal conductivity study is also of great technological interest. Materials with both very high and very low thermal conductivities are technologically important. High-thermal-conductivity materials like diamond or silicon have been extensively studied in part because of their potential applications in thermal management of electronics.[1–8] Low-thermal-conductivity materials like skutterudites, clathrates, half-heuslers, chalcogenides, and novel oxides are the focus of the recent quest for high-efficiency thermoelectric materials.[9-34]

The aim of this chapter is to review and explain the main mechanisms and models that govern heat conduction in solids. More detailed theoretical treatments can be found elsewhere.[35-38] Furthermore, only heat conduction by carriers (electrons) and lattice waves (phonons) at low temperatures will be discussed. It is in this temperature region that most of the theoretical models can be compared against experimental results.

2. SIMPLE KINETIC THEORY

Thermal conductivity is defined as

$$\kappa = -\frac{\vec{Q}}{\vec{\nabla}T}, \qquad (2)$$

where $\vec{Q}$ is the heat flow rate (or heat flux) vector across a unit cross section perpendicular to $\vec{Q}$ and T is the absolute temperature. For the kinetic formulation of thermal conduction in gases, let us assume that c is the heat capacity of each particle and n is the concentration of the particles. In the presence of a temperature gradient $\vec{\nabla}T$, for a particle to travel with velocity $\vec{v}$ its energy must change at a rate of

$$\frac{\partial E}{\partial t} = c\vec{v}\cdot\vec{\nabla}T. \qquad (3)$$

The average distance a particle travels before being scattered is $v\tau$, where τ is the relaxation time. The average total heat flow rate per unit area summing over all particles is therefore

$$\vec{Q} = -nc\tau\langle\vec{v}\cdot\vec{v}\rangle\vec{\nabla}T = -\frac{1}{3}nc\tau v^2\vec{\nabla}T. \qquad (4)$$

The brackets in Eq. (4) represent an average over all particles. Combining Eqs. (2) and (4), we have

$$\kappa = \frac{1}{3}nc\tau v^2 = \frac{1}{3}Cvl, \qquad (5)$$

where $C = nc$ is the total heat capacity and $l = v\tau$ is the particle mean free path. In solids the same derivation can be made for various excitations (electrons, phonons, photons, etc.). Equation (5) can then be generalized to

$$\kappa = \frac{1}{3}\sum_\alpha C_\alpha v_\alpha l_\alpha, \qquad (6)$$

where the summation is over all excitations, denoted by α. In general, Eq. (6) gives a good phenomenological description of the thermal conductivity, and it is practically very useful for order of magnitude estimates.

Like most of the nonequilibrium transport parameters, thermal conductivity cannot be solved exactly. Calculations are usually based on a combination of perturbation theory and the Boltzmann equation, which are the bases for analyzing the microscopic processes that govern the heat conduction by carriers and lattice waves.

3. ELECTRONIC THERMAL CONDUCTION

The free electron theory of electron conduction in solids in the first instance considers each electron as moving in a periodic potential produced by the ions and other electrons without disturbance, and then regards the deviation from the periodicity due to the vibrations of the lattice as a perturbation. The possible values of the electron wave vector $\vec{k}$ depend on the periodicity and the size of the crystal. The $\vec{k}$-space is separated into Brillouin zones. The electron energy $E_{\vec{k}}$ depends on the form of the potential and is a continuous function of $\vec{k}$ in each zone, but it is discontinuous at the zone boundaries. The values of $E_{\vec{k}}$ in a zone trace out a "band" of energy values.

The distribution function that measures the number of electrons in the state $\vec{k}$ and at the location $\vec{r}$ is $f_{\vec{k}}$. In equilibrium, the distribution function is $f_{\vec{k}}^0$, given by

$$f_{\vec{k}}^0 = \frac{1}{\exp\left(\frac{E_{\vec{k}} - E_F}{k_B T}\right) + 1}, \tag{7}$$

where E_F and k_B are the Fermi energy and Boltzmann's constant, respectively. According to the Boltzmann equation, in the presence of an electrical field $\vec{E}$ and a temperature gradient $\vec{\nabla}T$, the steady state that represents a balance between the effects of the scattering processes and the external field and temperature gradient can be described as

$$\frac{f_{\vec{k}} - f_{\vec{k}}^0}{\tau(k)} = -\vec{v}_{\vec{k}} \cdot \left(\frac{\partial f_{\vec{k}}^0}{\partial T}\vec{\nabla}T + e\frac{\partial f_{\vec{k}}^0}{\partial E_{\vec{k}}}\vec{E}\right), \tag{8}$$

where $\tau(k)$, $\vec{v}_{\vec{k}}$, and e are the relaxation time, the electron velocity, and the electron charge, respectively. Taking into account the spatial variation of the Fermi energy and the explicit expression of $f_{\vec{k}}^0$, Eq. (8) can be written in the form

$$\frac{f_{\vec{k}} - f_{\vec{k}}^0}{\tau(k)} = -\vec{v}_{\vec{k}} \cdot \left(-\frac{E_{\vec{k}} - E_F}{T}\frac{\partial f_{\vec{k}}^0}{\partial E_{\vec{k}}}\vec{\nabla}T + e\frac{\partial f_{\vec{k}}^0}{\partial E_{\vec{k}}}\left(\vec{E} - \frac{\vec{\nabla}E_F}{e}\right)\right). \tag{9}$$

The effective field acting on the electrons is $\vec{E}_{\text{eff}} = \vec{E} - \frac{\vec{\nabla}E_F}{e}$. Equation (9) can be entered into the expressions for the electrical current density $\vec{J}$ and the flux of energy $\vec{Q}$:

$$\vec{J} = \int e\vec{v}_{\vec{k}} f_{\vec{k}} d\vec{k} \tag{10}$$

and

$$\vec{Q} = \int (E_{\vec{k}} - E_F)\vec{v}_{\vec{k}} f_{\vec{k}} d\vec{k}. \tag{11}$$

If we define a general integral K_n as

$$K_n = -\frac{1}{3}\int (\vec{v}_{\vec{k}})^2 \tau(k)(E_{\vec{k}} - E_F)^n \frac{\partial f_{\vec{k}}^0}{\partial E_{\vec{k}}} d\vec{k}, \tag{12}$$

then Eqs. (10) and (11) can be written as

$$\vec{J} = e^2 K_0 \vec{E}_{\text{eff}} - \frac{e}{T} K_1 \vec{\nabla} T \tag{13}$$

and

$$\vec{Q} = e K_1 \vec{E}_{\text{eff}} - \frac{1}{T} K_2 \vec{\nabla} T, \tag{14}$$

respectively. The electronic thermal conductivity can be found with $\vec{J} = 0$ such that

$$\kappa_e = -\left[\frac{\vec{Q}}{\vec{\nabla} T}\right]_{\vec{J}=0} = \frac{1}{T}\left(K_2 - \frac{K_1^2}{K_0}\right). \tag{15}$$

The electrical conductivity σ can be derived from Eq. (13) as

$$\sigma = e^2 K_0. \tag{16}$$

Since $\partial f_{\vec{k}}^0 / \partial E_{\vec{k}}$ is approximately a delta function at the Fermi surface with width $k_B T$, K_n can be evaluated by expansion:

$$K_n = \left[(E_{\vec{k}} - E_F)^n \frac{\sigma(E_{\vec{k}})}{e^2} + \frac{\pi^2}{6} k_B^2 T^2 \frac{\partial^2}{\partial E^2}\left\{(E_{\vec{k}} - E_F)^n \frac{\sigma(E_{\vec{k}})}{e^2}\right\} + \ldots \right]_{E_{\vec{k}} = E_F}. \tag{17}$$

Therefore[39]

$$K_2 = \frac{\pi^2}{3} \frac{k_B^2 T^2}{e^2} \sigma(E_F) + O\left(k_B T / E_F\right)^2, \tag{18}$$

and

$$\frac{K_1^2}{K_0} \sim O\left(k_B T / E_F\right)^2. \tag{19}$$

Equation (15) through (19) lead to the Wiedemann–Franz law with the standard Lorentz number L_0 as

$$L_0 = \frac{\kappa_e}{\sigma T} = \frac{\pi^2}{3} \frac{k_B^2}{e^2}. \tag{20}$$

The numerical value of L_0 is 2.4453×10^{-8} W-Ω/K^2. This shows that all metals have the same electronic thermal conductivity to electrical conductivity ratio, and this ratio is proportional to the absolute temperature. Furthermore, for strong degenerate electron gases, Eq. (20) is independent of the scattering mechanism and the band structure for the electrons as long as the scatterings are elastic. The Wiedemann–Franz law is generally well obeyed at high-temperatures. In the low– and intermediate–temperature regions, however, the law fails due to the inelastic scattering of the charge carriers.[36, 37, 39]

Typically one needs to calculate the relaxation times for the various relevant electron scattering processes and use Eqs. (15) and (16) to determine the electronic thermal conductivity and the electrical conductivity. Over a wide temperature range, scattering of electrons by phonons is a major factor for determining the electrical and electronic thermal conductivities. The resistance due to this type of scattering is called the ideal resistance. The ideal electrical and electronic thermal resistances can be approximately written as[36]

Sec. 3 · ELECTRONIC THERMAL CONDUCTION

$$\rho_i = \frac{1}{\sigma_i} = A\left(\frac{T}{\theta_D}\right)^5 J_5\left(\frac{\theta_D}{T}\right) \tag{21}$$

and

$$W_i = \frac{1}{\kappa_i} = \frac{A}{L_0 T}\left(\frac{T}{\theta_D}\right)^5 J_5\left(\frac{\theta_D}{T}\right)\left\{1 + \frac{3}{\pi^2}\left(\frac{k_F}{q_D}\right)^2\left(\frac{\theta_D}{T}\right)^2 - \frac{1}{2\pi^2}\frac{J_7(\theta_D/T)}{J_5(\theta_D/T)}\right\}, \tag{22}$$

respectively, where

$$J_n\left(\frac{\theta_D}{T}\right) = \int_0^{\theta_D/T} \frac{x^n e^x}{(e^x - 1)^2} dx, \tag{23}$$

$$A = \frac{3\pi\hbar q_D^6 (G')^2}{4e^2 (m^*)^2 n_c k_B \theta_D k_F^2 v_F^2}, \tag{24}$$

θ_D is the Debye temperature for phonons (to be discussed below), k_F is the electron wave number at the Fermi surface, q_D is the phonon Debye wave number, $\hbar$ is the Planck constant, G' is the constant representing the strength of the electron–phonon interaction, m^* is the electron effective mass, n_c is the number of unit cells per unit volume, and v_F is the electron velocity at the Fermi surface. According to Eqs. (21) through (24), at high-temperatures ($T \gg \theta_D$)

$$\rho_i = \frac{AT}{4\theta_D} \tag{25}$$

and

$$W_i = \frac{A}{4\theta_D L_0} = \frac{\rho_i}{L_0 T}. \tag{26}$$

The Wiedemann–Franz law is obeyed; W_i is a constant. At low temperatures ($T \ll \theta_D$)

$$\rho_i = 124.4 A \left(\frac{T}{\theta_D}\right)^5 \tag{27}$$

and

$$W_i = 124.4 \frac{A}{L_0 T}\left(\frac{T}{\theta_D}\right)^3 \frac{3}{\pi^2}\left(\frac{k_F}{q_D}\right)^2 \propto T^2. \tag{28}$$

In addition to the electron–phonon interaction, the electron-defect interaction contributes to the electrical resistivity and electronic thermal resistivity. One can approximately write the resistivity ρ and electronic thermal resistivity W_e as (Matthiessen's Rule)

$$\rho = \frac{1}{\sigma} = \rho_0 + \rho_i \tag{29}$$

and

$$W_e = \frac{1}{\kappa_e} = W_0 + W_i. \tag{30}$$

Here; ρ_0 and W_0 are the residual electrical resistivity and electronic thermal resistivity, respectively, caused by electron scattering due to impurities and defects, and ρ_0 is independent of temperature. If we assume that all defects scatter electrons elastically, ρ_0 and W_0 should be related by the Wiedemann–Franz law:

$$W_0 = \rho_0/L_0 T \propto 1/T. \tag{31}$$

At very low temperature, since W_0 increases and W_i decreases with decreasing temperature,

$$\kappa_e \approx \frac{1}{W_0} = \frac{L_0}{\rho_0} T. \tag{32}$$

The electronic thermal conductivity increases linearly with increasing temperature. As temperature increases, W_i becomes relatively more important. If W_i increases to values comparable to that of W_0 at sufficiently low temperature, then κ_e will pass a maximum, decrease, and eventually attain the high-temperature constant described by Eq. (26). In alloys or metals with a high concentration of defects, W_0 and W_i only become comparable at high-temperatures. In this case, there is no maximum in the κ_e, versus, T curve, and κ_e will approach the high-temperature constant monotonically as T increases. Both cases are illustrated in Fig. 1, which shows the low-temperature thermal conductivity of silver measured by White.[40] In Fig. 1, sample (a) is obtained by annealing sample (b). Low, temperature thermal conductivity is therefore dramatically affected by the residual resistance. By combining Eqs. (27) through (31), we have

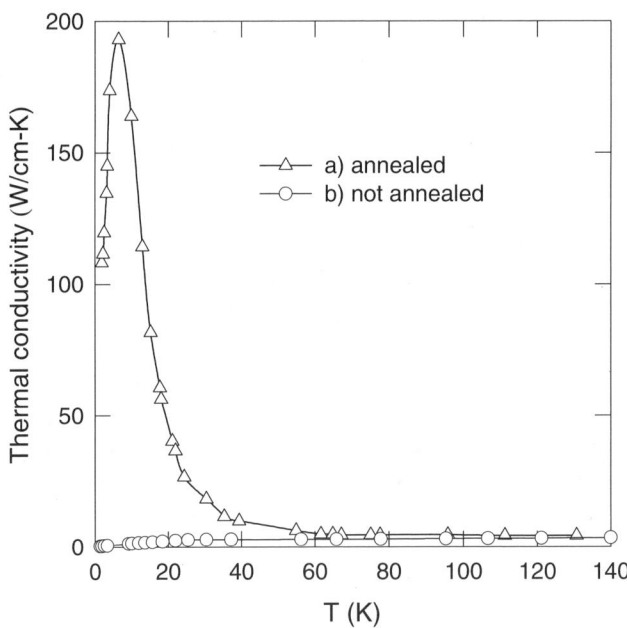

FIGURE 1 Thermal conductivity of two silver samples with (a) very low and (b) moderately high residual resistivities from Ref. 40, showing the profound influence of the residual resistivity on the low-temperature thermal conductivity. Sample (a) was obtained by annealing sample (b). Reprinted with permission from IOP Publishing Limited.

Sec. 3 · ELECTRONIC THERMAL CONDUCTION

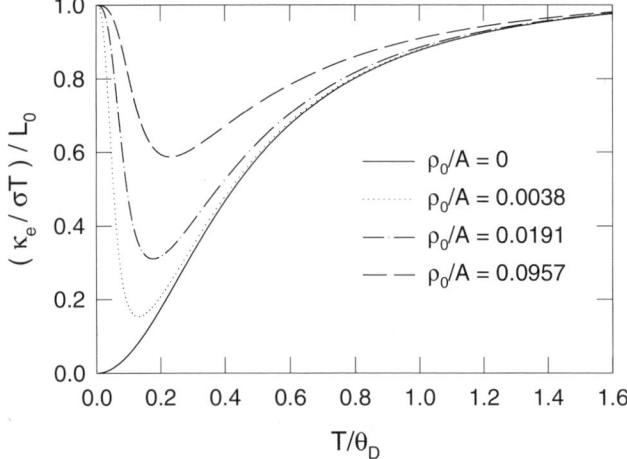

FIGURE 2 Calculated $(\kappa_e/\sigma T)/L_0$ versus T/θ_D curves for various values of ρ_0/A, and for monovalent metals. The calculations are based on Eq. (33) using $(k_F/q_D)^2 = \sqrt[3]{2}/2$ for monovalent metals.

$$\frac{\kappa_e/\sigma T}{L_0} = \frac{\frac{\rho_0}{A} + \left(\frac{T}{\theta_D}\right)^5 J_5\left(\frac{\theta_D}{T}\right)}{\frac{\rho_0}{A} + \left(\frac{T}{\theta_D}\right)^5 \{[1 + \frac{3}{\pi^2}\left(\frac{k_F}{q_D}\right)^2\left(\frac{\theta_D}{T}\right)^2] J_5\left(\frac{\theta_D}{T}\right) - \frac{1}{2\pi^2} J_7\left(\frac{\theta_D}{T}\right)\}}. \tag{33}$$

This quantity versus T/θ_D is shown in Fig. 2 for various values of $\frac{\rho_0}{A}$, and for monovalent metals ($(k_F/q_D)^2 = \sqrt[3]{2}/2$). Figure 2 shows that the Wiedemann–Franz law holds at very high and very low temperatures except for very pure metals, but $\kappa_e/\sigma T$ underestimates L_0 at intermediate temperatures in a manner that strongly depends on the amount of impurity present.

It should be stressed that Eqs. (21) and (22) were derived without electron–phonon umklapp processes, and the strength of the electron–phonon interaction was assumed to be a constant. Close agreement between the model and experimental data should not be expected in most cases. Modifications to the model are discussed elsewhere.[37]

The discussion so far has been focused on degenerate electron gases (like those in metals). In semiconductors, because of the energy gap E_G between the top of the valence band and the bottom of the conduction band and the consequence that the Fermi level lies in the gap, the electron distribution may no longer be degenerate. Also, there may exist positively charged particles, the "holes," that may also contribute to the transport. In this case, the distribution functions for electrons and holes may be written as

$$f_e^0 \approx \exp\left[-\frac{E_e + E_F^e}{k_B T}\right] \tag{34}$$

and

$$f_h^0 \approx \exp\left[-\frac{E_h + E_F^h}{k_B T}\right], \tag{35}$$

where f_e^0 and f_h^0 are the equilibrium distribution functions for electrons and holes, respectively, E_e and E_h are the electron and hole energies, respectively, and E_F^e and E_F^h are the distances of the Fermi level below the bottom of the conduction band

and above the top of the valence band, respectively. If one wishes to use the general integral defined in Eq. (12), the conductivities can be written similarly to those in Eqs. (15) and (16):

$$\kappa_e = \frac{1}{T}\left\{(K_2 + K_2') - \frac{(K_1 + K_1')^2}{K_0 + K_0'}\right\} \quad (36)$$

and

$$\sigma = e^2(K_0 + K_0'), \quad (37)$$

where K_n and K_n' denote integrals for electrons and holes, respectively, and the distribution functions for the integrals should be those listed in Eqs. (34) and (35). In general, the ratio $\kappa_e/\sigma T$ is rather complicated to work out. For simplicity, assume that the energy bands are parabolic and the energy dependences of the carriers' relaxation times are the same such that

$$\tau_{e,h}(E) \propto E^\alpha, \quad (38)$$

where α is a constant, and the subscripts e and h denote the electrons and holes. With some algebra one finds[36]

$$\frac{\kappa_e}{\sigma T} = \left(\frac{k_B}{e}\right)^2\left[\left(\frac{5}{2}+\alpha\right) + \left(5 + 2\alpha + \frac{E_G}{k_BT}\right)^2 \frac{n_e\mu_e n_h\mu_h}{(n_e\mu_e + n_h\mu_h)^2}\right]. \quad (39)$$

In Eq. (39), n and μ are the carrier concentration and mobility, and again the subscripts e and h denote electrons and holes. The first term in Eq. (39) is the standard Lorentz number for nondegenerate Fermi distributions. When carrier-acoustic phonon scattering dominates, $\alpha = -1/2$, and if $n_e = n_h$, Eq. (39) can be written as

$$\kappa_e = \sigma T\left(\frac{k_B}{e}\right)^2\left[2 + \left(4 + \frac{E_G}{k_BT}\right)^2 \frac{\mu_e/\mu_h}{(1 + \mu_e/\mu_h)^2}\right]. \quad (40)$$

The second term in Eq. (39) or (40) is called the bipolar diffusion term. This type of energy transport in addition to that carried by the electrons and the holes is due to the creation of electron–hole pairs that extract an amount of energy E_G at the high-temperature end; on recombination this energy is given up to the cold end. The bipolar diffusion may be a noticeable contribution to the electronic thermal conductivity when both the carrier concentration and the mobility are about equal for electrons and holes, and their mobilities have reasonably high values.[41, 42] This should be the case for intrinsic semiconductors. In the case of doped semiconductors only one of the carriers has high mobility; the low-mobility carrier will not be fast enough to accompany the high-mobility carrier for the recombination at the cold end; therefore, the bipolar contribution may not be noticeable. Anomalously, high thermal conductivity at high-temperatures for InSb, Ge, Si, and Bi may be explained by this bipolar diffusion.[43, 44, 45]

Figure 3 shows the temperature dependence of $(W - W_I)/T$ for pure Ge measured by Glassbrenner and Slack.[44] The thermal resistance due to isotopes W_I is almost independent of temperature and is subtracted from the total thermal resistance W. The term $(W - W_I)/T$ increases linearly as a function of T for $T < 700$ K, reaches a maximum at about 700 K, and decreases with increasing T for $T >$

Sec. 3 · ELECTRONIC THERMAL CONDUCTION

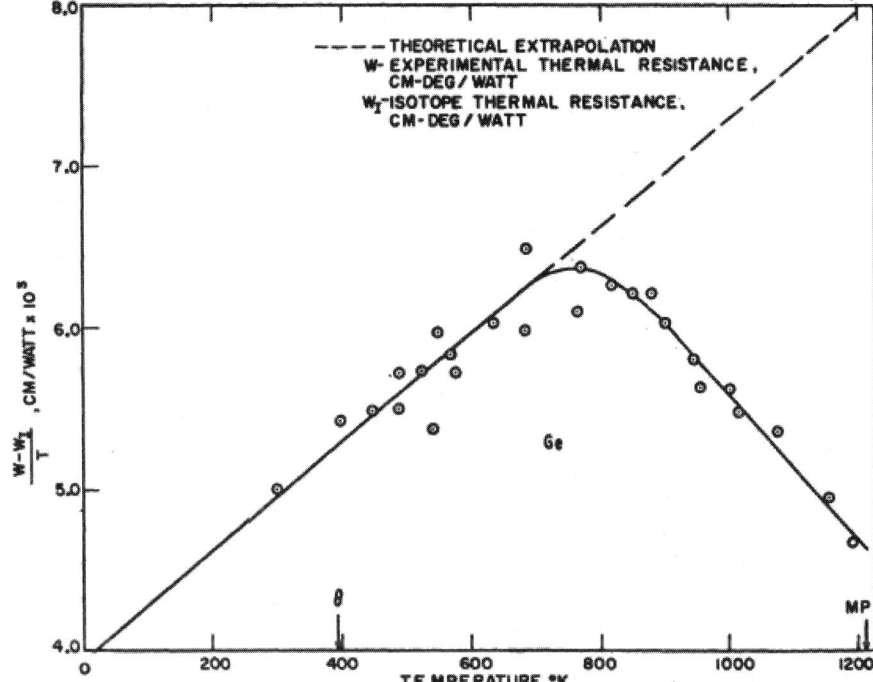

FIGURE 3 The temperature dependence of $(W - W_I)/T$ for pure Ge from Ref. 44. W is the total measured thermal resistivity, and W_I is the calculated thermal resistance due to isotopes. The dashed line is a theoretical extrapolation for thermal resistivity induced by phonon–phonon interactions.

700 K. The theoretical extrapolation (dashed line) represents T and T^2 components of the thermal resistivity due to phonon–phonon interactions. The former is due to three-phonon umklapp processes (to be discussed), and the latter is attributed to four-phonon processes.[46] Deviation from the dashed line at high-temperatures is ascribed to the electronic thermal resistance. It was found that the contribution to the total thermal conductivity from the first term of Eq. (40) is less than 10% up to the melting point. The majority of the deviation between $(W - W_I)/T$ and the dashed line at high-temperatures is due to dipolar diffusion. The values of E_G can be estimated by subtracting the lattice component and the first term of Eq. (40) from the total thermal conductivity and comparing it against the second term in Eq. (40). The results are in agreement with the published results.[44]

3.1. Lattice Thermal Conductivity

Lattice thermal conduction is the dominant thermal conduction mechanism in nonmetals, if not the only one. Even in some semiconductors and alloys, it dominates a wide temperature range. In solids atoms vibrate about their equilibrium positions (crystal lattice). The vibrations of atoms are not independent of each other, but are rather strongly coupled with neighboring atoms. The crystal lattice vibration can be characterized by the normal modes, or standing waves. The quanta of the crystal vibrational field are referred to as "phonons." In the presence of a temperature gradient, the thermal energy is considered as propagating by means of wave packets consisting of various normal modes, or phonons. Derivation of phonon dispersion curves ($\omega_{\vec{q}}$ versus q curves, where $\omega_{\vec{q}}$ and q are the phonon frequency and wave

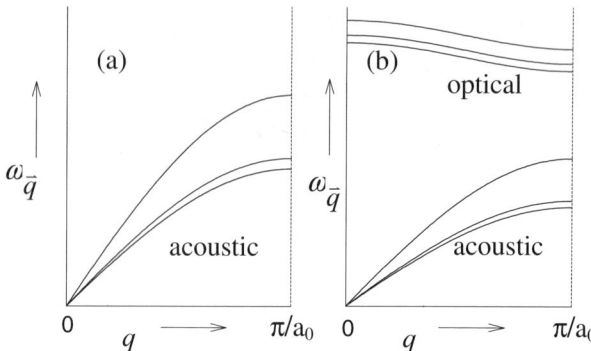

FIGURE 4 Schematic phonon dispersion curves for a given direction of $\vec{q}$ of (a) monatomic lattice and (b) diatomic lattice. The lattice parameter is denoted a_0.

vector, respectively) can be found in a standard solid-state physics book.[47] Schematic phonon dispersion curves for monatomic and diatomic lattices are shown in Fig. 4. Phonon dispersion curves for solids normally consist of acoustic and optical branches. The low-frequency acoustic branches correspond to atoms in a unit cell moving in same phase, whereas the high-frequency optical branches represent atoms in a unit cell moving in opposite phases. Normally optical phonons themselves are not effective in transporting heat energy because of their small group velocity $\partial\omega_{\vec{q}}/\partial q$, but they may affect the heat conduction by interacting with the acoustic phonons that are the main heat conductors.

The phonon distribution function, which represents the average number of phonons with wave vector $\vec{q}$, is $N_{\vec{q}}$. In equilibrium, the phonon distribution function can be written as

$$N_{\vec{q}}^0 = \frac{1}{\exp(\hbar\omega_{\vec{q}}/k_B T) - 1}. \tag{41}$$

The Boltzmann equation assumes that the scattering processes tend to restore a phonon distribution $N_{\vec{q}}$ to its equilibrium form $N_{\vec{q}}^0$ at a rate proportional to the departure of the distribution from equilibrium, such that

$$\frac{N_{\vec{q}} - N_{\vec{q}}^0}{\tau_q} = -(\vec{v}_g \cdot \vec{\nabla} T)\frac{\partial N_{\vec{q}}^0}{\partial T}, \tag{42}$$

where $\vec{v}_g$ is the phonon group velocity and τ_q is the phonon scattering relaxation time. The heat flux due to a phonon mode $\vec{q}$ is the product of the average phonon energy and the group velocity. Therefore the total heat flux carried by all phonon modes can be written as

$$\vec{Q} = \sum_{\vec{q}} N_{\vec{q}} \hbar \omega_{\vec{q}} \vec{v}_g. \tag{43}$$

Substituting Eq. (42) into Eq. (43) yields

$$\vec{Q} = -\sum_{\vec{q}} \hbar\omega_{\vec{q}} v_g^2 \langle\cos^2\theta\rangle \tau_q \frac{\partial N_{\vec{q}}^0}{\partial T}\vec{\nabla} T = -\frac{1}{3}\sum_{\vec{q}} \hbar\omega_{\vec{q}} v_g^2 \tau_q \frac{\partial N_{\vec{q}}^0}{\partial T}\vec{\nabla} T, \tag{44}$$

where θ is the angle between $\vec{v}_g$ and $\vec{\nabla} T$; the lattice thermal conductivity is

Sec. 3 · ELECTRONIC THERMAL CONDUCTION

$$\kappa_L = -\frac{\vec{Q}}{\vec{\nabla}T} = \frac{1}{3}\sum_{\vec{q}} \hbar\omega_{\vec{q}} v_g^2 \tau_q \frac{\partial N_{\vec{q}}^0}{\partial T}. \tag{45}$$

At this point approximations need to be made to Eq. (45) in order to obtain meaningful results. Furthermore, evaluations of various phonon scattering relaxation times are usually difficult to make precise. It is not worthwhile to try to calculate Eq. (45) for the precise phonon frequency spectrum and dispersion curve of a real solid. Assumptions of Debye theory should be used: an average phonon velocity v (approximately equal to the velocity of sound in solids) is used to replace v_g, $\omega_{\vec{q}} = v \cdot q$ for all the phonon branches, and the phonon velocities are the same for all polarizations. The summation in Eq. (45) can be replaced by the integral

$$\kappa_L = \frac{1}{3}\int \hbar\omega_{\vec{q}} v_g^2 \tau_q \frac{\partial N_{\vec{q}}^0}{\partial T} f(q) d\vec{q}, \tag{46}$$

where $f(q)d\vec{q} = (3q^2/2\pi^2)\,dq$, and therefore $f(\omega)d\omega = (3\omega^2/2\pi^2 v^3)\,d\omega$. Using the Debye assumptions and Eqs. (41) and (45) leads to

$$\kappa_L = \frac{1}{2\pi^2 v}\int_0^{\omega_D} \hbar\omega^3 \tau_q(\omega) \frac{(\hbar\omega/k_B T^2)\exp(\hbar\omega/k_B T)}{[\exp(\hbar\omega/k_B T) - 1]^2} d\omega, \tag{47}$$

where ω_D is the Debye frequency such that

$$3N = \int_0^{\omega_D} f(\omega) d\omega \tag{48}$$

is the total number of all distinguishable phonon modes. If we make the substitution $x = \hbar\omega/k_B T$ and define the Debye temperature $\theta_D = \hbar\omega_D/k_B$, Eq. (47) becomes

$$\kappa_L = \frac{k_B}{2\pi^2 v}\left(\frac{k_B}{\hbar}\right)^3 T^3 \int_0^{\theta_D/T} \tau_q(x) \frac{x^4 e^x}{(e^x - 1)^2} dx. \tag{49}$$

Within the Debye approximation the differential lattice specific heat is

$$C(x)dx = \frac{3k_B}{2\pi^2 v^3}\left(\frac{k_B}{\hbar}\right)^3 T^3 \frac{x^4 e^x}{(e^x - 1)^2} dx. \tag{50}$$

If we define the mean free path of the phonons as $l(x) = v\tau_q(x)$, the lattice thermal conductivity can be written as

$$\kappa_L = \frac{1}{3}\int_0^{\theta_D/T} v^2 \tau_q(x) C(x) dx = \frac{1}{3}\int_0^{\theta_D/T} C(x) v l(x) dx. \tag{51}$$

This is analogous to the thermal conductivity formula [Eq. (6)] derived from simple kinetic theory.

Equation (49) is usually called the Debye approximation for the lattice thermal conductivity. If one can calculate the relaxation times $\tau_i(x)$ for various phonon scattering processes in solids and add the scattering rates such that

$$\tau_q^{-1}(x) = \sum_i \tau_i^{-1}(x), \tag{52}$$

Eq. (49) should be sufficient to obtain the lattice thermal conductivity. It is indeed then adequate for analyzing and predicting a lot of the experimental data especially when a large concentration of lattice defects prevails in the solid. Sometimes, than even in the case of some chemically pure crystals, because of the existence of appreciable amounts of isotopes, the lattice thermal conductivity can be represented satisfactorily by the Debye approximation.

The phonon scattering processes included in the Debye approximate are resistive and are called umklapp processes or U-processes. The total crystal momentum is not conserved for U-processes. Because such processes tend to restore non-equilibrium phonon distribution to the equilibrium distribution described by Eq. (42), they give rise to thermal resistance. There exist, however, other nonresistive and total crystal-momentum-conserving processes that do not contribute to the thermal resistance but may still have profound influence on the lattice thermal conductivity of solids. Such processes are called normal processes or N-processes. Even though N-processes themselves do not contribute to thermal resistance directly, they have the great effect of transferring energy between different phonon modes, thus preventing large deviations from the equilibrium distribution.

Since the N-processes themselves do not tend to restore the phonon equilibrium distribution, they cannot be simply added to Eq. (52). The Callaway model is the most widely used in analyzing the effect of N-processes on the lattice thermal conductivity.[48] Callaway's model assumes that N-processes tend to restore a non-equilibrium phonon distribution to a displaced phonon distribution of the form[49]

$$N_{\vec{q}}(\vec{\lambda}) = \frac{1}{\exp[(\hbar\omega - \vec{q}\cdot\vec{\lambda})/k_BT] - 1} = N_{\vec{q}}^0 + \frac{\vec{q}\cdot\vec{\lambda}}{k_BT}\frac{\exp(\hbar\omega/k_BT)}{[\exp(\hbar\omega/k_BT) - 1]^2}, \quad (53)$$

where $\vec{\lambda}$ is some constant vector (in the direction of the temperature gradient) that determines the anisotropy of the phonon distribution and the total phonon momentum. If the relaxation time for the N-processes is τ_N, Eq. (42) can be modified to

$$\frac{N_{\vec{q}} - N_{\vec{q}}^0}{\tau_q} + \frac{N_{\vec{q}} - N_{\vec{q}}(\vec{\lambda})}{\tau_N} = -(\vec{v}\cdot\vec{\nabla}T)\frac{\partial N_{\vec{q}}^0}{\partial T}. \quad (54)$$

If we define a combined relaxation time τ_c by

$$\tau_c^{-1} = \tau_q^{-1} + \tau_N^{-1}, \quad (55)$$

and define

$$n_1 = N_{\vec{q}} - N_{\vec{q}}^0, \quad (56)$$

the Boltzmann equation [Eq. (54)] can be written as

$$-\frac{\hbar\omega}{k_BT^2}(\vec{v}\cdot\vec{\nabla}T)\frac{\exp(\hbar\omega/k_BT)}{[\exp(\hbar\omega/k_BT) - 1]^2} + \frac{\vec{q}\cdot\vec{\lambda}}{\tau_N k_BT}\frac{\exp(\hbar\omega/k_BT)}{[\exp(\hbar\omega/k_BT) - 1]^2} - \frac{n_1}{\tau_c} = 0. \quad (57)$$

If we express n_1 as

$$n_1 = -\alpha_q\frac{\hbar\omega}{k_BT^2}(\vec{v}\cdot\vec{\nabla}T)\frac{\exp(\hbar\omega/k_BT)}{[\exp(\hbar\omega/k_BT) - 1]^2}, \quad (58)$$

then Eq. (57) can be simplified to

Sec. 3 · ELECTRONIC THERMAL CONDUCTION

$$\frac{\hbar\omega\alpha_q}{\tau_c T}\vec{v}\cdot\vec{\nabla}T + \frac{\vec{q}\cdot\vec{\lambda}}{\tau_N} = \frac{\hbar\omega}{T}\vec{v}\cdot\vec{\nabla}T. \tag{59}$$

Since $\vec{\lambda}$ is in the direction of the temperature gradient, it is convenient to define a parameter β with the same dimension as the relaxation time:

$$\vec{\lambda} = -\frac{\hbar}{T}\beta v^2\vec{\nabla}T. \tag{60}$$

Because $\vec{q} = \vec{v}\omega/v^2$, Eq. (59) can be further simplified to

$$\alpha_q = \tau_c(1 + \beta/\tau_N). \tag{61}$$

From Eq. (58) it is straightforward [the same procedure as Eqs. (42) through (49)] to show that the lattice thermal conductivity can be expressed as

$$\kappa_L = \frac{k_B}{2\pi^2 v}\left(\frac{k_B}{\hbar}\right)^3 T^3 \int_0^{\theta_D/T} \alpha_q(x)\frac{x^4 e^x}{(e^x - 1)^2}dx$$

$$= \frac{k_B}{2\pi^2 v}\left(\frac{k_B}{\hbar}\right)^3 T^3 \int_0^{\theta_D/T} \tau_c(x)\left(1 + \frac{\beta}{\tau_N(x)}\right)\frac{x^4 e^x}{(e^x - 1)^2}dx. \tag{62}$$

The task now is to determine β. Because the total crystal momentum is conserved for the N-processes, the rate of phonon momentum change is zero. Therefore

$$\int \frac{N_{\vec{q}}^{\vec{\lambda}} - N_{\vec{q}}}{\tau_N}\vec{q}d\vec{q} = 0. \tag{63}$$

Substituting Eqs. (53) and (58) into Eq. (63), we have

$$\int \frac{\exp(\hbar\omega/k_B T)}{[\exp(\hbar\omega/k_B T) - 1]^2}\left[\frac{\hbar\omega}{k_B T^2}\alpha_q(\vec{v}\cdot\vec{\nabla}T) + \frac{\vec{q}\cdot\vec{\lambda}}{k_B T}\right]\frac{\vec{q}}{\tau_N}d\vec{q} = 0. \tag{64}$$

This can be further simplified by using Eqs. (60) and (61), so that

$$\int \frac{\exp(\hbar\omega/k_B T)}{[\exp(\hbar\omega/k_B T) - 1]^2}\frac{\hbar\omega}{k_B T^2}(\vec{v}\cdot\vec{\nabla}T)(\alpha_q - \beta)\frac{\vec{v}\omega}{\tau_N v^2}d\vec{q} = 0. \tag{65}$$

By inserting Eq. (63) into Eq. (65) and using the dimensionless x as defined earlier, we can solve for β:

$$\beta = \int_0^{\theta_D/T} \frac{\tau_c}{\tau_N}\frac{e^x x^4}{(e^x - 1)^2}dx \bigg/ \int_0^{\theta_D/T} \frac{\tau_c}{\tau_N \tau_q}\frac{e^x x^4}{(e^x - 1)^2}dx. \tag{66}$$

Therefore the lattice thermal conductivity can be written as

$$\kappa_L = \kappa_1 + \kappa_2, \tag{67}$$

where

$$\kappa_1 = \frac{k_B}{2\pi^2 v}\left(\frac{k_B}{\hbar}\right)^3 T^3 \int_0^{\theta_D/T} \tau_c\frac{x^4 e^x}{(e^x - 1)^2}dx \tag{67a}$$

and

$$\kappa_2 = \frac{k_B}{2\pi^2 v}\left(\frac{k_B}{\hbar}\right)^3 T^3 \frac{(\int_0^{\theta_D/T} \frac{\tau_c}{\tau_N} \frac{x^4 e^x}{(e^x-1)^2} dx)^2}{\int_0^{\theta_D/T} \frac{\tau_c}{\tau_N \tau_q} \frac{x^4 e^x}{(e^x-1)^2} dx}. \tag{67b}$$

When the impurity level is significant and all phonon modes are strongly scattered by the resistive processes in a solid, then $\tau_N \gg \tau_q$ and $\tau_c \approx \tau_q$. Under this circumstance $\kappa_1 \gg \kappa_2$ and κ_L is given by Eq. (67a), the same as Eq. (49) since the N-processes do not exist. In the opposite extreme, when N-processes are the only phonon scattering processes, we have $\tau_N \ll \tau_q$ and $\tau_c \approx \tau_N$. The denominator of κ_2 then approaches 0, leading to infinite lattice thermal conductivity as expected because the N-processes do not give rise to thermal resistance.

Now that we have derived the formula for the lattice thermal conductivity, the problem is to calculate the relaxation times. Phonon scatterings have been treated by numerous authors. Here we list the main conclusions.

For phonon–phonon normal scattering the relaxation rate

$$\tau_N^{-1} = B\omega^a T^b \tag{68}$$

is the general form suggested by best fits to experimental thermal conductivity data; B is a constant independent of ω and T, $(a, b) = (1, 3)$ was recommended for LiF and diamond[50,51], and $(a, b) = (1, 4)$ and $(2, 3)$ were used for some group IV and III–V semiconductors.[8]

Peierls suggested the form

$$\tau_U^{-1} \propto T^n \exp(\theta_D/mT) \tag{69}$$

for the phonon–phonon umklapp scattering with constants n and m on the order of 1.[52] Based on the Leibfried and Schlömann model,[53] Slack et al. proposed the following form for the Grüneisen constant γ and the average atomic mass of M in the crystal:[54]

$$\tau_U^{-1} \approx \frac{\hbar \gamma^2}{Mv^2 \theta_D}\omega^2 T \exp(-\theta_D/3T). \tag{70}$$

Other empirical n and m values were also used,[50,55–57] all of which were based again on best fits to experimental data. At sufficiently high-temperatures $\kappa_L \propto 1/T$ if phonon–phonon umklapp scattering is the dominant process.

Klemens was the first to calculate the relaxation rate for phonon–point-defect scattering where the linear dimensions of the defects are much smaller than the phonon wavelength.[58] The corresponding phonon–point-defect scattering rate is

$$\tau_{PD}^{-1} = \frac{V}{4\pi v^3}\omega^4 \sum_i f_i \left(\frac{\bar{m} - m_i}{\bar{m}}\right)^2, \tag{71}$$

where V is the volume per atom, m_i is the mass of an atom, f_i is the fraction of atoms with mass m_i, and $\bar{m}$ is the average mass of all atoms. A strain field modification to Eq. (71) has been described by Abeles.[59]

The phonon-boundary scattering rate is independent of phonon frequency and temperature and can be written as

$$\tau_B^{-1} = v/d,$$

Sec. 3 · ELECTRONIC THERMAL CONDUCTION

where d is the sample size for a single crystal or the grain size for a polycrystalline sample.

For phonon-dislocation scattering, Nabarro separated the effects of the core from the surrounding strain field.[60] The corresponding relaxation rates are

$$\tau_{Core}^{-1} \propto N_D \frac{r^4}{v^2} \omega^3 \tag{72}$$

and

$$\tau_{Str}^{-1} \propto N_D \frac{\gamma^2 B_D^2 \omega}{2\pi}, \tag{73}$$

where N_D is the number of dislocation lines per unit area, r is the core radius, and B_D is the Burgers vector of the dislocation.

Pohl suggested an empirical nonmagnetic phonon-resonance scattering relaxation rate of

$$\tau_{Res}^{-1} = \frac{C\omega^2}{(\omega^2 - \omega_0^2)^2}, \tag{74}$$

where C is a constant proportional to the concentration of the resonant defects and ω_0 is the resonance frequency.[61] This formula accounted well for the observed low-temperature dip in the thermal conductivity of KNO_2-containing KCl crystals. It has also been used for fitting experimental data for clathrates and skutterudites.[22,23]

According to Ziman, the relaxation time for the scattering of phonons by electrons in the conduction state is given by[62]

$$\tau_{EPC}^{-1} = \frac{\varepsilon^2 (m^*)^3 v}{4\pi \hbar^4 d} \left(\frac{k_B T}{\frac{1}{2} m^* v^2} \right)$$

$$\times \left\{ \frac{\hbar \omega}{k_B T} - \ln \frac{1 + \exp[(\frac{1}{2} m^* v^2 - E_F)/k_B T + \hbar^2 \omega^2/8 m^* v^2 k_B T + \hbar \omega/2 k_B T]}{1 + \exp[(\frac{1}{2} m^* v^2 - E_F)/k_B T + \hbar^2 \omega^2/8 m^* v^2 k_B T - \hbar \omega/2 k_B T]} \right\}, \tag{75}$$

where ε is the electron–phonon interaction constant or deformation potential and d is the mass density. The relaxation time for scattering of phonons by electrons in a bound state, as given by Griffin and Carruthers,[63] is

$$\tau_{EPB}^{-1} = \frac{G\omega^4}{[\omega^2 - (4\Delta/\hbar)^2]^2} \frac{1}{[1 + r_0^2 \omega^2/4v^2]^8}, \tag{76}$$

where G is a proportionality constant containing the number of scattering centers, Δ is the chemical shift related to the splitting of electronic states, and r_0 is the mean radius of the localized state. It should be pointed out that these formulas for electron–phonon interaction are based on the adiabatic principle and perturbation theory. As argued by Ziman,[36] the theory is only valid if the mean free path of electrons L_e satisfies the condition

$$1 < qL_e. \tag{77}$$

Since the phonon wave vector $q = 2\pi/\lambda_p$, this means that the wavelength of a phonon λ_p must not be longer than the mean free path of the electron it scatters. The model originally developed by Pippard for explaining the ultrasonic attenuation in metals is, however, applicable over the entire range of qL_e.[64] Pippard's relaxation times are

$$\tau_{EP}^{-1} = \frac{4nm^*v_F L_e \omega^2}{15 d v^2} \quad \text{for} \quad qL_e \ll 1 \tag{78}$$

and

$$\tau_{EP}^{-1} = \frac{\pi n m^* v_F \omega}{6 d v} \quad \text{for} \quad qL_e \gg 1 \tag{79}$$

where n is the electron concentration and v_F is the Fermi velocity. There have been several cases in which Pippard's theory was applied to lattice thermal conductivity.[21,65,66]

Typically it is necessary to use the full Callaway model [Eqs. (67), (67a), and (67b)] to interpret experimental data on the effect of isotopes when one starts with an isotopically pure crystal. The addition of a small amount of defects will rapidly suppress κ_2, which becomes negligible for an impure sample. The decrease of κ_1 upon increasing defect concentration is much slower. For samples with an appreciable amount of defects (even isotopic defects) it is sufficient to use the Debye approximation [Eq. (49)] for examining the low-temperature lattice thermal conductivity. The measured thermal conductivity versus temperature curve is usually fit by trial and error for one sample, where an appropriate relaxation rate is carefully chosen for each scattering mechanism believed to be present. The same curves for samples with additional defects are then fit by using suitable relaxation rates to reflect scattering by additional defects. Much interesting physics has been revealed by this method. Different phonon scattering processes usually dominate in different temperature ranges. Figure 5 plots the lattice thermal conductivity of a binary

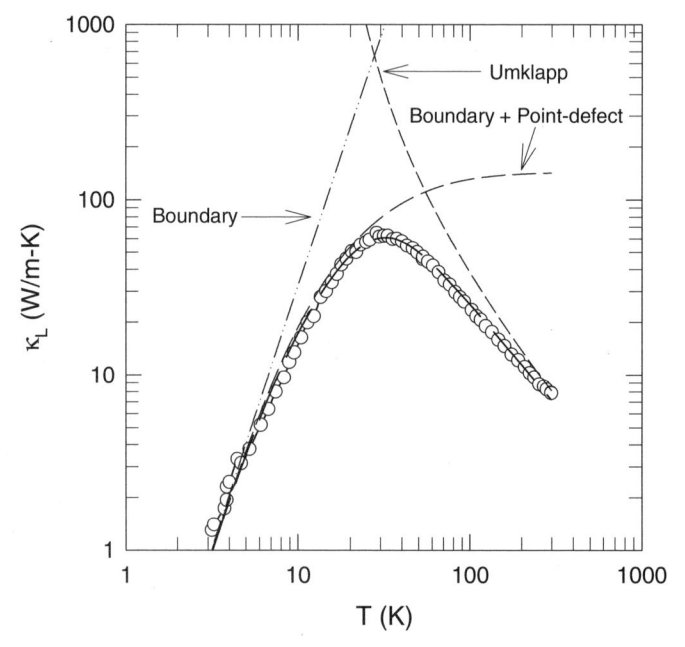

FIGURE 5 The lattice thermal conductivity versus temperature of a $CoSb_3$ sample from Ref. 21. The dots and the solid line represent the experimental data and the theoretical fit. The dashed curves are the theoretical limits imposed on the phonon heat transport by boundary scatterings, by a combination of boundary plus point-defect scatterings, and by umklapp scatterings.

skutterudite compound $CoSb_3$.[16] The dots and the solid line represent experimental data and a theoretical fit using the Debye approximation, respectively. The possible phonon scattering mechanisms are phonon-boundary, phonon-point defect, and phonon–phonon umklapp scatterings. The dashed lines in Fig. 5 correspond to the theoretical limits on the lattice thermal conductivity set by boundary scattering, a combination of boundary plus point-defect scattering, and umklapp scattering. At high-temperatures (close to $\theta_D = 300$ K), umklapp is the dominant phonon scattering mechanism, while boundary and point-defect scattering dominate at low and intermediate temperatures, respectively.

In practice, the dominant phonon method works quite well in predicting the effect of phonon scattering processes. The dominant phonon method assumes that at a given temperature all phonons are concentrated about a particular dominant frequency $\omega_{\text{dom}} \sim k_B T/\hbar$. An empirical power law can then be deduced as follows. If one particular defect can be described by $1/\tau \propto \omega^a$ or, equivalently, by $1/\tau \propto x^a T^a$, taking $C \propto T^3$ (for low T) and employing the simple kinetic formula [Eq. (5)] yields $\kappa_L \propto T^{3-a}$. Even though the dominant phonon method is not mathematically justified, the power law is usually valid at low temperatures. For example, one should have $\kappa_L \propto T^3$ at low temperatures if boundary scattering is the dominant phonon scattering mechanism.

4. SUMMARY

In this chapter theoretical treatments of the electronic and the lattice thermal conductivities at low temperatures have been reviewed. The electronic thermal resistance for degenerate electron gases is the sum of residual and ideal components. At low temperatures $W_e \approx AT + B/T^2$, where A and B are constants, while at high-temperatures W_e approaches a constant. Except for very pure metals, the Wiedemann–Franz law holds well at low and high-temperatures with standard Lorenz number L_0. For intermediate temperatures, the Wiedemann–Franz law breaks down in a way strongly dependent on the amount of the impurity. For intrinsic semiconductors, the bipolar diffusion process enhances the electronic thermal conductivity. The N- and U-processes are both important for analyzing and predicting the lattice thermal conductivity of solids. The lattice thermal conductivity for isotopically pure crystals can be well described by the full Callaway model. When impurity concentrations are appreciable, the Debye approximation is adequate for modeling experimental data. The different resistive phonon scattering processes in the Debye approximation dominate in different temperature ranges. At low temperatures, the power law predicted by the dominant phonon method is often useful in identifying the characteristics of the phonon scattering processes.

5. REFERENCES

1. D. G. ONN, A. WITEK, Y. Z. QIU, T. R. ANTHONY, and W. F. BANHOLZER *Some Aspects of the Thermal Conductivity of Isotopically Enriched Diamond Single Crystals* Phys. Rev. Lett. **68**, 2806–2809 (1992).
2. T. R. ANTHONY, W. F. BANHOLZER, J. F. FLEISCHER, L. WEI, P. K. KUO, R. L. THOMAS, and R. W. PRYOR *Thermal Diffusivity of Isotopically Enriched ^{12}C Diamond* Phys. Rev. B **42**, 1104-1111 (1990).
3. J. R. OLSON, R. O. POHL, J. W. VANDERSANDE, A. ZOLTAN, T. R. ANTHONY, and W. F. BANHOLZER

Thermal Conductivity of Diamond between 170 and 1200 K and the Isotope Effect Phys. Rev B **47**, 14850-14856 (1993).
4. L. WEI, P. K. KUO, R. L. THOMAS, T. R. ANTHONY, and W. F. BANHOLZER *Thermal Conductivity of Isotopically Modified Single Crystal Diamond* Phys. Rev. Lett. **70**, 3764-3767 (1993).
5. K. C. HASS, M. A. TAMOR, T. R. ANTHONY, and W. F. BANHOLZER *Lattice Dynamics and Raman Spectra of Isotopically Mixed Diamond* Phys. Rev. B **45**, 7171-7182 (1992).
6. R. BERMAN *Thermal Conductivity of Isotopically Enriched Diamonds* Phys. Rev B **45**, 5726-5728 (1992).
7. T. RUF, W. HENN, M. ASEN-PALMER, E. GMELIN, M. CARDONA, H.-J. POHL, G. G. DEVYATYCH, and P. G. SENNIKOV *Thermal Conductivity of Isotopically Enriched Silicon* Solid State Commun. **115**, 243 (2000).
8. D. T. MORELLI, J. P. HEREMANS, and G. A. SLACK *Estimation of the Isotope Effect on the Lattice Thermal Conductivity of Group IV and Group III-V Semiconductors* Phys. Rev B **66**, 195 304 (2002).
9. T. M. TRITT, ed., *Semiconductors and Semimetals*, Vol. **69-71** (Academic Press, San Diego, CA, 2000).
10. D. T. MORELLI and G. P. MEISNER *Low Temperature Properties of the Filled Skutterudite $CeFe_4Sb_{12}$* J. Appl. Phys. **77**, 3777-3781 (1995).
11. J.-P. FLEURIAL, A. BORSHCHEVSKY, T. CAILLAT, D. T. MORELLI, and G. P. MEISNER *High Figure of Merit in Ce-Filled Skutterudites* Proc. 16th Intl. Conf. on Thermoelectrics, (IEEE, Piscataway, NJ, 1996), p. 91-95.
12. B. C. SALES, D. MANDRUS, and R. K. WILLIAMS *Filled Skutterudite Antimonides: A New Class of Thermoelectric Materials* Science **272**, 1325-1328 (1996).
13. T. M. TRITT, G. S. NOLAS, G. A. SLACK, A. C. EHRLICH, D. J. GILLESPIE, and J. L. COHN *Low-Temperature Transport Properties of the Filled and Unfilled $IrSb_3$ Skutterudite System* J. Appl. Phys. **79**, 8412-8418 (1996).
14. G. S. NOLAS, G. A. SLACK, D. T. MORELLI, T. M. TRITT, and A. C. EHRLICH *The Effect of Rare-Earth Filling on the Lattice Thermal Conductivity of Skutterudites* J. Appl. Phys. **79**, 4002-4008 (1996).
15. G. P. MEISNER, D. T. MORELLI, S. HU, J. YANG, and C. UHER *Structure and Lattice Thermal Conductivity of Fractionally Filled Skutterudites: Solid Solutions of Fully Filled and Unfilled End Members* Phys. Rev. Lett. **80**, 3551-3554 (1998).
16. J. YANG, G. P. MEISNER, D. T. MORELLI, and C. UHER *Iron Valence in Skutterudites: Transport and Magnetic Properties of $Co_{1-x}Fe_xSb_3$* Phys. Rev. B **63**, 014410 (2000).
17. G. S. NOLAS, M. KAESER, R. T. LITTLETON IV, and T. M. TRITT *High Figure of Merit in Partially Filled Ytterbium Skutterudite Materials* Appl. Phys. Lett. **77**, 1855-1857 (2000).
18. B. C. SALES, B. C. CHAKOUMAKOS, and D. MANDRUS *Thermoelectric Properties of Thallium-Filled Skutterudites* Phys. Rev. B **61**, 2475-2481 (2000).
19. L. D. CHEN, T. KAWAHARA, X. F. TANG, T. HIRAI, J. S. DYCK, W. CHEN, and C. UHER *Anomalous Barium Filling Fraction and n-type Thermoelectric Performance of $Ba_yCo_4Sb_{12}$* J. Appl. Phys. **90**, 1864-1868 (2001).
20. G. A. LAMBERTON, JR., S. BHATTACHARYA, R. T. LITTLETON IV, M. A. KAESER, R. H. TEDSTROM, T. M. TRITT, J. YANG, and G. S. NOLAS *High Figure of Merit in Eu-filled $CoSb_3$-based Skutterudites* Appl. Phys. Lett. **80**, 598-600 (2002).
21. J. YANG, D. T. MORELLI, G. P. MEISNER, W. CHEN, J. S. DYCK, and C. UHER *Influence of Electron-Phonon Interaction on the Lattice Thermal Conductivity of $Co_{1-x}Ni_xSb_3$* Phys. Rev. B **65**, 094115 (2002).
22. J. YANG, D. T. MORELLI, G. P. MEISNER, W. CHEN, J. S. DYCK, and C. UHER *Effect of Sn Substituting Sb on the Low Temperature Transport Properties of Ytterbium-Filled Skutterudites*, Phys. Rev. B 67, 165207 (2003)
23. J. L. COHN, G. S. NOLAS, V. FESSATIDIS, T. M. METCALF, and G. A. SLACK *Glasslike Heat Conduction in High-Mobility Crystalline Semiconductors* Phys. Rev. Lett. **82**, 779-782 (1999).
24. J. S. TSE, K. UEHARA, R. ROUSSEAU, A. KER, C. I. RATCLIFFE, M. A. WHITE, and G. MACKAY *Structural Principles and Amorphouslike Thermal Conductivity of Na-Doped Si Clathrates* Phys. Rev. Lett. **85**, 114-117 (2000).
25. G. S. NOLAS and G. A. SLACK *Thermoelectric Clathrates* Am. Sci. **89**, 136-141 (2001).
26. J. DONG, O. F. SANKEY, AND W. MYLES *Theoretical Study of the Lattice Thermal Conductivity in Ge Framework Semiconductors* Phys. Rev. Lett. **86**, 2361-2364 (2001).
27. B. C. SALES, B. C. CHAKOUMAKOS, R. JIN, J. R. THOMPSON, and D. MANDRUS *Structural, Magnetic,*

Sec. 5 · REFERENCES

Thermal, and Transport Properties of $X_8Ga_{16}Ge_{30}$ *(X = Eu, Sr, Ba) Single Crystals* Phys. Rev. B **63**, 245113 (2001).
28. B. A. COOK, G. P. MEISNER, J. YANG, and C. UHER *High Temperature Thermoelectric Properties of MNiSn (M=Zr, Hf)* Proc. 18th Intl. Conf. on Thermoelectrics, (IEEE, Piscataway, NJ, 1999), pp. 64-67.
29. C. UHER, J. YANG, S. HU, D. T. MORELLI, and G. P. MEISNER *Transport Properties of Pure and Doped MNiSn (M=Zr, Hf)* Phys. Rev. B **59**, 8615-8621 (1999).
30. Q. SHEN, L. CHEN, T. GOTO, T. HIRAI, J. YANG, G. P. MEISNER and C. UHER *Effects of Partial Substitution of Ni by Pd on the Thermoelectric Properties of ZrNiSn-Based Half-Heusler Compounds* Appl. Phys. Lett. **79**, 4165-4167 (2001).
31. S. BHATTACHARYA, Y. XIA, V. PONNAMBALAM, S. J. POON, N. THADANI, and T. M. TRITT *Reductions in the Lattice Thermal Conductivity of Ball-Milled and Shock Compacted* $TiNiSn_{1-x}Sb_x$ *Half-Heusler Alloys* Mat. Res. Soc. Symp. Proc. **691**, G7.1 (2002).
32. I. TERASAKI, I. TSUKADA, Y. IGUCHI *Impurity-Induced Transition and Impurity-Enhanced Thermopower in the Thermoelectric Oxide* $NaCo_{2-x}Cu_xO_4$ Phys. Rev. B **65**, 195106 (2002).
33. I. TERASAKI, Y. ISHII, D. TANAKA, K. TAKAHATA and Y. IGUCHI *Thermoelectric Properties of* $NaCo_{2-x}Cu_xO_4$ *Improved by the Substitution of Cu for Co* Japanese J. Appl. Phys, Lett., **40**, L65-7 (2001).
34. R. FUNAHASHI and M. SHIKANO $Bi_2Sr_2Co_2O_y$ *Whiskers with High Thermoelectric Figure of Merit* Appl. Phys. Lett. **81**, 1459-1461 (2002).
35. P. G. KLEMENS *Thermal Conductivity and Lattice Vibrational Modes*, in *Solid State Physics*, Vol. 7, edited by F. Seitz and D. Turnbull, (Academic Press, New York, 1958), pp. 1-98.
36. J. M. ZIMAN *Electrons and Phonons* (Clarendon Press, Oxford, UK, 1960).
37. R. BERMAN *Thermal Conduction in Solids* (Clarendon Press, Oxford, UK, 1976).
38. G. A. SLACK *The Thermal Conductivity of Nonmetallic Crystals*, in *Solid State Physics*, Vol. **34**, edited by H. Ehrenreich, F. Seitz, and D. Turnbull (Academic Press, New York, 1979), pp. 1-71.
39. N. W. ASHCROFT and N. D. MERMIN, *Solid State Physics* (Saunders, Philadelphia, 1976).
40. G. K. WHITE *Thermal Conductivity of Silver at Low Temperatures* Proc. Phys. Soc. London **A66**, 844-845 (1953).
41. P. J. PRICE *Ambipolar Thermodiffusion of Electrons and Holes in Semiconductors* Phil. Mag. **46**, 1252-1260 (1955).
42. J. R. DRABBLE and H. J. GOLDSMID, *Thermal Conduction in Semiconductors* (Pergamon Press, Oxford, UK, 1961).
43. G. BUSCH and M. SCHNEIDER *Heat Conduction in Semiconductors* Physica **20**, 1084-1086 (1954).
44. C. J. GLASSBRENNER and G. A. SLACK *Thermal Conductivity of Silicon and Germanium from 3 K to the Melting Point* Phys. Rev **134**, A1058-A1069 (1964).
45. C. UHER and H. J. GOLDSMID *Separation of the Electronic and Lattice Thermal Conductivities in Bismuth Crystals* Phys. Stat. Sol. (b) **65**, 765-772 (1974).
46. I. POMERANCHUK *On the Thermal Conductivity of Dielectrics* Phys. Rev. **60**, 820-821 (1941).
47. J. S. BLAKEMORE *Solid State Physics*, 2nd ed., (Cambridge University Press, Cambridge, UK, 1985).
48. J. CALLAWAY *Model for Lattice Thermal Conductivity at Low Temperature* Phys. Rev. **113**, 1046-1051 (1959).
49. P. G. KLEMENS *Thermal Conductivity of Solids at Low Temperature*, in *Encyclopedia of Physics*, Vol. **14**, edited by S. Flügge (Springer-Verlag, Berlin, 1956), pp. 198-281.
50. R. BERMAN and J. C. F. BROCK *The Effect of Isotopes on Lattice Heat Conduction. I: Lithium Fluoride* Proc. R. Soc. **A289**, 46-65 (1965).
51. N. V. NOVIKOV, A. P. PODOBA, S. V. SHMEGARA, A. WITEK, A. M. ZIATSEV, A. B. DENISENKO, W. R. FAHRNER, and M. WERNER *Influence of Isotopic Content on Diamond Thermal Conductivity* Diamond Rel. Mat. **8**, 1602-1606 (1999).
52. R. PEIERLS *Kinetic Theory of Thermal Conduction in Dielectric Crystals* Ann. Phys., Leipzig **3**, 1055 (1929).
53. G. LEIBFRIED and E. SCHLÖMANN *Thermal Conductivity of Dielectric Solids by a Variational Technique* Nachr. Akad. Wiss. Göttingen II **a(4)**, 71 (1954).
54. G. A. SLACK and S. GALGINAITIS *Thermal Conductivity and Phonon Scattering by Magnetic Impurities in CdTe* Phys. Rev. **133**, A253-A268 (1964).
55. R. O. POHL *Influence of F Centers on the Lattice Thermal Conductivity in LiF* Phys. Rev. **118**, 1499-1508 (1960).
56. B. K. AGRAWAL and G. S. VERMA *Lattice Thermal Conductivity of Solid Helium* Phys. Rev. **128**, 603-605 (1962).

57. R. M. KIMBER and S. J. ROGERS *The Transport of Heat in Isotopic Mixtures of Solid Neon: An Experimental Study Concerning the Possibility of Second-Sound Propagation* J. Phys. C **6**, 2279-2293 (1973).
58. P. G. KLEMENS *Scattering of Low Frequency Phonons by Static Imperfections* Proc. Phys. Soc. London **A68**, 1113-1128 (1955).
59. B. ABELES *Lattice Thermal Conductivity of Disordered Semiconductors at High Temperatures* Phys. Rev. **131**, 1906-1911 (1963).
60. F. R. N. NABARRO *The Interaction of Screw Dislocations and Sound Waves* Proc. R. Soc. **A209**, 278-290 (1951).
61. R. O. POHL *Thermal Conductivity and Phonon Resonance Scattering* Phys. Rev. Lett. **8**, 481-483 (1962).
62. J. M. ZIMAN *The Effect of Free Electrons on Lattice Conduction* Phil. Mag. **1**, 191-198 (1956); corrected Phil. Mag. **2**, 292 (1957).
63. A. GRIFFIN and P. CARRUTHERS *Thermal Conductivity of Solids.IV: Resonance Fluorescence Scattering of Phonons by Electrons in Germanium* Phys. Rev. **131**, 1976-1992 (1963).
64. A. B. PIPPARD *Ultrasonic Attenuation in Metals* Phil. Mag. **46**, 1104-1114 (1955).
65. P. LINDENFELD and W. B. PENNEBRAKER *Lattice Thermal Conductivity of Copper Alloys* Phys. Rev. **127**, 1881-1889 (1962).
66. L. G. RADOSEVICH and W. S. WILLIAMS *Thermal Conductivity of Transition Metal Carbides* J. Am. Ceram. Soc. **53**, 30-33 (1970).

Chapter 1.2

THERMAL CONDUCTIVITY OF METALS

Ctirad Uher

Department of Physics, University of Michigan, Ann Arbor, MI 48109, USA

1. INTRODUCTION

Metals represent a vast number of materials that have been the backbone of industrial development during the past two centuries. The importance of this development upon future technological progress is undiminished and unquestionable. Whether in pure, elemental form or as new lightweight, high-strength alloys, metals are simply indispensable to modern industrial society. Metals are, of course, known for their lustrous and shiny appearance, for their malleability and ductility and, above all, for their ability to conduct electric current. Because of their myriad applications, it is highly desirable to be able to tailor the properties of metals to match and optimize their use for specific tasks. Often among the important criteria is how well a given metal or alloy conducts heat.

The physical parameter that characterizes and quantifies the material's ability to conduct heat is called thermal conductivity, often designated by κ. Understanding the nature of heat conduction process in metals and being able to predict how well a particular alloy will conduct heat are issues of scientific and technological interest. In this section we review the fundamental physical principles that underscore the phenomenon of heat conduction in metals, develop an understanding of why some metals are better than others in their ability to conduct heat, and illustrate the behavior of thermal conductivity on several specific examples encompassing pure metals and alloys.

1.1. Carriers of Heat in Metals

Metals are solids[*] and as such they possess crystalline structure where the ions (nuclei with their surrounding shells of core electrons) occupy translationally equivalent positions in the crystal lattice. We know see Chapter (1.1) that crystalline lattices support heat flow when an external thermal gradient is imposed on the structure. Thus, like every other solid material, metals possess a component of heat conduction associated with the vibrations of the lattice called lattice (or phonon) thermal conductivity, κ_p. The unique feature of metals as far as their structure is concerned is the presence of charge carriers, specifically electrons. These point charge entities are responsible not just for the transport of charge (i.e., for the electric current) but also for the transport of heat. Their contribution to the thermal conductivity is referred to as the electronic thermal conductivity κ_e. In fact, in pure metals such as gold, silver, copper, and aluminum, the heat current associated with the flow of electrons by far exceeds a small contribution due to the flow of phonons, so, for all practical purposes (and essentially for the entire regime of temperatures from sub-Kelvin to the melting point), the thermal conductivity can be taken as that due to the charge carriers. In other metals (e.g., transition metals and certainly in alloys) the electronic term is less dominant, and one has to take into account the phonon contribution in order to properly assess the heat conducting potential of such materials.

In discussions of the heat transport in metals (and in semiconductors), one makes an implicit and essential assumption that the charge carriers and lattice vibrations (phonons) are independent entities. They are described by their respective unperturbed wave functions, and any kind of interaction between the charge carriers and lattice vibrations enters the theory subsequently in the form of transitions between the unperturbed states. This suggests that one can express the overall thermal conductivity of metals (and other solids) as consisting of two independent terms, the phonon contribution and the electronic contribution:

$$\kappa = \kappa_e + \kappa_p. \qquad (1)$$

These two—electrons and phonons—are certainly the main heat carrying entities. However, there are other possible excitations in the structure of metals, such as spin waves, that may, under certain circumstances, contribute a small additional term to the thermal conductivity. For the most part, we shall not consider these small and often conjectured contributions.

It is important to recognize that the theory of heat conduction, whether in nonmetallic or metallic systems, represents an exceptionally complex many-body quantum statistical problem. As such, it is unreasonable to assume that the theory will be able to describe all the nuances in the heat conduction behavior of any given material or that it will predict a value for the thermal conductivity that exactly matches that obtained from experiment[†]. What one really hopes for is to capture the general trend in the behavior of the thermal conductivity, be it among the group of materials or in regard to their temperature dependence, and have some reasonably reliable guidelines and perhaps predictive power as to whether a particular

---

[*] We consider here metals in their solid form only.

[†] Measurements of thermal conductivity are among the most difficult of the transport studies, and the accuracy of the data itself is usually no better than 2%.

Sec. 1 · INTRODUCTION

class of solids is likely to be useful in applications where heat carrying ability is an important concern or consideration.

The theory of heat conduction has been outlined in Chapter 1.1. There are numerous other monographs and review articles where this topic is treated in detail. I am particularly fond of the following texts: Berman's *Thermal Conduction in Solids*,[1] Ashcroft and Mermin's *Solid State Physics*,[2] Ziman's *Electrons and Phonons*,[3] Smith and Jensen's *Transport Phenomena*,[4] and Mahan's *Many-Particle Physics*.[5] The level of coverage varies from an introductory treatment to a comprehensive quantum statistical description. On one hand, a too elementary description is unlikely to provide much more beyond an outline of the phenomenon; on the other hand, a many-body quantum statistical exposure is likely to overwhelm most readers and, more often than not, its results are very difficult to apply in practice or use as a guide. I will therefore settle here on a treatment that captures the essential physics of the problem, provides the formulas that describe the behavior of thermal conductivity in metallic systems, and has some predictive power regarding the choice of materials in various applications where heat conduction is an important issue.

Because of the obvious practical relevance of noble metals such as copper, gold and silver and of various alloys containing copper, there has always been a strong interest in comparing the current and heat conducting characteristics of these materials. It is therefore no surprise that certain important empirical relations were discovered nearly 50 years before the concept of an electron was firmly established and some 80 years before the notion of band structure emerged from the quantum theory of solids. Such is the case of the Wiedemann–Franz law,[6] which states that, at a given temperature, the thermal conductivity of a reasonably pure metal is directly related to its electrical conductivity; in other words, by making a simple measurement of electrical resistivity (conductivity σ equals inverse resistivity ρ^{-1}), we essentially know how good such a metal will be as a heat conductor,

$$\frac{\kappa}{\sigma} \equiv \kappa\rho = LT. \qquad (2)$$

In many respects, the Wiedemann–Franz law and the constant L that relates the thermal conductivity of pure metals to their electrical conductivity, called the Lorenz number,[7] are the fundamental tenets in the theory of heat conduction in metals. Most of our discussion will focus on the conditions under which this law is valid and on deviations that might arise. Such knowledge is of great interest: on one hand it allows a fairly accurate estimate of the electronic thermal conductivity without actually performing the rather tedious thermal conductivity measurements; on the other hand, deviations from the Wiedemann–Franz law inform on the presence and strength of particular (inelastic) scattering processes that might influence the carrier dynamics.

I wish to stress right here that the Wiedemann–Franz law strictly compares the *electronic thermal conductivity* with the electrical conductivity. In metals that have a substantial phonon contribution to the overall thermal conductivity (impure metals and alloys), forming the ratio with the measured values of the thermal and electrical conductivities will naturally lead to a larger magnitude of the constant L because the overall thermal conductivity will contain a significant contribution from phonons.

1.2. The Drude Model

The first important attempt in the theoretical understanding of transport processes in metals is due to Paul Drude.[8] This comes merely three years after J. J. Thomson[9] discovered the electron, an elementary particle that clearly fits the role of a carrier of charge in metals regardless of the fact that the convention for the positive current direction would much prefer a positively charged carrier.

The classical free-electron model of metals developed by Drude builds on the existence of electrons as freely moving noninteracting particles that navigate through a positively charged background formed by much heavier and immobile particles. At his time, Drude had no idea of the shell structure of atoms—nowadays we might be more specific and say that the gas of electrons is formed from the conduction or valence electrons that are stripped from atoms upon the formation of a solid while the positively charged ions (nuclei surrounded by the core electrons) are the immobile particles located at the lattice sites. In any case, the classical electron gas is then described using the language of kinetic theory. In this picture, the straight-line thermal motion of electrons is interrupted by collisions with the lattice ions. Collisions are instantaneous events in which electrons abruptly change their velocity and completely "forget" what their direction of motion was just prior to the collision. Moreover, it is assumed that thermalization occurs only in the process of collisions; i.e., following a collision, although having a completely random direction of its velocity, the speed of the electron corresponds to the temperature of the region where the collision occurred.

In defining conductivities, it is advantageous first to introduce the respective current densities. Since electrons carry both charge and energy, their flow implies both electric and heat currents. The term *electric current density*, designated $\mathbf{J}_e$, is understood to represent the mean electric charge crossing a unit area perpendicular to the direction of flow per unit time. Let us assume a current flows in a wire of cross-sectional area A as a result of voltage applied along the wire. If n is the number of electrons per unit volume and all move with the same average velocity v, then in time dt the charge crossing the area A is $-nevAdt$. Hence, the current density is

$$J_e = -ne\bar{v}. \tag{3}$$

In the absence of driving forces (electric field or thermal gradient), all directions of electron velocity are equally likely and the average velocity is zero; hence there is no flow of charge or energy.

If an electric field ε is applied to an electron gas, electrons accelerate between the collisions and acquire a mean (drift) velocity directed opposite to the field:

$$\mathbf{a} \equiv \frac{d\mathbf{v}}{dt} = -\frac{e\mathbf{\varepsilon}}{m}. \tag{4}$$

Integrating Eq. (4), one obtains

$$\mathbf{v}(t) = -\frac{e\mathbf{\varepsilon}}{m}t + \mathbf{v}(0) \tag{5}$$

where $\mathbf{v}(0)$ is the velocity at $t=0$, i.e., right after the collision at which time the electric field has not yet exerted its influence. The apparently unlimited rise in the velocity $\mathbf{v}(t)$ with time is checked by a subsequent collision in which the electron is restored to a local thermal equilibrium. There will be a range of times between

Sec. 1 · INTRODUCTION

collisions, so to find out the mean value of the velocity over all possible times t between collisions, we need to know how these times are distributed. The probability that an electron which has not collided with an ion during the time t will now suffer a collision in the time interval $t + \Delta t$ is

$$P(t)dt = \frac{e^{-t/\tau}}{\tau} dt. \tag{6}$$

The mean velocity then follows from

$$\bar{v} = -\frac{e\varepsilon}{m}\bar{t} = -\frac{e\varepsilon}{m}\int t\frac{e^{-t/\tau}}{\tau}dt = -\frac{e\tau}{m}\varepsilon; \tag{7}$$

i.e., the mean time $\bar{t}$ between collisions is equal to τ, the parameter frequently called the collision time or relaxation time. After Eq. (7) is substituted into Eq. (3), the electric current density becomes

$$\boldsymbol{J}_e = \frac{ne^2\varepsilon\tau}{m} = \sigma\varepsilon, \tag{8}$$

from which one obtains the familiar expression for the electrical conductivity:

$$\sigma = \frac{ne^2\tau}{m}. \tag{9}$$

Thus, the Drude Theory predicts the correct functional form for Ohm's law. By defining the mean free path between the collisions l_e as

$$l_e = \bar{v}\tau, \tag{10}$$

We can write Eq. (9) as

$$\sigma = \frac{ne^2 l_e}{m\bar{v}}. \tag{11}$$

In Drude's time, the mean speed of electrons was calculated by using the equipartition theorem, namely

$$\frac{1}{2}m(\bar{v})^2 = \frac{3}{2}k_B T. \tag{12}$$

Thus, with the obtained mean velocity and using the experimental values of the electrical conductivity, the mean free path of metals at room temperature invariably came out in the range 1–5Å, seemingly providing an excellent support for Drude's model that assumed frequent collisions of electrons with ions on the lattice sites.

Similar to the electric current density, one can define the thermal current density $\boldsymbol{J}_Q$ as a vector parallel to the direction of heat flow with a magnitude equal to the mean thermal energy per unit time that crosses a unit area perpendicular to the flow. Because the speed of an electron relates to the temperature of the place where the electron suffered the most recent collision, the hotter the place of collision the more energetic the electron. Thus, electrons arriving at a given point from the hotter region of the sample will have higher energy than those arriving at the same point from the lower-temperature region. Hence, under the influence of the thermal gradient there will develop a net flux of energy from the higher-temperature end to the lower-temperature side. If we take $n/2$ as both the number of electrons per unit volume arriving from the higher-temperature side and the same density of electrons arriving from the lower-temperature region, and assuming that the thermal gradient is small (i.e., the temperature change over a distance equal to the collision length is negligible), it is easy to show that the kinetic theory yields for the heat current density the expression

$$J_Q = \frac{1}{3}\ell_e \bar{v} c_v (-\nabla T) = -\kappa_e \nabla T. \tag{13}$$

Here ℓ_e is the mean free path of electrons, c_v is their electronic specific heat per unit volume, and κ_e is the thermal conductivity of electrons. Equation (13) is a statement of the Fourier law with the electronic thermal conductivity given as

$$\kappa_e = \frac{1}{3}\ell_e \bar{v} c_v. \tag{14}$$

Relying again on the classical description of the electron gas, Drude writes the specific heat of electrons per unit volume c_v in terms of the molar specific heat c_m:

$$c_v = c_m \frac{n}{N_A} = \frac{3}{2} N_A k_B \left(\frac{n}{N_A}\right) = \frac{3}{2} k_B n. \tag{15}$$

Substituting into Eq. (14) yields the classical expression for the thermal conductivity of electrons

$$\kappa_e = \frac{1}{3}\ell_e \bar{v} \frac{3}{2} k_B n = \frac{n k_B \ell_e \bar{v}}{2}. \tag{16}$$

By forming the ratio κ_e/σ, one arrives at the Wiedemann–Franz law:

$$\frac{\kappa_e}{\sigma} = \frac{n k_B \ell_e \bar{v}/2}{n e^2 \ell_e / m \bar{v}} = \frac{k_B m (\bar{v})^2}{2 e^2} = \frac{3}{2}\left(\frac{k_B}{e}\right)^2 T. \tag{17}$$

In the last step in Eq. (17) the mean square velocity is taken in its Maxwell–Boltzmann form $3k_B T/m$.

Equation (17) is a crowning achievement of the Drude theory—it accounts for an empirically-well-established relation between thermal and electrical conductivities of metals, and the numerical factor $\frac{3}{2}(k_B/e)^2 = 1.1 \times 10^{-8} \text{V}^2/\text{K}^2$ is only a factor of 2 or so smaller than the experimental data[*].

Overall, the success of the Drude theory appears impressive, especially when viewed from the perspective of the early years of the 1900s, and for the next three decades it formed the basis of understanding of the transport properties of metals. In retrospect, the success of this theory is quite fortuitous and stems from a lucky cancellation of a couple of large and grossly erroneous parameters that have their origin in the use of classical statistics. The most blatant error comes from the classically calculated value of the electronic specific heat, which grossly overestimates the actual electronic contribution to the heat capacity of metals. With its value of $3R/2$ per mole, the classical electronic term is comparable to that due to the lattice and would result in far too large a specific heat of metals. Experimentally, no such classical electronic contribution was ever seen. This outstanding puzzle in the classical theory of metals lasted until it was recognized that electrons

---

[*] In his original paper, Drude makes a mistake in his evaluation of the electrical conductivity, which comes out only one-half of what Eq. (11) implies. With such an erroneous result his value for the Lorenz constant is 2.2×10^{-8} V^2/K^2 in excellent agreement with experimental values.

are fermions and their properties must be accounted for with Fermi–Dirac statistics.

In the expression for the Wiedemann–Franz law, Eq. (17), this large error is compensated by a comparably large error arising from the mean square electronic speed which, in the classical Maxwell–Boltzmann form, is a factor of 100 smaller than the actual (Fermi) velocity of electrons. The fortuitous cancellation of these two large errors leaves the Lorenz constant approximately correct except for its numerical factor of 3/2. With Fermi–Dirac statistics this numerical factor becomes equal to $\pi^2/3$, in which case it is often stated that the Lorenz constant assumes its Sommerfeld value:

$$L_0 = \frac{\pi^2}{3}\left(\frac{k_B}{e}\right)^2 = 2.44 \times 10^{-8}\, \text{V}^2\text{K}^{-2}. \tag{18}$$

Such fortuitous cancellation does not extend to the electrical conductivity in Eq. (11) and classical statistics is responsible for underestimating the mean free path by a factor of 100. In reality, l_e should be on the order of several hundred angstroms rather than a few angstroms, as obtained by Drude. Actually there is another reason why Drude's viewpoint of very frequent scattering of electrons on ions is not a good physical picture. With the advent of quantum mechanics it was soon recognized that, because the electrons have a wave nature (apart from their corpuscular character), a perfectly periodic lattice presents no resistance to the current of electrons and the electrical conductivity in that case should be infinite. It is only because of deviations from perfect periodicity that the electrical conductivity attains its finite value. Imperfections can be of either dynamic nature and intrinsic to the structure, such as displacements of ions about the lattice sites due to thermal vibrations, or they can have a static character, such as impurities, vacancies, interstitials, and other structural defects.

Another less than satisfactory outcome of the Drude theory that is a direct consequence of the classical treatment concerns the predicted temperature dependence of electrical resistivity. With the Maxwell–Boltzmann mean electron velocity, $v = (3k_BT/m)^{1/2}$, the temperature dependence of resistivity is expected to be proportional to

$$\rho = \frac{1}{\sigma} = \frac{m\bar{v}}{ne^2\ell_e} \sim T^{1/2}. \tag{19}$$

Experimentally, the resistivity of metals at temperatures above the liquid nitrogen temperature is typically a linear function of temperature. Again, the discrepancy with the classical Drude model is eliminated when quantum mechanics is used, namely when the mean velocity is taken as the Fermi velocity.

The fact that we talk about the Drude model some 100 years after its inception indicates that, in spite of its shortcomings, the model provided convenient and compact expressions for electrical and thermal conductivities of metals and, with properly calculated parameters, offered a useful measure of comparison between important transport properties. Table 1 presents thermal and electrical conductivities of metals together with their Lorenz ratio, all referring to a temperature of 273 K.

TABLE 1 Thermal Conductivity of Pure Metals at 273 K.[a]

Metal	κ (W/m-K)	ρ (10^{-8} Ωm)	L (10^{-8} V^2/K^2)	Ref.
Ag	436	1.47	2.34	11
Al	237	2.43	2.10	12
Au	318	2.03	2.39	13
Ba	23.3	29.8	2.55	14
Be	230	2.8	2.36	15
Ca	186	3.08	2.13	16
Cd	100	6.80	2.49	15
Ce	11.2 (291 K)	80.0	~ 3.5	17
Co	99 (300 K)	5.99	1.98 (300 K)	18
Cr	95.7 (280 K)	11.8	4.11	19
Cs	37 (295 K)	18.0	2.5 (295 K)	20
Cu	402 (300 K)	1.73 (300 K)	2.31 (300 K)	21
Dy	10.4 (291 K)	93	3.75 (291 K)	22
Er	13.8 (291 K)	79	3.75 (291 K)	22
Fe	80.2 (280 K)	8.64	2.57 (280 K)	23
Ga ($\parallel c$)	16.0	50.3	2.95	19
($\parallel a$)	41.0	16.1	2.41	19
($\parallel b$)	88.6	7.5	2.43	24
Gd	9.1 (291 K)	128	4.2 (291 K)	22
Hf	22.4 (293 K)	31.0 (293 K)	2.45 (293 K)	25
Hg ($\parallel$)	34.1 (197 K)	14.6 (197 K)	2.53 (197 K)	26
($\perp$)	25.9 (197 K)	19.3 (197 K)	2.55 (197 K)	26
Ho	11.8 (300 K)	78.0 (300 K)	3.2 (300 K)	27
In	81.0 (280 K)	8.25 (280 K)	2.39 (280 K)	28
Ir	149 (277 K)	4.70	2.57 (277 K)	29
K	98.5	6.20	2.24	30
La	14.0 (291 K)	59	2.9	22
Li	65	8.5	2.05	31
Lu	16.2 (291 K)	~ 50	3.3 (291 K)	22
Mg	153 (301 K)	4.5 (301 K)	2.29 (301 K)	32
Mn (α)	7.8 (291 K)	137	4.0 (291 K)	22
Mo	143	4.88	2.56	33
Na	142	4.29	2.23	34
Nb	51.8 (280 K)	13.3	2.53 (280 K)	35
Nd	16.5 (291 K)	58	3.7 (291 K)	22
Ni	93 (280 K)	6.24	2.19 (280 K)	36
Os	87 (323 K)	8.3	2.7 (323 K)	29
Pb	35.5	19.2	2.50	37
Pd	71.7	9.74	2.57	38
Pr	12.8	65	3.1 (280 K)	39
Pt	71.9 (280 K)	9.82	2.59 (280 K)	23
Pu	5.2 (298 K)	~ 130	2.48 (298 K)	40
Rb	55.8	11.3	2.30	41
Re	49	16.9	3.05	42
Rh	153 (280 K)	4.35	2.46 (280 K)	43
Ru	110 (280 K)	6.7	2.72 (280 K)	43
Sc	21.8	44	4.3	44
Sm	13.4 (291 K)	90	4.3 (291 K)	22
Sn	64	10.6	2.48	15
Sr	51.9	11.0	2.18	14
Ta	57.7 (280 K)	12.1	2.56 (280 K)	45
Tb	10.4 (291 K)	110	4.25 (291 K)	22
Tc	51 (300 K)	16.7	~ 3.4 (300 K)	46

Sec. 2 · SPECIFIC HEAT OF METALS

Metal	κ (W/m-K)	ρ (10^{-8} Ωm)	L (10^{-8} V^2/K^2)	Ref.
Th	49.3	13.9	2.56	47
Ti	22.3	40	3.25	48
Tl	50.6	15	2.8	49
U	28 (278 K)	24	2.8 (278 K)	50
V	35 (260 K)	18.9	2.4 (260 K)	51
W	183 (280 K)	4.85	3.27 (280 K)	23
Y	15.9 (291 K)	~ 52	2.9 (291 K)	22
Zn	114.5 (283 K)	5.5	2.31 (283 K)	52
Zr	20.5 (323 K)	39	3.4 (323 K)	53
As (*pc*)	27 (298 K)	31.7 (298 K)	2.87 (298 K)	54
($\perp$ *c*-axis)	51.0 (300 K)	28.4 (300 K)	4.83 (300 K)	55
($\parallel$ *c*-axis)	29.0 (300 K)	38.6 (300 K)	3.73 (300 K)	55
Bi (*pc*)	7.8 (298 K)	148 (298 K)	3.87 (298 K)	56
($\perp$ *c*-axis)	9.8 (300 K)	112 (300 K)	3.65 (300 K)	57
($\parallel$ *c*-axis)	6.0 (300 K)	135 (300 K)	2.69 (300 K)	57
Sb	18.2	43	2.87	58

[a.] The data also include values of electrical resistivity and the Lorenz number. Most of the data are taken from the entries in the extensive tables in Landolt-Börnstein, New Series[10].

2. SPECIFIC HEAT OF METALS

One of the major problems and an outstanding puzzle in the Drude picture of metals was a much too large contribution electrons were supposed to provide toward the specific heat of metals. While the classical treatment invariably suggested an additional $3R/2$ contribution to the molar specific heat of metals (on top of the $3R$ term due to the lattice vibrations), the experimental evidence was unequivocal: the specific heat of simple solids at and above room temperature did not distinguish between insulators and metals, and the measurements yielded values compatible with the Dulong–Petit law; i.e., the molar specific heat of $3R = 6$ cal mole^{-1}K^{-1} = 25 J mole^{-1}K^{-1}. This puzzle was solved when electrons were treated as quantum mechanical objects and described in terms of Fermi–Dirac statistics. We briefly outline here the key steps.

Having half-integral spin, electrons are classified as Fermi particles (fermions) and are subject to the Pauli exclusion principle. As intrinsically indistinguishable particles, symmetry requirements imposed on the wave function of electrons are such that no two electrons are allowed to share the same quantum state, i.e., to possess identical sets of quantum numbers. This fundamental restriction has far-reaching consequences as to how to treat the electrons, how to "build up" a metal, and, for that matter, how to assemble elements in the periodic table. The statistics that guarantee that the Pauli principle is obeyed were developed by Fermi and Dirac, and the distribution of fermions governed by these statistics is referred to as the Fermi–Dirac distribution $f(E, T)$:

$$f(E, T) = \frac{1}{e^{(E-\zeta)/k_B T} + 1}. \tag{20}$$

Here ζ is the chemical potential, often referred to as the Fermi level or Fermi energy and written as E_F instead of ζ. Equation (20) expresses the probability of finding an electron in a particular energy state E and has a very interesting property. At $T=0$ K, the function $f(E,T)$ is a step function that sharply divides the occupied states from the unoccupied states, with the Fermi energy playing the demarcation line between the two. At $E < E_F$, all available single-electron states are fully occupied while for $E > E_F$ all states are empty. The Fermi energy is thus the highest occupied energy state. As the temperature increases, a certain degree of "smearing" takes place at the interface between the occupied and unoccupied states, with states close to but below E_F now having a finite probability of being unoccupied, while states close to but slightly above E_F have a finite chance to be occupied. The range of energies over which the distribution is smeared out is fairly narrow and amounts to only about $4k_BT$ around the Fermi level. Since the Fermi energy of typical metals is several electron–volts (eV) while the value of $4k_BT$ at 300 K is only about 100 meV, the smearing of the distribution is indeed small on the fundamental energy scale of metals. Nevertheless, this lack of sharpness in the distribution of electrons is all-important for their physical properties. Because the deep lying electron states are fully occupied with no empty states in the vicinity, electrons occupying such states cannot gain energy nor can they respond to external stimuli such as an electric field or thermal gradient. Only electrons within the smeared-out region of the distribution have empty states accessible to them and can thus absorb the energy. Of the entire number of free electrons constituting a metal, it is thus only a fraction $4k_BT/E_F$ (~0.01) of them that are involved in the specific heat. This is the reason why one does not detect electronic contribution to the specific heat of metals at and above ambient temperatures; the contribution of electrons is simply too small on the scale of the Dulong–Petit law.

To calculate the specific heat due to electrons exactly within the single-electron approximation, apart from their distribution function, one also needs to know the number of energy states per unit volume in a unit energy interval, i.e., the density of states, $D(E)$. This quantity is model dependent. Regardless of the model chosen for $D(E)$, its product with the distribution function integrated over all energies of the system must yield the density of electrons per unit volume:

$$n = \int_0^\infty D(E)f(E,T)dE. \qquad (21)$$

When electrons are heated from absolute zero temperature to some finite temperature T, they absorb energy and increase their internal energy by

$$U(T) = \int_0^\infty ED(E)f(E,T)dE - \int_0^{E_F^0} ED(E)dE. \qquad (22)$$

The second integral in Eq. (22) represents the internal energy of electrons $U(0)$ at $T=0$ K. Electrons have this non-zero internal energy at absolute zero purely because they obey the Pauli exclusion principle, and only two electrons out of n electrons (one with spin up and one with spin down) can occupy the lowest energy state. All other electrons have to be accommodated on the progressively higher energy levels, hence a significant internal energy even at $T=0$ K. By differentiating with respect to temperature and employing a trick of subtracting a term the value of which is zero but the presence of which allows one to express the result in a

Sec. 2 · SPECIFIC HEAT OF METALS

convenient form,* we obtain

$$c_v = \int_0^\infty (E - E_F)D(E)\frac{\partial f(E,T)}{\partial T}dE. \tag{23}$$

Assuming that the density of states does not vary much over the narrow temperature range where the distribution function is smeared out around the Fermi energy, (i.e., taking the density of states at its value of the Fermi energy E_F and carrying out the partial derivative of the distribution function), one obtains

$$c_v = k_B^2 TD(E_F) \int_{-\infty}^{+\infty} \frac{x^2 e^x}{(e^x + 1)^2}dx = \frac{\pi^2}{3}k_B^2 TD(E_F). \tag{24}$$

This is a general result for the specific heat per unit volume of electrons, independent of a particular form of the density of states. For a free-electron gas, it is easy to show that the density of states becomes

$$D(E) = \frac{(2m)^{3/2}}{2\pi^2\hbar^3}E^{1/2}. \tag{25}$$

In that case, the density of electrons per unit volume can be conveniently expressed as

$$n = \frac{2}{3}D(E_F)E_F, \tag{26}$$

and with the definition of the Fermi temperature $T_F = E_F/k_B$, the heat capacity per unit volume becomes

$$c_v = \frac{\pi^2}{2}nk_B\frac{T}{T_F}. \tag{27}$$

Comparing with Eq. (15), one sees that the effect of Fermi–Dirac statistics is to lower the specific heat of electrons by a factor of $(\pi^2/3)(T/T_F)$. Because the Fermi temperature of metals is very high, $\sim 10^4$–10^5 K, this factor is less than 0.01 even at room temperature.

To get from the specific heat per unit volume into a more practically useful quantity—specific heat per mole of the substance—Eq. (27) must be multiplied by the volume per mole, ZN_A/n, where ZN_A is the number of electrons in a mole of metal of valence Z and N_A stands for Avogadro's number:

$$C = \frac{\pi^2}{2}Zk_BN_A\frac{T}{T_F} = \frac{\pi^2}{2}ZR\frac{T}{T_F} = \gamma T. \tag{28}$$

In Eq. (28), R is the universal gas constant and γ, the coefficient of the linear specific heat of metals, is often called the Sommerfeld constant. Although the magnitude of the electronic specific heat is small, and certainly at ambient temperature it is negligible compared to the lattice contribution, it gains importance at low

---

* Because of the high degeneracy of electrons, the electron density n in Eq. (21) is a constant and by multiplying it by the value of the Fermi energy, the product EFn remains temperature independent. Differentiating it with respect to temperature gives $0 = \int E_F D(E)(F/T)dE$. This term is then subtracted from the derivative of Eq. (22) with respect to temperature to obtain Eq. (23).

temperatures. Because the lattice specific heat decreases with a much faster T^3 power law, there will be a crossover temperature (typically a few kelvins) below which a slower, linear dependence of the electronic contribution will start to be manifested and eventually will become the dominant contribution.

It is interesting to compare experimental values of the coefficient γ of selected metals with the corresponding free-electron values given by Eq. (28). In essence, such a comparison (see Table 2) informs how appropriate it is to describe the electrons of each respective metal as being free electrons. We note that for the noble metals and the alkaline metals the free-electron description is a reasonable starting point, while for most of the transition metals the free-electron picture is grossly inadequate.

The large deviations between the experimental and free-electron γ-values observed in the case of transition metals are believed to arise from a very high density of states of the d-electrons near the Fermi energy. Transition metals are characterized by their incomplete d-shells. When transition metals form by bringing together atoms, the two outer s-electrons split off and go into a wide conduction band. The more tightly bound d-electrons also form a band, but this is a rather narrow band because the overlap between the neighboring d-orbitals is small. Nevertheless, this band must be able to accommodate 10 electrons per atom, hence its large density of states. On the other hand, semimetals such as Sb and Bi possess very low values of the γ parameter because their carrier density is several orders of magnitude less than that of a typical metal.

3. THE BOLTZMANN EQUATION

When one talks about transport phenomena, one means processes such as the flow

TABLE 2 Comparison of Experimental and Calculated (free-electron) Values of the γ Coefficient of the Electronic Specific Heat for Selected Metals.[a.]

Metal	Valence	Free-electron γ (10^{-3} J mole^{-1}K^{-2})	Experimental γ (10^{-3} J mole^{-1} K^{-2})
Na	1	1.09	1.7
K	1	1.67	1.7
Ag	1	0.64	0.66
Au	1	0.64	0.67
Cu	1	0.50	0.69
Ba	2	1.95	2.72
Ca	2	1.51	2.72
Sr	2	1.80	3.64
Co	2	0.61	4.98
Fe	2	0.64	5.02
Ni	2	0.61	7.02
Al	3	0.91	1.30
Ga	3	1.02	0.63
Sn	4	1.39	1.84
Pb	4	1.50	2.93
As	5	1.30	–
Sb	5	1.63	0.628
Bi	5	1.79	0.08

a. Experimental values are taken from Ref. 2.

Sec. 3 · THE BOLTZMANN EQUATION

of charge or flow of heat in solids. One is usually concerned about the steady-state flow, i.e., the flow established as a result of the influence of external driving forces (electric field or thermal gradient in our case) and the internal scattering processes tending to restore the system to equilibrium. A steady state therefore must be distinguished from the equilibrium state, and transport theory is thus a special branch of nonequilibrium statistical mechanics. The deviations from the truly equilibrium state, however, are usually small. The questions are: how are the driving forces and the scattering processes interrelated? and how does the population of the species under investigation evolve as a function of time? The answers are provided by the Boltzmann equation. Although there are other approaches describing how to treat the problem and arrive at formulas for the transport parameters, the Boltzmann equation has proven itself time and again as the most versatile approach and the one that is reasonably easy to track. Moreover, it yields a form of the expressions for the transport parameters that are intuitive and that can readily be compared with the experimental results. In this section we show how one uses the Boltzmann equation to arrive at useful formulas for the transport parameters of metals.

Metals are crystalline solids. The periodic potential of ions located at the crystal lattice sites is an essential feature of the system and must be incorporated in a realistic description of metals. Lattice periodicity has the following consequences: the momentum of a free electron is replaced by a quantity $\hbar\mathbf{k}$, where $\mathbf{k}$ is the Bloch wave vector of an electron. The wave vector contains information about the translational symmetry of the lattice. The electron velocity, which in the free-electron model is simply $v = p/m$, is replaced by the group velocity $(1/\hbar)\partial E(\mathbf{k})/\partial \mathbf{k}$, where $E(\mathbf{k})$ is the Bloch energy of an electron. Wave functions, which in the free-electron picture were simply plane waves $\exp(i_{\mathbf{k}\cdot\mathbf{r}})$, must now reflect the periodicity of the lattice. This is accomplished by introducing Bloch waves $\psi_k(\mathbf{r}) = u_k(\mathbf{r})\exp(i_{\mathbf{k}\cdot\mathbf{r}})$, i.e., plane waves modulated by a function $u_k(\mathbf{r})$ that has the periodicity of the lattice.

The fundamental issue concerns the distribution function of electrons in a metal, i.e., how the occupation number of electrons changes as a result of the influence of an electric field and thermal gradient, and the effect of various scattering processes that the electrons undergo. The equilibrium distribution of electrons is governed by the Fermi–Dirac function [Eq. (20)], which is independent of the spatial coordinate $\mathbf{r}$ because of the assumed homogeneity (in equilibrium, the temperature is the same everywhere). Away from equilibrium, the distribution may depend on the spatial coordinate $\mathbf{r}$, since we assume that local equilibrium extends only over the region larger than the atomic dimensions, and of course on the time t. Moreover, because the energy of electrons is a function of the wave vector, we also assume that the distribution function depends on $\mathbf{k}$. We therefore consider explicitly $f(\mathbf{r},\mathbf{k},t)$.

Temporal changes in $f(\mathbf{r},\mathbf{k},t)$ arise because of three influences:

1. Electrons may move in or out of the region in the vicinity of point $\mathbf{r}$ as a result of diffusion.
2. Electrons are acted upon by driving forces such as an electric field or thermal gradient.
3. Electrons are deflected into and out of the region near point $\mathbf{r}$ because they scatter on lattice vibrations or on crystal imperfections. This is the scattering influence on the distribution function, and the respective partial derivative is called the collision term.

The first two influences on $f(\mathbf{r},\mathbf{k},t)$ have their origin in the Liouville theorem concerning the invariance of the volume occupied in phase space. Thus, the number of electrons in the neighborhood of point $\mathbf{r}$ at time t must equal the number of electrons in the neighborhood of point $\mathbf{r} - \mathbf{v}(\mathbf{k})dt$ at the earlier time $t - dt$. Since the electrons are also subjected to an external electric field E, they accelerate and in time dt change their momentum $\hbar\mathbf{k}$ by $(d\mathbf{k}/dt)dt = e\,\varepsilon\,dt/\hbar$. (We assume that there is no magnetic field present.) Therefore, in analogy with the Liouville theorem for the spatial coordinate $\mathbf{r}$, we consider the same volume invariance but this time for k-space; i.e., the electrons that are at point $\mathbf{k}$ at time t must have been at a location $\mathbf{k} - (-e)\,\varepsilon\,dt/\hbar$ at time $t - dt$. In the absence of any kind of collisions, we therefore can write

$$f(\mathbf{r},\mathbf{k},t) = f\left[\mathbf{r} - \mathbf{v}(\mathbf{k})dt, \mathbf{k} + e\varepsilon\frac{dt}{\hbar}, t - dt\right]. \tag{29}$$

However, collisions cannot be neglected since they change the population of electrons. Some electrons will be deflected away from the stream of electrons during the time interval dt as the electrons proceed from $(\mathbf{r} - \mathbf{v}(t)dt, \mathbf{k} + e\varepsilon\,dt/\hbar)$ to $(\mathbf{r}, \mathbf{k})$. However, it is also possible that some electrons that arrive at $(\mathbf{r}, \mathbf{k})$ at time t do not belong to the stream of electrons we considered at $(\mathbf{r} - \mathbf{v}(t)dt, \mathbf{k} + e\varepsilon\,dt/\hbar)$ at time $t - dt$ because they were deflected into the stream by collisions in the neighboring regions during the time dt. It is customary to capture both the "out-deflected" and "in-deflected" electrons in a term $(\partial f/\partial t)_{\text{coll}}$. Thus, in time dt there will be a change of population due to scattering of magnitude $(\partial f/\partial t)_{\text{coll}}dt$. Putting it all together, we write

$$f(\mathbf{r},\mathbf{k},t) = f\left[\mathbf{r} - \mathbf{v}(\mathbf{k})dt, \mathbf{k} + e\varepsilon\frac{dt}{\hbar}, t - dt\right] + \left(\frac{df}{dt}\right)_{\text{coll}}dt. \tag{30}$$

Expanding the equation to terms linear in dt, one obtains the Boltzmann equation:

$$\frac{\partial f}{\partial t} + \mathbf{v}\cdot\nabla_r f - \frac{e\varepsilon}{\hbar}\cdot\nabla_k f = \left(\frac{\partial f}{\partial t}\right)_{\text{coll}}. \tag{31}$$

Equation (31) describes a steady state, not the equilibrium state. The terms on the left-hand side are frequently called the streaming terms. In principle, the collision term on the right-hand side contains all the information about the nature of the scattering. Using the quantum mechanical probability for transitions between the Bloch states ψ_k and $\psi_{k'}$, specifically $W_{kk'} \propto |\langle \mathbf{k}'|H'|\mathbf{k}\rangle|^2$, where H' is the perturbation Hamiltonian, one can write the collision term as

$$\left(\frac{\partial f(\mathbf{k})}{\partial t}\right)_{\text{coll}} = \frac{V}{(2\pi)^3}\int d\mathbf{k}'\{[1 - f(\mathbf{k})]W_{kk'}f(\mathbf{k}') - [1 - f(\mathbf{k}')]W_{k'k}f(\mathbf{k})\} \tag{32}$$

The two terms in braces represent transitions from the occupied state $\mathbf{k}'$ to an empty state $[1 - f(\mathbf{k})]$ and a corresponding transition from the occupied state $\mathbf{k}$ to an empty state $[1 - f(\mathbf{k}')]$. These terms are present to comply with the Pauli exclusion principle. Substituting Eq. (32) into Eq. (31), one gets a complicated integrodifferential equation for the distribution function $f(\mathbf{r},\mathbf{k},t)$.

One of the most successful approaches for solving the Boltzmann equation relies on the use of a *relaxation time approximation*. The essence of this approach is the assumption that scattering processes can be described by a relaxation time $\tau(\mathbf{k})$ that specifies how the system returns to equilibrium, i.e., how the distribution function

Sec. 3 · THE BOLTZMANN EQUATION

$f(\mathbf{r},\mathbf{k},t)$ approaches its equilibrium value $f^o(\mathbf{k})$. One therefore writes the collision term as

$$\left(\frac{\partial f}{\partial t}\right)_{\text{coll}} = -\frac{f(\mathbf{k}) - f^o(\mathbf{k})}{\tau(\mathbf{k})} = -\frac{g(\mathbf{k})}{\tau(\mathbf{k})}, \quad (33)$$

where $g(\mathbf{k})$ is the deviation of the distribution function $f(\mathbf{k})$ from its equilibrium value $f^o(\mathbf{k})$, and $\tau(\mathbf{k})$ is the relaxation time. Further considerable simplification of the transport problem is achieved by *linearization* of the Boltzmann equation. This is done simply by replacing the steady-state electron distribution in the gradients $\nabla_r f(\mathbf{r},\mathbf{k},t)$ and $\nabla_k f(\mathbf{r},\mathbf{k},t)$ by the equilibrium distribution function, i.e., by taking $\nabla_r f(\mathbf{r},\mathbf{k},t) = \nabla_r f^o(\mathbf{r},\mathbf{k})$ and $\nabla_k f(\mathbf{r},\mathbf{k},t) = \nabla_k f^o(\mathbf{r},\mathbf{k})$. This seems eminently reasonable provided that the external fields are not too strong so that the steady-state distribution is not too far from equilibrium. We can easily evaluate these two respective gradients (including specifically the dependence of the Fermi energy on temperature) and rewrite Eq. (31) as

$$\left(-\frac{\partial f^o}{\partial E}\right)\mathbf{v}(\mathbf{k})\left\{\frac{E - E_F}{T}\nabla_r T - e\left[\boldsymbol{\varepsilon} - \frac{1}{e}\nabla_r E_F\right]\right\} = \left(\frac{\partial f}{\partial t}\right)_{\text{coll}} \quad (34)$$

This is the linearized Boltzmann equation. After some algebra, we can write, for the perturbed distribution function $f(\mathbf{r},\mathbf{k},t)$,

$$f(\mathbf{k}) = f^o(\mathbf{k}) + \left(-\frac{\partial f^o}{\partial E}\right)\mathbf{v}(\mathbf{k})\tau(\mathbf{k})\left\{e\left(\boldsymbol{\varepsilon} - \frac{1}{e}\nabla_r E_F\right) + \frac{E - E_F}{T}(-\nabla_r T)\right\}. \quad (35)$$

This is a general form of the distribution function describing electron population perturbed by a weak electric field and a small temperature gradient. Before we proceed further it is important to comment on the quantity we call the relaxation time $\tau(\mathbf{k})$.

It is a major consideration whether $\tau(\mathbf{k})$ is a true relaxation time in the sense of representing something like the time between collisions and, therefore, whether it faithfully represents the way the equilibrium state is approached. We can, of course, always formally introduce the parameter $\tau(\mathbf{k})$ for any process, but if this quantity turns out to depend on a particular form of the field that created the out-of-equilibrium state, then we would need different functional forms of $\tau(\mathbf{k})$ for different external fields and the meaning we associate with the relaxation time would be lost. Thus, whether the true relaxation time exists (i.e., whether the relaxation time approximation is valid) is a critical issue, and we shall see that it has a decisive influence on the discourse concerning the thermal conductivity at low temperatures.

3.1. Transport Coefficients

Knowing how a weak electric field and a thermal gradient perturb the population of electrons, we may now proceed to inquire what currents arise when the system is allowed to reach steady state. To do that, it is customary to define current densities. If we consider a flow of particles, we know that a volume element $d\mathbf{k}$ around point $\mathbf{k}$ in reciprocal space will contribute a flux of particles of magnitude $(1/8\pi^3)v(\mathbf{k})d\mathbf{k}$, where $v(\mathbf{k})$ is the particle velocity. Since we consider electrons, there are two spin states for every k-state and the electrons are distributed according to the Fermi–Dirac function. Summing over all available states and assigning a charge e to each electron, we obtain the current density:

Chap. 1.2 · THERMAL CONDUCTIVITY OF METALS

$$\mathbf{J}_e = \frac{2e}{8\pi^3} \int \mathbf{v}(\mathbf{k}) f(\mathbf{k}) d\mathbf{k}. \tag{36}$$

Similarly for the heat current density, we write

$$\mathbf{J}_Q = \frac{2}{8\pi^3} \int \mathbf{v}(\mathbf{k})[E - \zeta] f(\mathbf{k}) d\mathbf{k}. \tag{37}$$

Equation (37) contains a term $E - \zeta$, where ζ is the chemical potential (free energy, which in metals can be taken as the Fermi energy) rather than just energy E. This is necessary because thermodynamics tells us that heat is internal energy minus free energy.

Since equilibrium distributions do not generate currents (there is no spontaneous current flow), substituting the perturbed distribution function in Eq. (35) into Eq. (36) causes the integral containing the equilibrium distribution to vanish. The remaining term is

$$\mathbf{J}_e = \frac{2e}{8\pi^3} \int \mathbf{v}(\mathbf{k})\mathbf{v}(\mathbf{k}) \tau \left(-\frac{\partial f^o}{\partial E}\right) \left[e\boldsymbol{\varepsilon} - \nabla\zeta + \frac{E-\zeta}{T}(-\nabla T)\right] d\mathbf{k}$$

$$= \frac{e^2 \tau}{4\pi^3 \hbar} \int \mathbf{v}(\mathbf{k})\mathbf{v}(\mathbf{k}) \left(-\frac{\partial f^o}{\partial E}\right) \left[\boldsymbol{\varepsilon} - \frac{1}{e}\nabla\zeta\right] \frac{dS}{|\mathbf{v}|} dE$$

$$+ \frac{e\tau}{4\pi^3 \hbar} \int \mathbf{v}(\mathbf{k})\mathbf{v}(\mathbf{k}) \left(-\frac{\partial f^o}{\partial E}\right) \left[\frac{E-\zeta}{T}\right] (-\nabla T) \frac{dS}{|\mathbf{v}|} dE. \tag{38}$$

Similarly, substituting Eq. (35) into Eq. (37) results in

$$\mathbf{J}_Q = \frac{2}{8\pi^3} \int \mathbf{v}(\mathbf{k})\mathbf{v}(\mathbf{k}) \tau \left(-\frac{\partial f^o}{\partial E}\right) \left[e\boldsymbol{\varepsilon} - \nabla\zeta + \frac{E-\zeta}{T}(-\nabla T)\right] (E-\zeta) d\mathbf{k}$$

$$= \frac{e\tau}{4\pi^3} \int \mathbf{v}(\mathbf{k})\mathbf{v}(\mathbf{k}) \left[\boldsymbol{\varepsilon} - \frac{1}{e}\nabla\zeta\right] (E-\zeta) \left(-\frac{\partial f^o}{\partial E}\right) \frac{dS}{|\mathbf{v}|} dE$$

$$+ \frac{\tau}{4\pi^3} \int \mathbf{v}(\mathbf{k})\mathbf{v}(\mathbf{k}) \frac{E-\zeta}{T}(-\nabla T) \left(-\frac{\partial f^o}{\partial E}\right) \frac{dS}{|\mathbf{v}|} dE, \tag{39}$$

where, with the aid of $d\mathbf{k} = dS d\mathbf{k}_\perp = dS dE/|\nabla_k E|$, we converted an integral over a volume in k-space into an integral over surfaces of constant energy. Defining now the integral

$$K_n = \frac{1}{4\pi^3} \frac{\tau}{\hbar} \iint \mathbf{v}(\mathbf{k})\mathbf{v}(\mathbf{k})(E-\zeta)^n \left(-\frac{\partial f^o}{\partial E}\right) \frac{dS}{|\mathbf{v}|} dE, \tag{40}$$

we can express Eqs. (38) and (39) in terms of K_n:

$$\mathbf{J}_e = e^2 K_o \boldsymbol{\varepsilon} + \frac{eK_1}{T}(-\nabla T), \tag{41}$$

$$\mathbf{J}_Q = eK_1 \boldsymbol{\varepsilon} + \frac{K_2}{T}(-\nabla T). \tag{42}$$

Sec. 3 · THE BOLTZMANN EQUATION

Equations (41) and (42) reveal an important point—electric current $\mathbf{J}_e$ can be generated not just by an external electric field but also by a thermal gradient. Likewise, heat current $\mathbf{J}_Q$ arises as a result of an electric field and a thermal gradient. In essence, there are interactions between the electric and heat currents.

Integrals K_n; defined in Eq. (40), can be evaluated, using a general theorem for integrals of the Fermi function over energy that are of the form

$$\int_{-\infty}^{+\infty} \Psi(E) f(E) dE = \int_{-\infty}^{+\infty} \Phi(E) \left(-\frac{\partial f^o}{\partial E}\right) dE, \tag{43}$$

where $\psi(E) = d\Phi(E)/dE$. Because $-\partial f^o/\partial E$ has an appreciable value only within a few $k_B T$ of ζ, a smoothly varying function of energy $\Phi(E)$ can be expanded about $E = \zeta$, where hopefully only the first few terms suffice for accuracy:

$$\Phi(E) = \Phi(\zeta) + \sum_{n=1}^{\infty} \left[\frac{(E-\zeta)^n}{n!}\right] \left[\frac{d^n \Phi(E)}{dE^n}\right]_{E=\zeta} \tag{44}$$

The integral in Eq. (43) then becomes

$$\int \Phi(E) \left(-\frac{\partial f^o}{\partial E}\right) dE = \Phi(\zeta) + \frac{\pi^2 (k_B T)^2}{6} \left[\frac{\partial^2 \Phi(E)}{\partial E^2}\right]_{E=\zeta} + \dots, \tag{45}$$

and only terms even in n contribute because $\partial f^o/\partial E$ is an even function of $E - \zeta$.

Equations (41) and (42) serve to define transport coefficients in zero magnetic field. The usual isothermal electrical conductivity follows from a condition that the thermal gradient is zero:

$$\mathbf{J}_e = e^2 K_o \boldsymbol{\varepsilon}. \tag{46}$$

With the explicit form of K_o, this can be written as

$$\mathbf{J}_e = \frac{e^2}{4\pi^3} \frac{\tau(E_F)}{\hbar} \int_{E=E_F} \mathbf{v}(\mathbf{k})\mathbf{v}(\mathbf{k}) \frac{dS_F}{|\mathbf{v}(\mathbf{k})|} \boldsymbol{\varepsilon} = \sigma \cdot \boldsymbol{\varepsilon}, \tag{47}$$

where σ is the electrical conductivity. In a general crystal structure the current density need not be parallel to the electric field, and the electrical conductivity is a tensor,

$$\sigma_{ij} = \frac{e^2 \tau(E_F)}{4\pi^3 \hbar} \int_{E=E_F} \frac{v_i(\mathbf{k}) v_j(\mathbf{k}) dS_E}{|\mathbf{v}(\mathbf{k})|}. \tag{48}$$

For cubic metals and for isotropic (i.e., polycrystalline metals), the tensor reduces to a scalar. Placing now the electric field $\boldsymbol{\varepsilon}$ along, say, the x-direction, we obtain, making use of $v_x^2 = (1/3)v^2$:

$$\sigma = \frac{e^2 \tau(E_F)}{12\pi^3 \hbar} \int_{E=E_F} v \, dS_E = \frac{e^2}{12\pi^3 \hbar} \int_{E=E_F} \ell_e \, dS_E. \tag{49}$$

In this equation, we introduced the mean free path of electrons as $|_e = v.\tau(E_F)$.

Equation (49) is the basic formula for the electrical conductivity. It states clearly that only electrons near the Fermi level can contribute to the transport process since the integral is over the Fermi surface $E(\mathbf{k}) = E_F$. This, of course, is in full agreement with the Pauli exclusion principle.

One might think that the thermal conductivity is obtained simply by taking it as a coefficient of the thermal gradient in Eq. (42), i.e., under the conditions that an external electric field is zero. However, this is not how thermal conductivity is measured. Although it is difficult to control electric fields, it is a straightforward matter to ensure that no electric current passes through the sample while the thermal conductivity is measured. We simply leave the sample in open-circuit configuration. Therefore, to obtain the desired coefficient of thermal conductivity, we set the current $\mathbf{J}_e = 0$ in Eq. (41) and express an electric field in terms of a thermal gradient:

$$\varepsilon = \frac{K_o^{-1} K_1}{eT} \nabla T. \tag{50}$$

Substituting Eq. (50) into Eq. (42) yields

$$\mathbf{J}_Q = \frac{1}{T} \left[K_2 - K_1 K_o^{-1} K_1 \right] (-\nabla T) = \kappa(-\nabla T). \tag{51}$$

The coefficient κ is the electronic thermal conductivity. The term $K_1 K_o^{-1} K_1$ in Eq. (51) is very small for metals and, for present purposes, can be neglected. The thermal conductivity is therefore

$$\kappa = \frac{1}{T} \left[K_2 - K_1 K_o^{-1} K_1 \right] \cong \frac{K_2}{T}. \tag{52}$$

If we define a function $\Phi(E) = (E - E_F)^2 K_o(E)$, it can be shown, using Eq. (45) and noting that the term $[\Phi(E)]_{E=E_F} = 0$, that K_2 is related to K_o via

$$K_2 = \frac{\pi^2 (k_B T)^2}{6} 2 K_o(E_F) = \frac{\pi^2 (k_B T)^2}{3} K_o(E_F) \tag{53}$$

Substituting Eq. (53) into Eq. (52), we obtain the thermal conductivity as

$$\kappa = \frac{K_2}{T} = \frac{\pi^2 k_B^2 T}{3} K_o(E_F) = \frac{\pi^2}{3} \left(\frac{k_B}{e} \right)^2 T \sigma \tag{54}$$

We recognize the ratio κ/σ as the Lorenz number of the Wiedemann–Franz law. Equation (54) is to be compared to the Drude result [Eq. (18)], where the only difference is in the factor $\pi^2/3 = 3.29$ in place of $3/2$, arising because Eq. (54) is derived with the proper statistical description of the electrons. Eq. (54) is quite a general result except for the fact that scattering processes must be elastic.

Now that the coefficients governing the flow of charge, σ, and the flow of heat, κ, have been established, two other coefficients remain to be determined based on Eqs. (41) and (42). These represent the interference terms between the electric and thermal fields.

Let us assume that we set up a temperature gradient across a sample that is in open-circuit configuration; i.e., there is no electric current $\mathbf{J}_e$. This gradient gives rise to an electric field ε:

$$\varepsilon = \frac{K_o^{-1} K_1}{eT} \nabla T = S \nabla T. \tag{55}$$

The coefficient S is the Seebeck coefficient, often, but somewhat unfortunately, called thermoelectric power:

$$S = \frac{K_o^{-1} K_1}{eT}. \tag{56}$$

Using a relation between the integrals K_1 and K_o,

$$K_1 = \frac{\pi^2}{3}(k_B T)^2 \left[\frac{\partial K_o(E)}{\partial E}\right]_{E=E_F} \tag{57}$$

one can write the Seebeck coefficient in the form

$$S = \frac{\pi^2}{3}\frac{ek_B^2 T}{\sigma}\left[\frac{\partial K_o(E)}{\partial E}\right]_{E=E_F} = \frac{\pi^2 k_B^2 T}{3e}\frac{1}{\sigma}\left[\frac{\partial \sigma(E)}{\partial E}\right]_{E=E_F} = \frac{\pi^2}{3}\left(\frac{k_B}{e}\right) k_B T \left[\frac{\partial \ln \sigma(E)}{\partial E}\right]_{E=E_F}. \tag{58}$$

The Seebeck coefficient is an important transport parameter and has a considerable bearing on technological applications such as thermocouple sensors and thermoelectric power conversion. The study of thermoelectric phenomena is an interesting topic but clearly beyond the purview of this chapter. A reader interested in thermoelectric effects might find it useful to consult monographs dedicated to this subject, among them Goldsmid,[59] Blatt et al.,[60] and Tritt.[61]

The remaining term in Eqs. (41) and (42) is obtained by setting $\nabla T = 0$, i.e., imposing isothermal conditions and relating the heat current to the electric current:

$$\mathbf{J}_Q = \frac{K_o^{-1} K_1}{e}\mathbf{J}_e = \Pi \mathbf{J}_e, \tag{59}$$

where Π is called the Peltier coefficient. It relates heat generation or absorption in junctions of dissimilar metals to the electric current passed through the circuit. The Peltier effect should not be confused with Joule heating, which is a quadratic function of current and is always dissipative. The Peltier effect is linear in current and is at the heart of an important but niche technology of thermoelectric cooling. The materials of choice are, however, small-gap semiconductors rather than metals.

For completeness, we also mention that it follows from Eqs. (56) and (59) that the Seebeck and Peltier effects are closely related, namely

$$\Pi = ST. \tag{60}$$

Having defined the transport coefficient and having established its form in terms of the transport integrals, we now address the key issue: the actual mechanism of transport. We will inquire about the dynamics of electrons and about the processes that limit both the electric and heat currents. Since the Wiedemann–Franz law plays such a pivotal role in the theory of transport and in the practical assessment of heat conducting ability of metals, we also inquire about the validity of this law in different regimes of transport. To address these issues, we rely on a quantum-mechanical description of the electron motion in a crystalline lattice and on the interactions of electrons with both intrinsic and extrinsic lattice imperfections. We also consider the influence of electrons interacting with each other. We should state right at the outset that a theoretical description of the transport mechanism is a difficult and challenging topic. As much as one would like to capture all nuances of the band structure of various metals and make the computations and analysis truly metal specific, one soon finds out that the sheer complexity of the problem necessitates various approximations and simplifications that, in spite of valiant efforts, yield not much more than a qualitative solution. It is, indeed, unrealistic to expect that the theory will provide perfect agreement with the experimental data. What we hope for are correct predictions regarding the temperature dependence of various transport phenomena, capturing the trends among various classes of metals, and perhaps a factor of 2 consent regarding specific numerical values of the transport coefficients. Because electrical conductivity of metals is so closely tied to their electronic thermal conductivity, we consider its behavior first. This will serve as an excellent preparation for discussions of heat transport in metals.

3.2. Electrical Conductivity

In Eq. (49) we have a basic formula for electrical conductivity. To solve the integral in this equation, we must consider specific models. The most amenable one assumes that electrons are in a band that is strictly parabolic (nearly free electrons). In this case the velocity of electrons at the Fermi surface is simply $v(E_F) = \hbar k_F/m^*$ (m^* being the effective mass), and the integral over the Fermi surface is $4\pi k_F^2$. Collecting these results, we obtain the electrical conductivity

$$\sigma = \frac{e^2 \tau(E_F)}{4\pi^3 \hbar} \frac{1}{3}\left(\frac{\hbar k_F}{m^*}\right)\left(4\pi k_F^2\right) = \frac{e^2 \tau(E_F)}{3\pi^2 m^*} k_F^3 = \frac{e^2 \tau(E_F)}{3\pi^2 m^*} 3\pi^2 n = \frac{ne^2 \tau(E_F)}{m^*}. \quad (61)$$

In Eq. (61) we have recovered a result formally equivalent to the Drude formula, Eq. (9). This time, however, we are on a firm footing in terms of the theory. The relaxation time, which in Eq. (9) was purely a statistical quantity, is now a relaxation time of the electrons on the Fermi surface. Only electrons that are within a thermal layer on the order of $k_B T$ of the Fermi energy can respond to an external electric field because they are the ones that can find empty states in their vicinity. Electrons deep down in the distribution have no empty states as neighbors and thus are unaffected. Metals have high conductivity because a relatively small number of them—those in the neighborhood of the Fermi energy—have very high velocities and not because all electrons drift "sluggishly" in response to the electric field. This is also easily understood from the following illustration that depicts how the out-of-equilibrium electron distribution actually looks. From Eq. (35), taking both gradients $\nabla_r T$ and $\nabla_r E_F$ equal to zero, we write

$$f(\mathbf{k}) = f^o(\mathbf{k}) - \frac{e\tau(\mathbf{k})}{\hbar} \frac{\partial f^o}{\partial E} \frac{\partial E(\mathbf{k})}{\partial \mathbf{k}} \cdot \boldsymbol{\varepsilon} = f^o\left(\mathbf{k} - \frac{e\tau}{\hbar}\boldsymbol{\varepsilon}\right), \quad (62)$$

where we used Taylor's theorem to arrive at the second equality. Eq. (62) states that the steady-state distribution $f(\mathbf{k})$ in reciprocal space (k-space) is the same as the equilibrium distribution that is shifted by $-(e\tau/\hbar)\boldsymbol{\varepsilon}$. Therefore, depicting the equilibrium distribution of electrons by a solid curve in Fig. 1(a), we see that the effect of a constant electric field applied along the negative x-axis is to shift the entire distribution in the positive x-direction by $(e\tau/\hbar)\boldsymbol{\varepsilon}$. The displaced electron distribution is indicated in Fig. 1a by a dashed curve. It is clear from the figure that only electrons near the Fermi edge are affected by the electric field; electrons lying deep inside the distribution are completely ignorant of the presence of the field. If the external field is switched off, the steady-state distribution (the dashed curve) will try to relax back to the equilibrium form. The question is, how does the distribution in

Sec. 3 • THE BOLTZMANN EQUATION

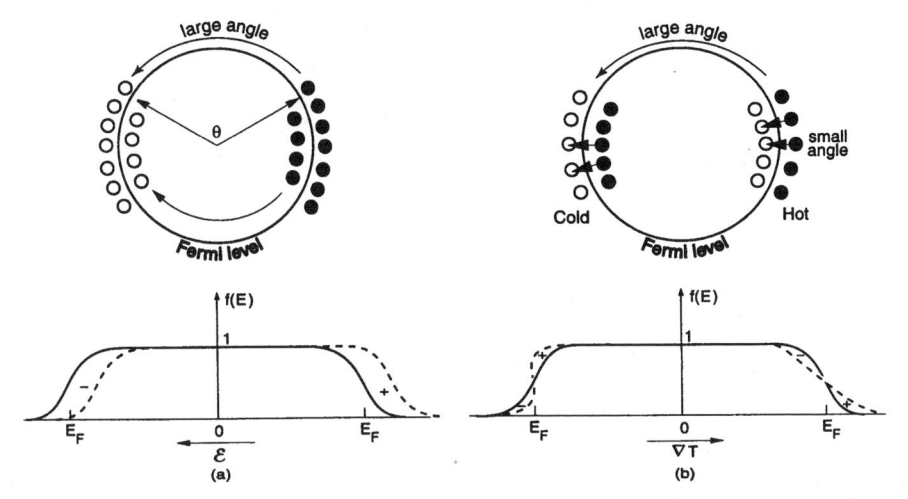

FIGURE 1 Schematic representation of the undisturbed (solid curves) and disturbed (dashed curves) Fermi distributions produced by (a) an electric field and (b) a temperature gradient. The overpopulated and underpopulated energy levels are marked with + and − signs, respectively. Solid and open circles in the upper panels represent the excess and deficiency of electrons relative to equilibrium distribution. An electric field shifts the entire distribution, whereas a temperature gradient creates asymmetry in the distribution function. The reader should note the distinction between the large-angle scattering (the so-called horizontal processes) and small-angle scattering but with a change in the electron's energy (the vertical processes).

Fig. 1a relax? Obviously electrons must be taken from the region on the right and moved to the opposite side of the Fermi surface. Whether this can be achieved in a single interaction involving a large angular change or whether many smaller steps are needed to bring electrons back determines how effective is the scattering process. The average time needed to reestablish equilibrium is a measure of relaxation time τ. This may turn out to be a different time than the average time between successive collisions, τ_o, and we must carefully distinguish these two times. Clearly, if the distribution can be relaxed in a single step, then $\tau = \tau_o$. However, if several collisions are needed to relax the distribution, the time τ is larger than τ_o and is given by

$$\frac{1}{\tau} \cong \frac{1}{\tau_o}(1 - \langle\cos\theta\rangle), \tag{63}$$

where $\langle\cos\theta\rangle$ stands for an average over the scattering angle. The point is that from the perspective of electrical resistivity[*] it is the change in the electron velocity in the direction of the electric field which gives rise to resistivity. If each collision changes the direction by only a small amount, many collisions will be needed to accomplish the task of relaxing the distribution. Because the factor $(1 - \langle\cos\theta\rangle)$ is sensitive to temperature, it will figure prominently in the discussion of electrical and thermal transport processes considered as a function of temperature.

In a strongly degenerate system such as metals the carrier density is essentially

---

[*] Thermally perturbed distributions, in addition to a possibility of being relaxed via the mechanism shown in Fig. 1a, i.e., using the so-called horizontal processes, have an alternative and often more effective means of relaxation through very small angle but with a small change in energy, the so-called vertical processes illustrated in Fig. 1b and discussed later.

temperature independent. It then follows from Eq. (61) that variations in the electrical resistivity with temperature come from the temperature dependence of the relaxation time τ. With the aid of quantum mechanics, it is, in principle, possible to calculate all scattering probabilities and, therefore, to determine the relaxation time. However, Fermi surfaces of real metals are quite complicated. One either does not have available all the input parameters, or, if one has them, the theory is not capable of incorporating all the fine nuances of the problem. In short, calculations of relaxation times are not a trivial task once one ventures beyond the comfort zone of spherical Fermi surfaces.

We have already mentioned that it is somewhat arbitrary to write the collision term in the form given in Eq. (33), and that only in situations where $\tau(\mathbf{k})$ is independent of the external fields is the relaxation time meaningful. Since in this chapter we only consider transport processes in zero magnetic field, we aim for a relaxation time that is common to both an electric field and a thermal gradient. Clearly, we would like to know when this is so; i.e., under what conditions does the relaxation time approximation yield a result equivalent to a rigorously calculated collision term in the Boltzmann equation? This is an important question that touches upon the validity of the Wiedemann–Franz law. It turns out that an electron population disturbed by an electric field and a thermal gradient will have the same relaxation time when electrons scatter *elastically*. Actually, as shown in Ref. 2, if the relaxation time approximation returns a solution for the distribution function of the form

$$f(\mathbf{k}) = f^o(\mathbf{k}) + \mathbf{A}(E) \cdot \mathbf{k}, \tag{64}$$

where a vector quantity $\mathbf{A}(E)$ depends on $\mathbf{k}$ only through its magnitude, then this solution is rigorous in the sense that the same result is obtained by solving the full Boltzmann equation.

We must not forget that by using the relaxation time approximation we have merely introduced the relaxation time as a parameter but we have not solved the scattering problem; this can be done only by having a general solution to the Boltzmann equation. Tackling such a formidable challenge, one must resort to more sophisticated techniques. Of these, the variational principle has proved to be the most successful. Developed by Köhler,[62] it is based on the linearized Boltzmann equation for an electric field and thermal gradient. It defines a function $\Phi(\mathbf{k})$ such that

$$f(\mathbf{k}) = f^o - \Phi(\mathbf{k}) \frac{\partial f^o}{\partial E(\mathbf{k})}. \tag{65}$$

The function Φ is then expressed as a linear combination of appropriately chosen functions with variable coefficients. The aim is to optimize the coefficients so as to produce a trial function that most closely approximates the true solution. The Boltzmann equation is thus reduced to an extremum problem. The better the trial function, the better is the approximation. An interesting upshot of this treatment is its thermodynamic interpretation. One can view the effect of an external electric field as aligning the velocities of electrons and thus decreasing the entropy. On the other hand, scattering tends to disperse electrons, i.e., decrease the degree of order in the system and thus increase the entropy. The Boltzmann equation therefore expresses a relation between the decrease in entropy (by external fields) and the increase in entropy (by collisions). Since variational calculations test for the max-

Sec. 3 • THE BOLTZMANN EQUATION

imum, the solution of the Boltzmann equation coincides with the maximum entropy production in the collision processes. For more information on the variational principle as applied to the Boltzmann equation readers might wish to consult Refs. 3 and 63.

For completeness, we mention that the Boltzmann equation is not the only plausible approach for the formulation of the transport problem. One may, for instance, use as a starting point linear response theory and derive the conductivity in terms of time correlation functions of current. This is the approach pioneered by Kubo[64] and by Greenwood:[65]

$$\sigma = \frac{1}{k_B T} \int_0^\infty \langle j_k(t) j_l(0) \rangle dt. \tag{66}$$

Other correlation functions (e.g., force–force correlations[66]) have been tried to arrive at an expression for the conductivity. A different approach, based on the density matrix formalism, was developed by Kohn and Luttinger.[67] Even the optical theorem was explored to derive an expression for conductivity.[68] Although these approaches capture the *a priori* microscopic quantum nature of the problem, none of these delivers a clearly superior solution. Yes, there is considerable comfort and satisfaction that one can start from the fundamentally atomistic perspective and use the rigorous field-theoretical techniques to arrive at a solution. However, the road to this solution is rather arduous, the treatment is less intuitive, and the results are often difficult to interpret. In spite of its shortcomings and a somewhat less than rigorous nature, the Boltzmann transport equation is not a bad starting point for the theoretical description of transport phenomena.

From Eqs. (33) and (61) it follows that the central problem in transport theory is a calculation of the relaxation time τ. In metals, three main scattering processes affect the electrical and thermal resistivities. (a) electrons can scatter on lattice defects such as foreign atoms (impurities) occupying the lattice sites; (b) electrons can be deflected via their interaction with lattice vibrations (phonons); and (c) electrons may interact with other electrons—after all, there are plenty of them around in a metal. We briefly consider key points on how to proceed in calculating the respective relaxation times for the three scattering scenarios. But first we mention a remarkable relation known for more than half a century prior to the development of quantum mechanics—Matthiessen's Rule. This empirical rule, described by Matthiessen in 1862,[69] asserts that if several distinct scattering mechanisms are at play the overall resistivity of metals is simply the sum of the resistivities one would obtain if each scattering mechanism alone were present. For example, for two distinct scattering mechanisms—scattering on impurities and on lattice vibrations—the resistivity can be written as

$$\rho = \frac{1}{\sigma} = \frac{m^*}{ne^2\tau} = \frac{m^*}{ne^2\tau_{(1)}} + \frac{m^*}{ne^2\tau_{(2)}} = \rho_{(1)} + \rho_{(2)}. \tag{67}$$

The essential point behind Matthiessen's Rule is the independence of scattering mechanisms; i.e., the overall collision rate is the sum of the collision rates of the participating scattering processes. In the context of the relaxation time approximation this immediately leads to a reciprocal addition of the relaxation times:

$$\frac{1}{\tau} = \frac{1}{\tau_{(1)}} + \frac{1}{\tau_{(2)}}. \tag{68}$$

Extensive studies (for a review see Ref. 70) have been done to test the validity of Matthiessen's Rule. Overall the rule seems to be reasonably robust, with deviations often less than 2%. However, there are situations where the rule breaks down, indicating that the scattering processes are not mutually independent after all. This happens when the relaxation time depends on the **k** vector; i.e., $\tau = \tau(\mathbf{k})$. As discussed in Ref. 2, the resistivity is then proportional to a reciprocal average of the relaxation time $1/\tau$, and Matthiessen's Rule implies

$$\frac{1}{\bar{\tau}} = \frac{1}{\bar{\tau}_{(1)}} + \frac{1}{\bar{\tau}_{(2)}}. \tag{69}$$

On the other hand, averages performed on Eq. (67) require

$$\overline{\left(\frac{1}{\tau}\right)} = \overline{\left(\frac{1}{\tau_{(1)}}\right)} + \overline{\left(\frac{1}{\tau_{(2)}}\right)}. \tag{70}$$

The formulas are not equivalent unless $\tau_{(1)}$ and $\tau_{(2)}$ are independent of **k**. Since there is a small but finite spread of electron wave vectors that can interact with phonons, $\tau_{(1)}$ and $\tau_{(2)}$ are somewhat interdependent and this, in general, leads to an inequality rather than an equality between the total and two partial resistivities:

$$\rho \geq \rho_{(1)} + \rho_{(2)}. \tag{71}$$

This inequality implies a positive deviation from Matthiessen's Rule; i.e., if we write for the total resistivity $\rho_{\text{tot}} = \rho_{\text{imp}} + \rho(T)$, the temperature-dependent resistivity always increases with increasing impurity resistivity ρ_{imp}. This old dogma has been challenged by measurements of Rowlands and Woods,[71] who observed negative deviations from Matthiessen's Rule in samples of Al, Ag, and Pd that were mechanically strained. Thus, it depends on how the resistivity ρ_{imp} is being increased. If the sample is more "dirty" (by stuffing more impurity into it), the deviation will always be positive. If, on the other hand, ρ_{imp} is increased by straining the sample, then the deviation will be negative. The origin of the negative deviations is believed to be electron-dislocation scattering. In mechanically deformed samples electron-dislocation scattering is dramatically enhanced and because this interaction process favors small-angle scattering in contrast to large-angle scattering predominant in the case of electron-impurity scattering, this leads to a reduction of $\rho(T)$ with increasing ρ_{imp}.

Although it is not perfectly obeyed, Matthiessen's Rule gives us a hope of being able to discuss electron scattering in metals in terms of more-or-less independent processes and focus on important characteristics and signatures of each relevant scattering mechanism. Before we do that, let us consider the form of an electron distribution function as it is perturbed by a thermal gradient.

3.3. Electronic Thermal Conductivity

An expression for the electronic thermal conductivity is given in Eq. (52). If we do not neglect the term $K_1 K_o^{-1} K_1$ with respect to the term K_2, we can write the thermal conductivity as

Sec. 3 · THE BOLTZMANN EQUATION

$$\kappa = \frac{1}{T}[K_2 - K_1 K_o^{-1} K_1] = \frac{1}{T}[K_2 - eSK_1 T] = \frac{K_2}{T} - eSK_1, \tag{72}$$

where we used Eq. (56), that defines the Seebeck coefficient S. Forming now a ratio $\kappa/\sigma T$ (i.e., defining the Lorenz function L), we obtain

$$\frac{\kappa}{\sigma T} \equiv L = \frac{\frac{K_2}{T} - eSK_1}{e^2 K_o T} = \frac{K_2}{e^2 T^2 K_o} - \frac{SK_o^{-1} K_1}{eT} = \frac{\pi^2}{3}\left(\frac{k_B}{e}\right)^2 - S^2 = L_o - S^2 \tag{73}$$

Thus, the Lorenz function L differs from the Sommerfeld value L_o by a term equal to the square of the Seebeck coefficient. In good metals at ambient temperatures the Seebeck coefficient is a few $\mu V/K$; i.e., the term S^2 is $\sim 10^{-11}$ V^2/K^2, which is three orders of magnitude smaller than L_o and can be omitted in most situations. Nevertheless, this correction term will be important when we discuss the high-temperature electron–electron contribution to the electronic thermal conductivity of metals.

It is instructive to illustrate the electron distribution that arises as a result of the presence of a thermal gradient. In particular, by comparing it with the distribution created by an electric field, we will be able to highlight the difference between the two and understand why thermal and electrical processes differ at low temperatures.

From Eq. (35), assuming only a thermal gradient is present, we can write for the distribution function of electrons

$$f(\mathbf{k}) = f^o(\mathbf{k}) + \left(-\frac{\partial f^o}{\partial E}\right)\left(\frac{E - \zeta}{T}\right)\tau \mathbf{v}(\mathbf{k})(-\nabla T) = f^o(T + \tau \mathbf{v}(\mathbf{k})(-\nabla T)), \tag{74}$$

where the second equality follows from the use of Taylor's theorem and the identity

$$\frac{\partial f^o}{\partial T} = \frac{E - \zeta}{T}\left(-\frac{\partial f^o}{\partial E}\right) \tag{75}$$

The distribution is shown in Fig. 1b. We note the following important points:

For $E > \zeta$ and for $\mathbf{v}(\mathbf{k})\cdot(-\nabla T) > 0$ we obtain $f(\mathbf{k}) > f^o(\mathbf{k})$
$\mathbf{v}(\mathbf{k})\cdot(-\nabla T) < 0$ $f(\mathbf{k}) < f^o(\mathbf{k})$

(76)

For $E < \zeta$ and for $\mathbf{v}(\mathbf{k})\cdot(-\nabla T) > 0$ we obtain $f(\mathbf{k}) < f^o(\mathbf{k})$
$\mathbf{v}(\mathbf{k})\cdot(-\nabla T) < 0$ $f(\mathbf{k}) > f^o(\mathbf{k})$

Therefore, for $E > \zeta$, electrons going in the direction of $-\nabla T$ (i.e., down the thermal gradient) are hotter and tend to spread the distribution, while electrons going in the direction of ∇T (i.e., up the thermal gradient) are colder and sharpen the distribution.

This is clearly a very different situation compared to that encountered in Fig. 1a, where the distribution was perturbed by an electric field with the result that the entire Fermi sphere shifted in response to the applied field.

The nature of the thermal distribution in Fig. 1b presents an alternative mechanism for the electron system to relax. While the electrons may achieve equilibrium by being scattered through large angles across the Fermi surface (the horizontal process effective in relaxing the electron population disturbed by an electric field in Fig. 1a), in Fig. 1b electrons may relax simply by undergoing collisions in which they change their energy by a small amount and move through the Fermi level, but with essentially no change in the wave vector. These types of processes—vertical processes—provide a very effective means of relaxing thermally driven distributions and govern thermal conductivity at low temperatures. At high-temperatures (usually including a room temperature) electrons scatter through large angles, the average of cos θ is zero, and a single collision may relax the distribution. Because scattering through large angles (horizontal processes) becomes less and less probable as the temperature decreases, electrons need many collisions before the direction of their velocity is changed substantially. However, if they can pick up or give up a bit of energy, no such limitation exists and they can relax readily. Thus, we are starting to appreciate that at low temperatures the relaxation time governing electrical conductivity will not be the same as the one governing thermal conductivity. The relaxation time in thermal processes will be considerably shorter because the electrons have an additional and very effective channel through which they can relax—the inelastic vertical process. This is the reason why the Wiedemann–Franz law breaks down at low temperatures—the electrical and thermal processes lack a common relaxation time.

4. SCATTERING PROCESSES

Electrons in a solid participate in three main interaction processes: they can scatter on lattice defects, they interact with lattice phonons, and they interact with each other. We have already noted that because of a small but finite spread of electron wave vectors, Matthiessen's Rule is not perfectly obeyed and interference effects, especially between the impurity and electron–phonon terms, lead to deviations from the rule. Such a slight nonadditive nature of scattering processes arises from the energy dependence of the electron–phonon interaction. As one would expect, deviations from Matthiessen's Rule also affect the electronic thermal resistivity and, in fact, might be more pronounced there than in the case of electrical resistivity. The reason is that the electrical resistivity is more immune to small changes in the electron wave vector than is the electronic thermal conductivity. Nevertheless, the statement must be qualified in the following sense: (a) As pointed out by Berman,[1] deviations from Matthiessen's Rule are usually largest in the regime where the scattering rates for the electrical and thermal processes differ most—at intermediate to low temperatures. Since in this regime phonons are much less effective in limiting electrical resistivity than the electronic thermal resistivity, the fractional influence of the same amount of impurity will be larger in the former case. (b) Experimentally, it is far more difficult to design and implement high–precision measurements of thermal conductivity than to do so for electrical resistivity. Thus, while there is no shortage of the data in the literature concerning deviations from Matthiessen's Rule in electrical resistivity, the corresponding data for the thermal conductivity are rather sparse.

We now consider how the different interaction mechanisms give rise to electrical and thermal resistivities.

4.1. Impurity Scattering

Impurities and lattice imperfections are present to a greater or lesser extent in all real crystals, and they are an impediment to the flow of electrons. Scattering of

Sec. 4 · SCATTERING PROCESSES

electrons on impurities is especially manifested at low temperatures, where the strength of the competing scattering mechanisms is weaker. In general, impurities are heavy objects on the scale of the electron mass, and, provided the impurity does not have an internal energy structure such that its energy levels would be closely spaced on the scale of $k_B T$, the impurity will not be able to absorb or give energy to a colliding electron. Consequently, impurity scattering is considered a purely elastic event. Since there is no change in the energy of an electron between its initial and final states, the collision probability $W_{kk'}$ is zero unless $E(\mathbf{k}) = E(\mathbf{k}')$. If U is the interaction potential, first-order perturbation theory gives for the collision probability:

$$W_{kk'} = \frac{2\pi}{\hbar} \delta[E(\mathbf{k}) - E(\mathbf{k}')] |\langle \mathbf{k}|U|\mathbf{k}'\rangle|^2. \tag{77}$$

During linearization of the Boltzmann equation, we invoke the principle of detailed balance, which requires that $W_{kk'} = W_{k'k}$. Thus, writing for the number of states of a particular spin in volume V, $V d\mathbf{k}/(2\pi)^3$, simplifies Eq. (32) to

$$\left(\frac{\partial f}{\partial t}\right)_{\text{coll}} = \frac{V}{(2\pi)^3} \int d\mathbf{k}' W_{kk'} [f(\mathbf{k}') - f(\mathbf{k})]. \tag{78}$$

The terms $[1 - f(\mathbf{k})]$ and $[1 - f(\mathbf{k}')]$ fall out and the exclusion principle does not affect the rate of change due to collisions. In order to relate the transition probability $W_{kk'}$ to the relaxation time τ, we must use a formula derived for the modified distribution function $f(\mathbf{r},\mathbf{k},t)$. The simplest case is that corresponding to an electric field, and let us say we take it along the x-axis; i.e.,

$$f = f^o + \tau e\, \boldsymbol{\varepsilon}\, v_x \frac{\partial f^o}{\partial E}. \tag{79}$$

From the definition of $(\partial f/\partial t)_{\text{coll}}$ we have

$$\left(\frac{\partial f}{\partial t}\right)_{\text{coll}} = \frac{f^o - f}{\tau} = -e\, \boldsymbol{\varepsilon}\, v_x \frac{\partial f^o}{\partial E}. \tag{80}$$

Substituting for $f(\mathbf{k})$ in Eq. (78), and equating the right-hand side with that of Eq. (80), we get for the relaxation time

$$\frac{1}{\tau} = \frac{V}{(2\pi)^3} \int d\mathbf{k}' W_{kk'} \left(1 - \frac{v'_x}{v_x}\right) = \frac{V m^* k_F}{2\pi \hbar^3} \int_0^\pi |\langle \mathbf{k}|U|\mathbf{k}'\rangle|^2 (1 - \cos\theta) \sin\theta\, d\theta. \tag{81}$$

In deriving the second equality in Eq. (81), we use the scattering geometry in Fig. 2. Equation (81) indicates that the scattering frequency is weighted with a factor $(1 - \cos\theta)$. This means that not all scattering processes are equally effective in altering the direction of the electron velocity. Much more weight is given to those processes that scatter electrons through a large angle.

Let us consider the case of a static charged impurity of valence Z that occupies a lattice site. The conduction electrons "feel" the presence of this impurity through a *screened Coulomb interaction*. This arises because the conduction electrons will try to respond to the electric field of the impurity and will distribute themselves so as to cancel the electric field at large distances. In the simplest case, modeling the electron gas as a free-particle system, the screened potential has the form of the Yukawa potential:

$$U(r) = -\frac{Ze^2}{4\pi\varepsilon_o r} e^{-k_{TF}r}, \tag{82}$$

where the parameter k_{TF} is called the Thomas–Fermi screening parameter and its inverse $(k_{TF})^{-1}$ is often referred to as the screening length. The Thomas–Fermi screening parameter is given solely by the electron density and can be written as

$$k_{TF} = \left(\frac{4k_F}{\pi a_o}\right)^{1/2}, \tag{83}$$

where k_F is the Fermi wave vector (or momentum) $[k_F = (3\pi^2 n)^{1/3}]$ and $a_o = \hbar^2/me^2 = 0.529 \times 10^{-10}$ m is the Bohr radius. For the electron densities in metals, the screening length is on the order of lattice spacing; i.e., electrons in metals are very effective in screening the electric field of impurity ions. The matrix element for this potential, $\langle \mathbf{k'}|U_{imp}|\mathbf{k}\rangle$, representing electron scattering from a state $\mathbf{k}$ to a state $\mathbf{k'}$ is the Fourier transform of the screened potential:

$$\langle \mathbf{k'}|U_{imp}|\mathbf{k}\rangle = -\frac{Ze^2}{\varepsilon_o\left(|\mathbf{k}-\mathbf{k'}|^2 + k_{TF}^2\right)V}. \tag{84}$$

Substituting this matrix element into Eq. (81), working out the integral, and introducing the density of impurities $n_i = N_i/V$, one obtains for the impurity resistivity, ρ_{imp}:

$$\rho_{imp} = \frac{m}{ne^2\tau} = \frac{2\pi m\, n_i}{3ne^2}\frac{E_F Z^2}{n}\frac{1}{\hbar}F\left(\frac{4k_F^2}{k_{TF}^2}\right), \tag{85}$$

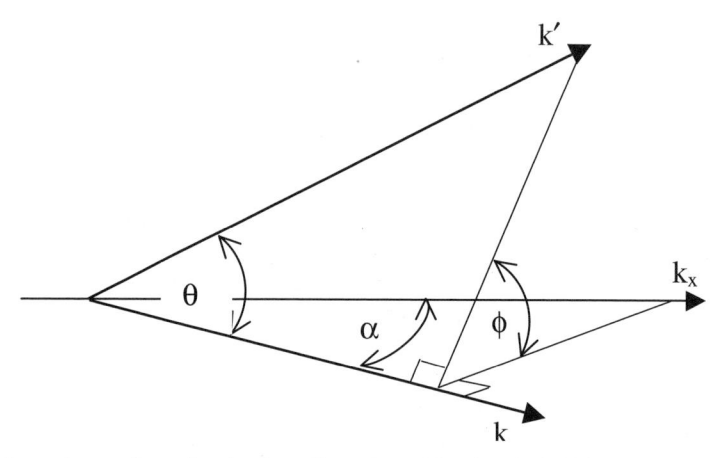

FIGURE 2 Scattering configuration in three dimensions. The electric field is assumed to be in the x-direction. Magnitudes of both initial and final electron wave vectors are k_F. A current change results from changes in the electron velocity with respect to the direction of the electric field. Assuming free electrons, the direction of the electron velocity coincides with the direction of its wave vector. Consequently, after averaging over ϕ, the change in the current is given by $(k_x - k'_x)/k_x = 1 - v'_x/v_x = (1-\cos\theta)$.

Sec. 4 · SCATTERING PROCESSES

where the function F is

$$F(x) = \frac{2[\ln(1+x) - x/(1+x)]}{x^2}. \tag{86}$$

It is of interest to estimate the impurity resistivity per 1% of impurity. As an example, we take copper where the Fermi energy is about 7 eV and the electron density is 8.5×10^{28} m^{-3}. Equations (83)–(86) then yield the impurity resistivity contribution of 1.8×10^{-8} Ω m per 1% of monovalent ($Z = 1$) impurity. This result is in a reasonable agreement with experimental values (see Table 3), although only two entries (Ag and Au) can be considered monovalent solutes.

The scattering cross section calculated with the Born approximation often tends to overestimate the cross section. The partial wave method is considered a better approach. In this scheme a charged impurity represents a spherically symmetric potential for which the Schrödinger equation has an asymptotic solution of the form

$$\psi_{k,l} \sim \frac{1}{r} \sin\left(kr - \frac{l\pi}{2} + \eta_l\right) P_l(\cos\theta). \tag{87}$$

Here $P_\ell(\cos\theta)$ is a spherical harmonic of order ℓ. By matching the functions $\psi_{k,\ell}$ to a plane wave (representing the initial state), one can write for the differential scattering cross section:

$$\sigma(\theta) = \frac{1}{k^2} \left| \sum_{l=0}^{\infty} (2l+1) e^{i\eta_l} \sin\eta_l P_l(\cos\theta) \right|^2. \tag{88}$$

TABLE 3 Resistivity increase in Cu per 1 at.% of Solute.[a]

Impurity	$\Delta\rho/c$ (10^{-8} Ω m/1 at.%)	Impurity	$\Delta\rho/c$ (10^{-8} Ω m/1 at.%)
Ag	0.14	Mn	4.3
Al	1.25	Ni	1.14
As	6.5	Pd	0.84
Au	0.53	Pt	1.86
B	1.4	Sb	5.4
Be	0.64	Se	10.6
Co	5.8	Si	3.25
Fe	14.5	Sn	2.8
Ge	3.7	V	5.8
In	1.2	Zn	0.25

[a] Data are taken from extensive tables in Landolt-Börstein, New Series.[10]

The phase shifts η_l must satisfy the Friedel sum rule,

$$Z = \frac{2}{\pi} \sum_l (2l+1)\eta_l(k_F), \quad (89)$$

where Z is the valence difference between the impurity and the solvent metal.

It is important to note that static lattice defects (such as charged impurities) scatter all electrons with equal effectiveness; consequently no temperature dependence arises in electrical resistivity due to static lattice defects. This contrasts with phonon transport, where the static defect scattering affects phonons of different frequencies differently, leading to a temperature dependence.

We conclude this section by recapitulating the main points: Because the electrons participating in transport are highly degenerate and are contained within a narrow band of width $\sim k_B T$ around the Fermi level, as they encounter a static lattice defect they will scatter elastically. A static lattice defect "looks" the same to all electrons (a very different situation from that of phonons), and the relaxation time will be temperature independent. Of course, different static crystalline imperfections (either point imperfections, such as interstitial atoms and vacancies, or extended imperfections, such as dislocations) scatter with different strength and effectiveness. Relaxation times relevant to each specific situation have been worked out and can be found in the literature.[3,72] The essential point concerning all of them is that, in each case, the relaxation time is independent of temperature. Collisions with static imperfections result in a large-enough change in the electron wave vector so that both electrical and thermal processes are affected to the same degree. Electrical resistivity that is given purely by electron scattering on static imperfections is often called the residual impurity resistivity. It is the resistivity that a metal would attain at very low temperatures. The common relaxation time for electrical and thermal transport when electrons scatter on static imperfections means that the Wiedemann–Franz law is valid. From this we glean that the impurity-dominated electronic thermal conductivity will be expected to follow linear temperature dependence and its inverse—thermal resistivity W_{imp}—will be inversely proportional to temperature:

$$W_{imp}(T) = \frac{\rho_{imp}}{L_o T} = \frac{B_{imp}}{T}, \quad (90)$$

where $B_{imp} = \rho_{imp}/L_o$.

4.2. Electron–Phonon Scattering

The most important process that limits the electrical and thermal currents has its origin in a particular kind of a crystal lattice imperfection—the disturbance of perfect periodicity created by a vibrating lattice ion. Just as an electron scatters when the lattice periodicity is interrupted by the presence of an impurity at a lattice site, the disturbance arising from vibrating ions (phonons in the language of quantum mechanics) leads to a deflection of an electron from its original path. The major difference between impurity scattering and electron–phonon scattering is that the former represents a purely elastic process, whereas the latter involves emission or absorption of a phonon, i.e., a small but finite change in the electron energy. From a practical perspective, one has some control over the impurity processes because one can, at least in principle, prepare samples of exceptional

Sec. 4 · SCATTERING PROCESSES

crystalline quality with an exceedingly small amount of impurity in the structure. In contrast, there is nothing one can do about a perturbation created by a vibrating lattice—it is there to stay whether we like it or not.

The topic of electron–phonon interaction is one of the pillars of solid–state physics. It presents a challenge that requires the full power of a quantum mechanical treatment of the problem. Rather than reproducing a lengthy and complex derivation that can be found in several texts, we will sketch the key steps and give results that are essential in the discussion of electrical and thermal transport in metals. Readers interested in a rigorous treatment of the electron–phonon interaction may find it useful to consult, among many others, Refs. 3, 5, 63, and 73.

In electron–phonon scattering we consider two distinct distributions, those of the electrons and phonons. We know how to treat each one separately–electrons are Bloch waves and obey Fermi–Dirac statistics, and vibrating metal ions show small displacements from their equilibrium positions and thus can be modeled as harmonic oscillators obeying Bose–Einstein statistics. In Chapter 1.1, in the discussion of heat transport by lattice vibrations, it was described how the coupled motions of atoms are transformed to form normal modes, each specified by its wave vector **q** and a frequency of vibration ω_q. These normal modes are quantized as harmonic oscillators, and the energy of a mode with frequency ω_q is $(n + 1/2)\hbar\omega_q$, n being an integer. Excitations of the lattice are energy quanta $\hbar\omega_q$, called phonons. The equilibrium distribution function, i.e., the mean number of thermally excited quanta $\hbar\omega_q$ at a temperature T (the average number of phonons), is given by the Planck distribution function (Bose–Einstein statistics):

$$N^o = \frac{1}{e^{\hbar\omega_q/k_B T} - 1}. \tag{91}$$

If the unit cell contains only one atom, there are only three acoustic modes (branches) of vibrations: one longitudinal with the atomic displacements parallel to the wave vector **q**, and two transverse, with the displacements perpendicular, to **q** and to each other. The dependence of the vibrational frequency ω_q on the wave vector **q**, $\omega_q(\mathbf{q})$, is called the dispersion relation. In the long-wavelength limit ($q \to 0$), $\omega_q = v_s q$, where v_s is the speed of sound. If more than one atom resides in the unit cell, more modes arise and they are called optical branches (the number of optical branches is given by $3(p - 1)$, where p is the number of atoms in the unit cell). Optical branches have non–zero frequency at $q = 0$. The group velocity of a mode, $v_g = d\omega_q/dq$, indicates the speed with which wave packets travel in the crystal and transport thermal energy. Because the optical branches are rather flat, the derivative $d\omega_q/dq$ is quite small, which explains why optical branches are substantially ineffective as carriers of heat; thus the task of transporting heat is assigned to acoustic phonon modes. Typical speeds of sound in metals are a few thousand meters per second compared to the speed of sound in air, 340 m/s. In describing the vibrational spectrum of solids, one often relies on the Debye model. A crystal is viewed as consisting of N lattice sites that give rise to $3N$ acoustic vibrational modes. Moreover, the vibrational spectrum is truncated so that the highest frequency that can be excited—the Debye frequency ω_D—is given by

$$\int_0^{\omega_D} D(\omega)d\omega = 3N. \tag{92}$$

In Eq. (92) $D(\omega)$ is the phonon density of states equal to

$$D(\omega) = \frac{3V_o\omega^2}{2\pi^2 v_s^3}, \tag{93}$$

where V_o is the sample volume. The question is how to describe the interaction between the electrons and phonons.

As pointed out, the feature that distinguishes the electron–phonon interaction from electron-impurity scattering is the quantum of energy $\hbar\omega_q$ that an electron absorbs or emits as it is scattered by a phonon. This is an important constraint because the electron's energy is changed in the interaction, i.e., the scattering is inelastic. We assume that only one quantum of energy (one phonon) is involved in the scattering process, and therefore the scattering satisfies the requirement

$$E(\mathbf{k}) \pm \hbar\omega = E(\mathbf{k}'). \tag{94}$$

If the wave vectors of the incoming and outgoing Bloch states are $\mathbf{k}$ and $\mathbf{k}'$, and the phonon wave vector is $\mathbf{q}$, the condition the wave vectors must satisfy (one can view it as a selection rule) is

$$\mathbf{k} \pm \mathbf{q} = \mathbf{k}' + \mathbf{G} \tag{95}$$

where $\mathbf{G}$ is a reciprocal lattice vector. Equation (95) delineates two distinct kinds of processes: interactions for which $\mathbf{G} = 0$ are called normal or N-processes; interactions where a nonzero vector $\mathbf{G}$ enters are known as umklapp or U-processes. In essence, Eq. (95) is a statement of the conservation law of momentum ($\mathbf{p} = \hbar\mathbf{k}$); in the case of U-processes the vector $\mathbf{k} + \mathbf{q}$ reaches over the edge of the first Brillouin zone, and the role of a reciprocal vector $\mathbf{G}$ is to bring it back to the first zone. As Fig. 3 illustrates, U-processes are exceptionally effective in generating resistance because they are capable of nearly reversing the direction of the electron wave vector. Of course, this necessitates a presence of phonons with a certain minimum wave vector which are more plentiful at high, rather than low, temperatures. Therefore, U-processes typically dominate at high-temperatures, and their influence weakens exponentially as the temperature decreases.

Now, because the lattice ions are much heavier than electrons ($M/m_e \sim 10^3$), it is assumed that the electrons quickly adjust to changes in the ionic positions. Consequently, within this concept, generally known as the Born–Oppenheimer approximation, the actual electron–phonon interaction is treated as a perturbation on the Bloch states that are calculated under the assumption that the ions are in equilibrium positions.

Considering electron–phonon interaction, one might think that the lower–energy cutoff (some kind of minimum electrical resistivity) would be given by the zero-point motion of the lattice ions, i.e., by the term $\hbar\omega(q)/2$ being emitted or absorbed by an electron. This, however, would be an incorrect viewpoint. To consider zero-point motion, we necessarily assume $T = 0$. But at absolute zero temperature neither phonon absorption nor phonon emission by electrons is possible. Absorption does not happen because $n_q = 0$, and phonon emission cannot take place because an electron would lose energy in such a process. Since at $T = 0$ the electron gas is perfectly degenerate, there is no available electron state below E_F to accommodate the electron.

Sec. 4 · SCATTERING PROCESSES

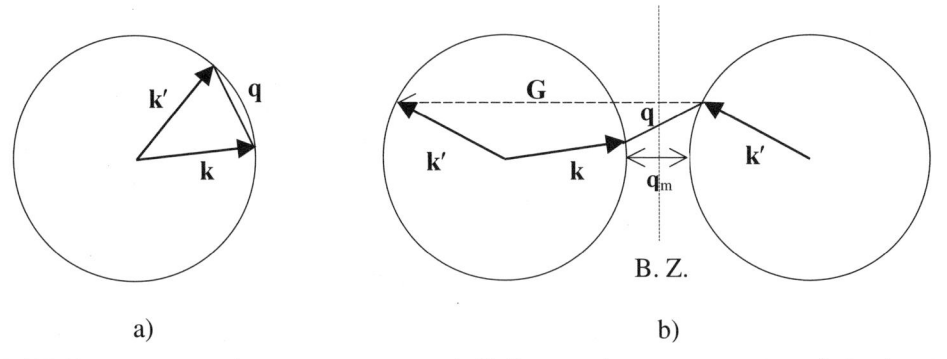

FIGURE 3 (a) Electron–phonon N-process $\mathbf{k}+\mathbf{q}=\mathbf{k}'$. (b) Electron–phonon U-process $\mathbf{k}+\mathbf{q}=\mathbf{k}'+\mathbf{G}$, where $\mathbf{G}$ is the reciprocal lattice vector. Note a minimum phonon wave vector $\mathbf{q}_m$ that is necessary for the U-process to take place. At low temperatures such wave vectors may not be available.

In order to describe the interaction between an electron and the lattice ions, let us assume that $g(\mathbf{k},\mathbf{k}',\mathbf{q})$ is a function that couples the Bloch states $\mathbf{k}$ and $\mathbf{k}'$ to the phonon of wave vector $\mathbf{q}$ via Eq. (95). We would like to find out the form of this coupling function, assuming a simple model that neglects any role of U-processes. This is the same simplification made by Bloch in the derivation of his Bloch–Grüneisen formula for the electrical conductivity. Consider an electron situated at point $\mathbf{r}$ that "senses" a lattice ion located at point $\mathbf{R}_i$ through a potential $U(\mathbf{r} - \mathbf{R}_i)$. The position of the ion $\mathbf{R}_i$ can be specified with respect to its equilibrium position $\mathbf{R}_i^o$ and the displacement $\delta\mathbf{R}_i$. The potential energy of an electron in the field of all ions of the lattice is the interaction Hamiltonian $\mathcal{H}_{\text{e-ion}}$:

$$\mathcal{H}_{\text{e-ion}} = \sum_i U\left[\mathbf{r} - \left(\mathbf{R}_i^o + \delta\mathbf{R}_i\right)\right], \tag{96}$$

where the summation runs over all ion sites. The potential $U[\mathbf{r} - (\mathbf{R}_i^o + \delta\mathbf{R}_i)]$ has an electrostatic origin except that electrons have an excellent ability to screen the long-range tail of the Coulomb potential. One of the simplest forms of screening—Thomas–Fermi theory—assumes a slowly varying potential as a function of $\mathbf{r}$ given in Eq. (82) together with its screening parameter k_{TF} in Eq. (83). The required smoothness of the potential must be on the scale of a Fermi wavelength, which, stated equivalently, requires $q \ll k_F$, i.e., the calculation using the Thomas–Fermi screening potential will be reliable only in the long-wavelength limit of lattice vibrations. Since we assume small displacements from the equilibrium position, we can expand the potential and, keeping only terms to first-order in $\delta\mathbf{R}_i$, we write

$$\mathcal{H}_{\text{e-ion}} = \sum_i U\left(\mathbf{r} - \mathbf{R}_i^o - \delta\mathbf{R}_i\right) = \sum_i U\left(\mathbf{r} - \mathbf{R}_i^o\right) - \sum_i \delta\mathbf{R}_i \nabla U\left(\mathbf{r} - \mathbf{R}_i^o\right). \tag{97}$$

The first term in the expansion is a constant and represents the potential to which the electrons are subjected when the ions are in their equilibrium lattice positions. This leads to the Bloch description of electrons in the crystalline lattice that we assume anyway. Thus, the electron–phonon interaction is described by the second term in Eq. (97). At this stage one can introduce the phonon coordinates $\mathbf{Q}_{q\alpha}$ via the displacements $\delta\mathbf{R}_i$:

$$\delta \mathbf{R}_i = N^{1/2} \sum_{q\alpha} \mathbf{Q}_{q\alpha} \hat{\mathbf{e}}_{q\alpha} e^{i\mathbf{q}\cdot\mathbf{R}_i^o}, \qquad (98)$$

where N is the number of lattice ions in a volume V_o and $\hat{\mathbf{e}}_{q\alpha}$ are unit polarization vectors chosen to be parallel or perpendicular to $\mathbf{q}$ when $\mathbf{q}$ is along a symmetry direction.

From this point, one would usually resort to using the second quantization formalism and express the phonon coordinates in terms of creation and annihilation operators. This is an elegant approach and one that keeps the terms easily tractable. Because not all readers may be familiar with this technique, we shall not follow this road. We simply recognize that we must calculate matrix elements of the form $<\mathbf{k}'|\delta \mathbf{R}_i \nabla U(\mathbf{r}-\mathbf{R}_i^o)|\mathbf{k}>$ and sum them up for all ions i. The actual computation is given in Ref. 4.

As noted, we only consider N-processes. If we further simplify the problem by treating electrons as free electrons (i.e., if we use a plane-wave $\exp(i\mathbf{k}\cdot\mathbf{r})/(V_o)^{1/2}$ to describe an electron in state $\mathbf{k}$), the coupling function for the electron–phonon interaction becomes a function of $\mathbf{q} = \mathbf{k}' - \mathbf{k}$ only; thus

$$C_q = \frac{ie^2 z}{(1 + k_{TF}^2/q^2)\varepsilon_o q^2 V_o} \sqrt{\frac{N\hbar}{2M\omega_q}} (\mathbf{q}\cdot\hat{\mathbf{e}}_q). \qquad (99)$$

Here, z is the charge state of an ion, and the subscript α is dropped because the polarization is assumed to be either longitudinal or transverse to the wave vector $\mathbf{q}$. We need the square of the modulus of C, which in the long-wavelength limit is

$$|C_q|^2_{q\to 0} = \frac{2\pi^2}{3} \frac{zE_F^2 \hbar \omega_q}{v_s^2 k_F^3 V_o M}. \qquad (100)$$

Eq. (100) is usually written in the form

$$|C_q|^2_{q\to 0} = \lambda \frac{\hbar \omega_q}{2N(0)V_o}, \qquad (101)$$

where

$$N(0) = \frac{mk_F}{2\pi^2 \hbar^2}, \qquad (102)$$

is the density of states per spin at the Fermi level, and the dimensionless parameter λ is

$$\lambda = \frac{2z}{3} \frac{E_F^2}{(k_B \Theta_s)^2} \frac{m_e}{M} = \frac{2z}{3} \frac{E_F^2}{(k_B \theta_D)^2} \left(\frac{q_D}{k_F}\right)^2 \left(\frac{c}{v_s}\right)^2 \frac{m_e}{M}. \qquad (103)$$

The temperature Θ_s in Eq. (103) is related to the longitudinal sound velocity v_s via the equation

$$\Theta_s = v_s \hbar k_F / k_B. \qquad (104)$$

The temperature Θ_s defined in Eq. (104) and the Debye temperature θ_D, in terms of which most of the asymptotic forms of the transport formuli and their integration limits are usually stated in the texts, are related by

Sec. 4 · SCATTERING PROCESSES

TABLE 4 Temperatures Θ_s and θ_D Together with the Parameter λ of Eq. (103) for Selected Metals.[a]

Metal	Θ_s (K)	θ_D (K)	θ_R (K)	λ (dimensionless)
Na	220	150	200	0.47
K	150	100	114	0.25
Cu	490	315	330	0.16
Ag	340	215	220	0.12
Al	910	394	395	0.90

[a] Values of Θ_s and λ are from Ref. 4, those of θ_D are from Ref. 74, and θ_R is from Ref. 75.

$$\frac{\Theta_s}{\theta_D} = \frac{v_s}{c}\frac{k_F}{q_D}. \quad (105)$$

Here c is the speed of sound in the Debye model, which is understood to be the average of the long-wavelength phase velocities of all three acoustic modes, not just the longitudinal one v_s. Since for free electrons $k_F = (2/Z)^{1/3}\mathbf{q}_D$, where Z is the nominal valence and $\mathbf{q}_D$ is the Debye wave vector, Θ_s could be larger (perhaps as much as a factor of 2, depending on Z) than the Debye temperature θ_D (see Table 4.).

Summarizing the approximations made to obtain the result in Eq. (101), we have:

a. Interaction processes were restricted to involve only N-processes
b. Thomas–Fermi screening theory was used
c. Electrons were treated as free electrons.

Such simplifications, of course, have consequences in terms of the validity of the calculation. Even in the simplest of metals (alkali and noble metals), U-processes cannot be completely neglected. Moreover, the coupling constant is appropriate for the long-wavelength limit only and not for phonons of higher frequency. Finally, Fermi surfaces of metals are, in general, quite complicated and electrons are not exactly free electrons. In spite of these limitations, the form of the electron–phonon coupling constant derived is used rather indiscriminately in most calculations. The reason is obvious—such a treatment yields a rather simple and compact result, and alternative approaches lead to very messy calculations, if possible at all.

With the coupling constant given by Eq. (101), one may proceed to calculate the collision integral $(\partial f(\mathbf{k})/\partial t)_{\text{coll}}$ and inquire about the form of the relaxation time describing the electron–phonon interaction:

$$\left(\frac{\partial f}{\partial t}\right)_{\text{coll}} = -\frac{2\pi}{\hbar}\sum_{k'\alpha}|C|^2\{[f_k(1-f_{k'})(1+N_{-q}) - f_{k'}(1-f_k)N_{-q}]\delta(E_{k'}-E_k+\hbar\omega_{-q})$$

$$+ [f_k(1-f_{k'})N_q - f_{k'}(1-f_k)(1+N_q)]\delta(E_{k'}-E_k-\hbar\omega_q)\}. \quad (106)$$

The factors such as $f_k(1-f_{k'})$ enter because of the Pauli exclusion principle, and N_q and N_{-q} stand for phonon distributions under the processes of phonon emission and absorption, respectively. Of course, when both electrons and phonons are in equilibrium, the collision term must vanish. When the distributions are disturbed by an electric field or thermal gradient, the deviations from the equilibrium population can be expressed with the aid of the deviation functions ψ_k for electrons and Φ_q for phonons:

$$f_k = f^o + f^o(1 - f^o)\psi_k, \tag{107}$$

$$N_q = N^o + N^o(1 + N^o)\phi_q. \tag{108}$$

Computations are considerably simplified if one makes the following approximation. Provided the phonon–phonon and phonon-impurity scattering processes are very frequent so that they relax the population of phonons effectively, the electron will interact with a phonon that is very close to equilibrium. Hence, one sets the deviation function of phonons to zero. With the identity equation that relates equilibrium distribution functions of electrons and phonons,

$$f^o(E)[1 - f^o(E + \hbar\omega)]N^o(\omega) = f^o(E + \hbar\omega)[1 - f^o(E)](1 + N^o(\omega)), \tag{109}$$

one can linearize Eq. (106). Introducing a relaxation time τ_k, one simplifies Eq. (106) to

$$\frac{1}{\tau_k} = \frac{2\pi}{\hbar} \sum_q |C|^2 \left[\delta(E_k - E_{k'} - \hbar\omega_q)\left(N_q^o + 1 - f_{k'}^o\right) + \delta(E_k - E_{k'} + \hbar\omega_q)\left(N_q^o + f_{k'}^o\right) \right]. \tag{110}$$

The coupling function depends only on the difference between the wave vectors of the outgoing and incoming electrons, $\mathbf{q} = \mathbf{k}' - \mathbf{k}$. Assuming again that the electrons are described by a free-electron model and converting from summation to integration, we obtain for the high- and low-temperature limits[*] the following temperature dependences for the electron–phonon relaxation rate:

$$\begin{aligned} 1/\tau_k &\propto T & \text{for} \quad T \geq \theta_D, \\ &\propto T^3 & \text{for} \quad T \ll \theta_D. \end{aligned} \tag{111}$$

The high-temperature result—the relaxation rate being a linear function of temperature—could have been anticipated based on a classical argument utilizing equipartition of energy, which is valid since we consider high-temperatures. One would expect the resistance to be proportional to the mean square displacement of an ion in a given direction. An ion of mass M with acceleration d^2x/dt^2 is subjected to a restoring force $-bx$ so as to satisfy the equation

$$M\ddot{x} + bx = 0. \tag{112}$$

Since $b/M = \omega^2 = 4\pi^2 f^2$, where f is the frequency, we can write for the mean potential energy of this oscillator at high-temperatures:

---

[*] The Debye temperature qD is the usual upper limit of the summation in Eq. (110) and, thus, the temperature delineating the asymptotic regions of transport. However, this is applicable only for highly degenerate systems such as metals. For semimetals a more appropriate criterion is $q/2 \leq k_F$. Moreover, if the model assumes that electrons couple only to longitudinal phonons, a somewhat higher temperature, such as Θ_s in Eq. (104) might be more appropriate.

Sec. 4 · SCATTERING PROCESSES

$$\frac{1}{2}b\overline{x^2} = \frac{1}{2}k_B T. \tag{113}$$

The resistance then becomes

$$R \propto \overline{x^2} = \frac{k_B T}{b} = \frac{k_B T}{4\pi^2 f^2 M} = \frac{h^2 T}{4\pi^2 M k_B \theta_D} \propto \frac{T}{M\theta_D^2}. \tag{114}$$

In order to appreciate what happens to an electron when it interacts with a phonon at high-temperatures, it is instructive to consider the largest possible change in the direction of an electron wave vector in such an electron–phonon encounter. If we assume a free-electron-like monovalent metal (one free electron per lattice ion) consisting of N atoms, the Brillouin zone must accommodate N distinct q-values (modes of vibrations), while the Fermi sphere must be large enough to fit ($N/2$) k-states ($N/2$ rather than N because each state can have two electrons, one with spin up and one with spin down). Thus, the volume of the Brillouin zone will be twice as large as the volume of the Fermi sphere, which implies that the ratio $q_D/k_F = 2^{1/3} \cong 1.26$. Therefore, the maximum change in the electron wave vector is $\Delta k = q_D = 1.26 k_F$. From Fig. 4 it follows that the largest possible angle θ is given by $\sin(\theta/2) = (q/2)/k_F$. Inverting this expression, we obtain the maximum scattering angle $\theta_{\max} = 2\sin^{-1}(q/2k_F) = 78°$. Of course, this is an idealized case not fully applicable even in alkali metals. Nevertheless, the estimate shows clearly that high-energy phonons (which are plentiful at high-temperatures) can, in a single collision, produce a large change in the direction of an electron with very little change in its energy. Because high-temperature phonons can relax out-of-equilibrium populations of electrons created by either an electric field or a thermal gradient in a single scattering event, there is one common relaxation time for both electrical and thermal processes.

As for the low-temperature limit of Eq. (111), it states that for temperatures well

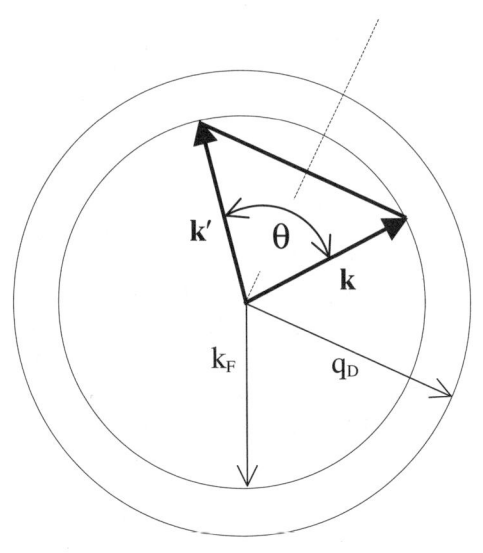

FIGURE 4 Relationship between the Fermi and Debye spheres for free electrons: $q_D = 2^{1/3} k_F \cong 1.26 k_F$. The maximum scattering angle $\theta_{\max}$ is given by $\theta_{\max} = 2\sin^{-1}(q_D/2k_F) \cong 78°$.

below θ_D the electron–phonon scattering rate declines as T^3. This must not be confused with the rate at which the current density diminishes because the two rates are not the same. What comes into play is the efficiency of scattering. It expresses the fact that, as the temperature decreases, phonon wave vectors become smaller and therefore the scattering angle decreases. Thus, more collisions are needed to make a substantial change in the direction of the electron velocity. We have seen in Eq. (63) that this is taken care of by introducing the efficiency factor $(1 - \cos\theta)$. From Fig. 4 it follows that

$$1 - \cos\theta = 2\sin^2\left(\frac{\theta}{2}\right) = 2\left(\frac{q}{2k_F}\right)^2 = \frac{1}{2}\left(\frac{q}{k_F}\right)^2. \tag{115}$$

Because $q \sim k_B T/\hbar v_s$, the factor $(1 - \cos\theta)$ introduces a temperature factor T^2. Thus, at low temperatures, the electrical resistivity is proportional to T^5. This result can be derived rigorously by using Kohler's variational technique to actually solve the Boltzmann equation. We will not reproduce the mathematical derivation, but merely state the final results.

$$\frac{1}{\tau_\sigma} = \frac{9\pi h^2 C^2}{8\sqrt{2}(m^*)^{1/2} \, ea^3 M k_B \theta_D E_F^{3/2}} \left(\frac{T}{\theta_D}\right)^5 J_5\left(\frac{\theta}{T}\right) = \frac{ne}{m^*} A \left(\frac{T}{\theta_D}\right)^5 J_5\left(\frac{\theta}{T}\right), \tag{116}$$

where

$$A = \frac{9\pi h^2 (m^*)^{1/2} C^2}{8\sqrt{2}ne^2 a^3 M k_B \theta_D E_F^{3/2}} \tag{117}$$

and

$$J_n\left(\frac{\theta}{T}\right) = \int_0^{\theta/T} \frac{x^n e^x}{(e^x - 1)^2} dx. \tag{118}$$

Inverting Eq. (61) and substituting for the relaxation time from Eq. (116), one obtains the famous Bloch–Grüneisen formula for the electrical resistivity:[76,77]

$$\rho_{\text{e-p}}(T) = \frac{m}{ne^2} \frac{1}{\tau_\sigma} = A\left(\frac{T}{\theta_D}\right)^5 J_5\left(\frac{\theta}{T}\right). \tag{119}$$

We indicate by subscript e-p that the resistivity arises as a result of the electron–phonon interaction. In most texts such a resistivity is referred to as the *ideal resistivity*—ideal in the sense that it would be the resistivity in the absence of impurities (i.e., the resistivity of an ideal solid). Note that different authors define the coefficient A differently. Apart from the form of Eq. (117), one often comes across a coefficient that is four times smaller, namely, $A' = A/4$. In that case, Eq. (119) is written with a prefactor of $4A'$.

It is of interest to look at the asymptotic solutions for low and high-temperatures:

$$\rho_{\text{e-p}}(T) = A\left(\frac{T}{\theta_D}\right)^5 \frac{(\theta_D/T)^4}{4} = \frac{A}{4}\left(\frac{T}{\theta_D}\right) \quad \text{for } T \geq \theta_D \tag{120}$$

and

Sec. 4 · SCATTERING PROCESSES

$$\rho_{e-p}(T) = 124.4A \left(\frac{T}{\theta_D}\right)^5 \text{ for } T \ll \theta_D. \tag{121}$$

In Eq. (120) we used the approximation $J_n(\theta/T) \cong (\theta/T)^{(n-1)/(n-1)}$, valid for high-temperatures, and in Eq. (121) we substituted $J_5(\infty) = 124.4$, applicable at $T \ll \theta_D$. Thus, at high-temperatures the resistivity is a linear function of temperature, while at low temperatures it should fall very rapidly following a T^5 power law. Note that the T^5 behavior of the resistivity is rarely observed. Apart from the obvious experimental challenges to measure resistivity at low temperatures with a precision high enough to ascertain temperature variations against a now large impurity contribution and a contribution due to electron–electron scattering (see the next section), we should keep in mind the simplifications introduced in the process of calculations, the main ones being spherical Fermi surfaces, N-processes only, Thomas–Fermi approximation, equilibrium phonon distribution. All of these do affect the final result. It is remarkable that with basically the same approach the theorists are able to calculate an electrical resistivity of even transition metals that agrees with the experimental data to within a factor of 2 (see Table 5).

We note that by fitting the experimental resistivity data to the Bloch-Grüneisen formula [Eq. (119)], we can obtain a characteristic temperature θ_R that plays a similar role as θ_D or the characteristic temperature associated with long-wavelength phonons, Θ_s. The temperature θ_R is usually much closer to θ_D (see Table 4) than to Θ_s, indicating that transverse phonons do play a role in electron scattering.

We now turn our attention to a relaxation time relevant to electronic thermal conductivity. This is obtained by performing similar variational calculations to those that led to Eq. (116) except that the trial function is now taken to be proportional to $E - E_F$:

$$\frac{1}{\tau_\kappa} = \frac{9\pi h^2 C^2}{2\sqrt{2}(m^*)^{1/2} a^3 M k_B \theta_D E_F^{3/2}} \left(\frac{T}{\theta_D}\right)^5 J_5\left(\frac{\theta}{T}\right) \times$$

$$\left\{1 + \frac{3}{\pi^2}\left(\frac{k_F}{q_D}\right)^2 \left(\frac{\theta_D}{T}\right)^2 - \frac{1}{2\pi^2} \frac{J_7(\theta/T)}{J_5(\theta/T)}\right\} \tag{122}$$

TABLE 5 Comparison of Calculated and Experimental Room Temperature Electrical Resistivities of Several Pure Metals.[a]

Metal	Ag	Au	Cu	Fe	Mo	Nb	Ni	Pd	Ta	W
Experiment	1.61	2.2	1.7	9.8	5.3	14.5	7.0	10.5	13.1	5.3
Calculation	1.5	1.9	1.9	11.4	6.1	18.4	5.2	5.2	13.2	7.2

[a] Calculated values are from Refs. 78 and 79, and the experimental data are from Ref. 80. Resistivities are in units of $10^{-8} \Omega$ m.

Electronic thermal resistivity arising from electron–phonon interaction (called ideal thermal resistivity) is then

$$W_{e-p}(T) = (\kappa_{e-p})^{-1} = \left(\frac{1}{3}C_v v_F^2 \tau_\kappa\right)^{-1}$$

$$= \frac{A}{L_o T}\left(\frac{T}{\theta_D}\right)^5 J_5\left(\frac{\theta}{T}\right)\left[1 + \frac{3}{\pi^2}\left(\frac{k_F}{q_D}\right)^2\left(\frac{\theta_D}{T}\right)^2 - \frac{1}{2\pi^2}\frac{J_7(\theta/T)}{J_5(\theta/T)}\right] \qquad (123)$$

where $C_v = (\pi^2 n k_B^2/2E_F)T$ is the electronic specific heat. The ideal thermal resistivity consists of three distinct contributions. The first contribution is recognized as having the same functional form as the one in Eq. (119) for electrical conductivity. This term arises from large-angle scattering and, therefore, satisfies the Wiedemann–Franz law. The second term is due to inelastic small-angle scattering (vertical processes) and has no counterpart in the electrical resistivity. The third term is a correction that accounts for situations where large-angle scattering can reverse the electron direction without actually assisting to restore the distribution back to equilibrium. This is due to the nature of the thermally perturbed distribution function, where an electron can be scattered through large angles between regions having similar deviations (for instance, regions marked + in Fig. 1b).

Frequently, asymptotic forms of Eq. (123) are required for low and high-temperatures, and they are

$$W_{e-p}(\text{H.T.}) = \frac{A}{4}\frac{1}{L_o T}\left(\frac{T}{\theta}\right) = \frac{\rho_{e-p}(\text{H.T.})}{L_o T} \quad \text{for } T \geq \theta_D \qquad (124)$$

$$W_{e-p}(\text{L.T.}) = \frac{A}{L_o T}\left(\frac{T}{\theta_D}\right)^5 J_5\left(\frac{\theta}{T}\right)\frac{3}{\pi^2}\left(\frac{k_F}{q_D}\right)^2\left(\frac{\theta_D}{T}\right)^2$$

$$= \frac{\rho_{e-p}(\text{L.T.})}{L_o T}\frac{3}{\pi^2}\left(\frac{k_F}{q_D}\right)^2\left(\frac{\theta_D}{T}\right)^2 = 37.8\frac{A}{L_o \theta_D}\left(\frac{k_F}{q_D}\right)^2\left(\frac{T}{\theta_D}\right)^2 \quad \text{for } T \ll \theta_D. \qquad (125)$$

Equation (124) is just the first term of Eq. (123) in the limit of high-temperatures. The second and third terms in the square brackets of Eq. (123) are both very small because at high-temperatures $J_7/J_5 \to (1/6)(\theta/T)^6/(1/4)(\theta/T)^4 \to (2/3)(\theta/T)^2 \to 0$. The second equality in this equation follows from using Eq. (120). Thus, at high-temperatures, we expect the ideal electronic thermal resistivity to be temperature independent and, because the relaxation times for the electrical and thermal processes are essentially identical, the Wiedemann–Franz law is obeyed, as Eq. (124) demonstrates.

Equation (125) was obtained by considering only the second term in the bracket of Eq. (123) because this term rises rapidly as temperature decreases, while the first and third terms are constant (1 and 5082/124.4, respectively). The second equality in Eq. (125) is obtained with the aid of Eq. (121). The third equality follows from approximating $J_5(\theta_D/T) \cong J_5(\infty) = 124.4$. We note an important feature—the thermal resistivity at low temperatures follows a quadratic function of temperature and its inverse, the thermal conductivity $\kappa_{e-p}(T)$, should be proportional to T^{-2}. This is indeed what is most frequently observed. Considering all the approximations made, the numerical factors should not be expected to be very precise, perhaps to

Sec. 4 · SCATTERING PROCESSES

within a factor of 2 or 3. We also observe from Eq. (125) that, at low temperatures, the Wiedemann–Franz law is clearly not obeyed.

Writing for the Lorenz ratio $L = \kappa/\sigma T$ and substituting from Eqs. (119) and (123), we obtain

$$L = \frac{\kappa}{\sigma T} = \left(\frac{\rho}{WT}\right)_{\text{e-p}} = \frac{A\left(\frac{T}{\theta_D}\right)^5 J_5\left(\frac{\theta}{T}\right)}{\frac{A}{L_o T}\left(\frac{T}{\theta_D}\right)^5 J_5\left(\frac{\theta}{T}\right)\left[1 + \frac{3}{\pi^2}\left(\frac{k_F}{q_D}\right)^2\left(\frac{\theta_D}{T}\right)^2 - \frac{1}{2\pi^2}\frac{J_7(\theta/T)}{J_5(\theta/T)}\right]T}$$

$$= \frac{L_o}{1 + \frac{3}{\pi^2}\left(\frac{k_F}{q_D}\right)^2\left(\frac{\theta_D}{T}\right)^2 - \frac{1}{2\pi^2}\frac{J_7(\theta/T)}{J_5(\theta/T)}}. \tag{126}$$

Asymptotic forms of the ratio L/L_o at high and low temperatures are

$$\frac{L}{L_o} = 1 \text{ for } T \geq \theta_D \tag{127}$$

$$\cong \frac{\pi^2}{3}\left(\frac{q_D}{k_F}\right)^2\left(\frac{T}{\theta_D}\right)^2 \text{ for } T \ll \theta_D. \tag{128}$$

Equations (127) and (128) state an important result. At high-temperatures, the Lorenz ratio attains its Sommerfeld value L_o. In other words, the relaxation times for electrical conductivity and thermal conductivity are identical, and the Wiedemann–Franz law holds. At these temperatures there are plenty of large–momentum (wave vector) phonons, and they scatter electrons elastically through large angles regardless of whether electron populations are perturbed by an electric field or a thermal gradient. At low temperatures, a very different scenario emerges. Here the Lorenz ratio rapidly decreases with decreasing temperatures and the Wiedemann–Franz law is obviously not obeyed. Phonon wave vectors decrease with temperature and they are incapable of scattering electrons through large angles. Thus, many collisions are necessary to change the velocity of electrons and therefore relax the electron distribution brought out of equilibrium by an electric field. In contrast, it is much easier to relax a thermally driven disturbance in the electron population. A very efficient route opens whereby electrons just slightly change their energy (inelastic processes) and pass through the thermal layer of width $\sim k_B T$ about the Fermi level, and the distribution is relaxed. This means that the relaxation time for thermal processes is considerably shorter than the one for electrical processes. The rate with which the two relaxation times depart as a function of temperature is given by Eq. (126), and a plot of L/L_o as a function of reduced temperature is shown in Fig. 1.10 in Sec. 6. At low temperatures the ratio attains a quadratic function of temperature. The decrease in the Lorenz ratio is particularly large in very pure metals because the temperature range where Eq. (128) is applicable extends to quite low temperatures before any influence of impurities is detected. Eventually, at some very low temperature, impurity scattering takes over and, since such processes are essentially elastic, the Lorenz ratio is restored.

4.3. Electron–Electron Scattering

Because of their high electron density, one might naively think that the electron–

electron interaction is a very important component of the electrical and thermal conductivities of metals. Just a rough estimate of the Coulomb energy [$e^2/4\pi\varepsilon_o r$] of two electrons separated by an interatomic distance comes out to be about 10 eV— an enormous amount of energy that exceeds even the Fermi energy. However, this is a very misleading and incorrect estimate. The point is that electrons have an exceptionally good ability to screen charged impurities. We have noted that, in metals, the screening length is typically on the order of an interatomic spacing. This fact, together with the constraints imposed by the Pauli exclusion principle, results in quite a surprising ineffectiveness of the electron–electron processes in metals, especially as far as electrical transport is concerned.

Interactions among electrons involve four electron states: two initial states that undergo scattering and two states into which the electrons are scattered. It should be clear that if only N-processes (processes in which the total electron momentum is conserved) are involved, the electron–electron interaction on its own would not lead to electrical resistivity because there is no change in the total momentum.[*] Thus, a possibility for a resistive process in interactions among electrons is tied to the presence of U-processes. But such processes are not too frequent because of the rather stringent conditions imposed by the Pauli exclusion principle. This is readily understood from an argument given in Ref. 2. Here we consider a metal at $T=0$ with fully occupied states up to the Fermi level and a single electron excited just above the Fermi level. Let us label this electron 1; its energy is $E_1 > E_F$. In order for this electron to interact, it must find a partner from among the electron states lying just below E_F because only these states are occupied. We label one such electron 2, and its energy is $E_2 < E_F$. The Pauli principle demands that after the interaction the two scattered electrons must go to unoccupied states and these are only available above the Fermi level; i.e., $E'_1 > E_F$ and $E'_2 > E_F$. Energy in this scattering process must be conserved,

$$E_1 + E_2 = E'_1 + E'_2, \tag{129}$$

and the wave vectors must satisfy

$$\mathbf{k}_1 + \mathbf{k}_2 = \mathbf{k}'_1 + \mathbf{k}'_2 + \mathbf{G}. \tag{130}$$

If E_1 is exactly equal to E_F, then E_2, E'_1, and E'_2 must also equal E_F. Thus, all four states would have to fall on the Fermi surface occupying zero volume in k-space and therefore yielding infinite relaxation time at $T=0$ K. With E_1 slightly larger than E_F (see Fig. 5), there is a very thin shell of k-space of thickness $|E_1-E_F|$ around the Fermi level available for final states E'_1 and E'_2. Because only two states (e.g., E_2 and E'_1) are independent rather than three [E'_2 is fixed by Eq. (129) once E_1, E_2, and E'_1 are chosen], the scattering rate is proportional to $(E_1 - E_F)^2$.

If we now assume a finite temperature, the electron distribution will be slightly smeared out (by an amount of thermal energy $k_B T$), and partially occupied levels will emerge in a shell of width $k_B T$ around E_F. Since each one of the two independent states now has this enlarged range of width $k_B T$ of possible values available, the scattering cross section is reduced by $(k_B T/E_F)^2 \sim 10^{-4}$ at ambient temperature. Thus, at room temperature, the contribution of the electron–electron processes to the transport effects in bulk metals is negligible in comparison to other

---

[*] This is strictly correct only for isotropic systems. Any source of anisotropy will give rise to a resistivity contribution associated with N-processes.

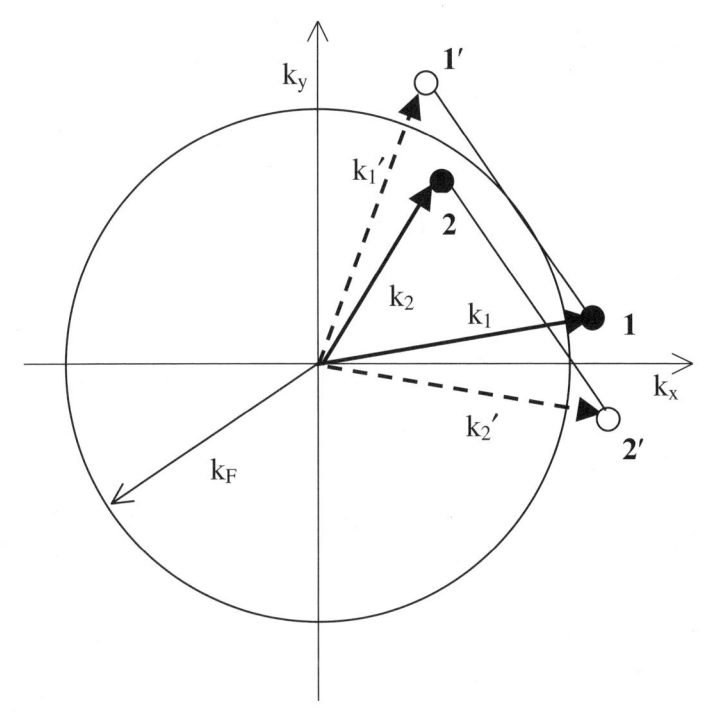

FIGURE 5 Schematic of a normal electron–electron interaction. Electrons 1 and 2 with initial wave vectors $\mathbf{k}_1$ and $\mathbf{k}_2$ scatter from one another to final states with wave vectors $\mathbf{k}_1'$ and $\mathbf{k}_2'$. The scattered electrons must go into unoccupied states (i.e., states above the Fermi energy).

scattering mechanisms. If one takes into account static screening, the collision frequency for a purely isotropic electron–electron scattering rate has been evaluated[81] and is

$$\frac{1}{\tau_{e-e}} = \frac{e^4(k_BT)^2}{16\pi\hbar^4\varepsilon_o^2 k_F v_F^3}\left[1 + \left(\frac{E_F}{\pi k_BT}\right)^2\right]\gamma\left(2k_F/k_{TF}\right), \quad (131)$$

where

$$\gamma(x) = \frac{1}{2}(V\varepsilon_o k_F^2/e^2)^2 \left\langle \frac{|w(k_1, k_2; k_1', k_2')|^2}{\cos(\theta/2)} \right\rangle$$

$$= \frac{x^3}{4}\left[\tan^{-1}x + \frac{x}{1+x^2} - \frac{1}{\sqrt{2+x^2}}\tan^{-1}x\sqrt{2+x^2}\right]. \quad (132)$$

The angular brackets denote an average over a solid angle, which is evaluated explicitly in Ref. 82.

The important point for our purposes is the T^2 dependence of the scattering rate. As the temperature decreases, the collision rate diminishes. The T^2 variation applies to bulk, three-dimensional metals. If a metallic structure has lower effective dimensionality, a different temperature dependence results. For instance, for a one-dimensional wire the electron–electron scattering rate is a linear function of temperature

TABLE 6 Calculated Values of the Coefficient A_{ee} (in units of $10^{-15}\,\Omega$ m K^{-2}) and the Effectivenes Parameter Δ (percent), Together with the Available Experimental values of A_{ee}.[a]

Metal	r_s/a_o	A_{ee} (calc)	Δ	A_{ee} (calc)/Δ	A_{ee} (exp)	Reference [b]
Li	3.25	2.1	0.054	40	30	94
Na	3.93	1.4	0.035	40	1.9	95
K	4.86	1.7	0.021	80	0.55–4.0	95–98
Rb	5.2	3.5	0.028	130		
Cs	5.62	6.7	0.028	240		
Cu	2.67	1.2	0.35	3	0.28–1.21	99–102
Ag	3.02	1.1	0.35	3	0.35±0.15	100, 103
Au	3.01	1.9	0.35	5	0.5±0.2	101
Al	2.07	4.1	0.4	10	2.8±0.2	104
Pb	2.3	16	0.4	40		
Fe	2.12				310	105
Co					98	106
Ni					340	107
Nb	3.07				23	108
Mo					12.6	109
Ru					27	110
Pd					350	111
W					4.8–6.4	112
Re					45	113
Os					20	113
Pt					140	114
Nb$_3$Sn					70,000	115
Bi	2.25	80,000			135,000	116
Graphite					50,000	117

[a] Also included is the electronic density normalized to the Bohr radius, r_s/a_o.
[b] Values of Δ for alkali metals are from Ref. 91; those for noble metals are from Ref. 92; those for the polyvalent metals are from Ref. 82. The calculated values A_{ee} (calc) include a phonon-mediated correction evaluated in Ref. 91. The theoretical estimate of A_{ee} for Bi and graphite is based on Ref. 93. References are provided for the experimental values of A_{ee}.

while for a two-dimensional thin metal film the electron–electron interaction yields a contribution proportional to $T^2\ln(T_F/T)$. Since our ultimate interest is the electronic thermal conductivity, and heat conduction measurements on lower-dimensional systems are exceedingly difficult to carry out, we limit our discussion to bulk metallic systems only.

Equation (132) implies a reduction with decreasing temperature of the already weak electron–electron scattering rate. This, however, does not mean that we can completely neglect the influence of electron–electron processes. At very low temperatures, where the electron–phonon processes rapidly disappear ($\rho_{e-p} \propto T^5$), and in very pure metals, where the impurity resistivity contribution is small, the relatively slow T^2 dependence may propel the electron–electron processes into a position of the dominant scattering mechanism.

4.3.1. Effect of e-e Processes on Electrical Resistivity

The earliest experimental evidence for a T^2 term in the resistivity due to electron–electron interaction is in the data on platinum measured in 1934 by de Haas and de

Sec. 4 · SCATTERING PROCESSES

Boer.[83] Shortly afterward, Landau and Pomeranchuk[84] and Baber[85] drew attention to the role of electron–electron scattering in the transport properties of metals. Prior to World War II, it was well established that, among the metals, transition elements are the ones that display an identifiable and measurable T^2 term in their low-temperature resistivity. A textbook explanation why transition metals behave this way was provided by Mott[86] in his model based on the *sd* scattering process. A characteristic feature of transition metals is an incomplete *d* band with high density of states, in addition to partially filled *s* and *p* bands of more or less ordinary (i.e., mobile) electrons. As the *s* electrons try to respond to an external field, they scatter preferentially into the high-density *d* states and acquire the character of the "sluggish" *d* electrons. This leads to a rather dramatic decrease in the electric current density. Electron–electron scattering of this kind does not even require U-processes; the ordinary N-processes perfectly suffice to generate the large resistivity observed in the transition metals (see Table 1). Because more recent studies indicate that the density of states in transition metals is not that much larger in comparison to the density of states in noble metals, an alternative explanation has been put forward[87] to explain the large T^2 term in the resistivity of transition metals. It stresses the importance of a realistic shape for the energy surfaces and the use of the *t* matrix rather than the Born approximation in calculations of the electron–electron scattering cross section. Nevertheless, the intuitive appeal of Mott's picture is hard to deny.

In the mid-1970s, following the development of high-sensitivity, SQUID-based detection systems that allowed for an unprecedented voltage resolution in resistivity measurements, major experimental efforts confirmed that the electron–electron interaction has its unmistakable T^2 imprint on the low-temperature resistivity of most metals (Table 6). Moreover, the interest in electron–electron processes has been hastened by the development of localization theories, and throughout the 1980s much has been learned about electron–electron interaction in lower-dimensional structures. Readers interested in these topics are referred to Kaveh and Wiser.[88]

Among the outcomes of the intense theoretical effort were two new perspectives on electron–electron scattering in metals. The first concerns situations where a metal is exceptionally impure, i.e., when the elastic (impurity) mean free path is reduced to a size comparable to the interatomic spacing. In this regime of transport, the usual Fermi liquid theory that is at the core of all our arguments breaks down and must be replaced by a theory that takes into account the effect of disorder on the electron–electron interaction.[89,90] Specifically, under the conditions of strong disorder, in addition to the Fermi liquid term T^2 of Eq. (131), there appears a new term

$$\frac{1}{\tau_{e-e}} = C(\ell_e)(k_B T)^{d/2}, \tag{133}$$

where C depends on the mean free path l_e and d is the dimensionality of the metallic system. Thus, the overall scattering rate due to electron–electron processes in very "dirty" metals is

$$\frac{1}{\tau_{e-e}} = A_{ee}T^2 + C(\ell_e)(k_B T)^{d/2} = A_{ee}T^2 + \frac{1}{2}\left(\frac{3}{E_F}\right)^{1/2}\left(\frac{k_B}{k_F \ell_e}\right)^{3/2} T^{3/2}, \tag{134}$$

where in the second equality the constant $C(l_e)$ is written explicitly. For reasonably pure metals ($k_F l_e \gg 1$) the second term in Eq. (134) is negligible in comparison to the first term. However, for a very impure metal for which the product $k_F l_e$ approaches the Yoffe–Regel criterion, $k_F l_e \sim 1$ (near a metal–insulator transition), the second term in Eq. (134) dominates. We shall not consider such highly disordered metals. Rather, we focus our attention on systems that can be treated in the spirit of the Fermi liquid theory—essentially all bulk metals. Therefore, we take it for granted that the scattering rate for electron–electron processes is simply just the first term in Eq. (134).

The second important feature emerged from a rigorous many-body calculation of MacDonald[118] and relates to a strong enhancement in the electron–electron collision rate observed in the low-temperature electrical resistivity of metals, which ultimately undergo a transition to a superconducting state. This enhancement arises from the phonon–exchange term in the "effective" electron–electron interaction. The reader may recall that in a conventional picture of superconductivity (the BCS theory) the mechanism of electron pairing is an attractive interaction between electrons via an exchange of virtual phonons. In nonsuperconducting metals (e.g., alkali metals and noble metals) this attractive interaction between electrons is smaller than their Coulomb repulsion. In metals that do superconduct, the attractive interaction is the dominant interaction. This effect leads to a spectacular enhancement (one to two orders of magnitude) in the electron–electron scattering rate of the polyvalent metals such as aluminum or lead. However, at high-temperatures this enhancement is substantially washed out. The limited range of temperatures where the effect shows up implies that one should be cautious when comparing the influence of electron–electron processes on the electrical resistivity to its influence on the electronic thermal conductivity with the latter accessible only at high-temperatures, as we discuss shortly.

With the electron–electron scattering term given by Eq. (131), the electrical resistivity ρ_{e-e} and the electronic thermal resistivity W_{e-e} (the subscript e-e refers to the electron–electron processes) can be written as

$$\rho_{e-e}(T) = A_{ee}T^2 \qquad (135)$$

and

$$W_{e-e} = B_{ee}T. \qquad (136)$$

The form of Eq. (136) is consistent with the Wiedemann–Franz law, in which case the ratio A_{ee}/B_{ee} plays the role of L. There are indications[119] that the value of L for electron–electron processes is close to 1.1×10^{-8} V^2/K^2.

An experimenter faces a daunting task when measuring the coefficients A_{ee} and B_{ee}. Writing for the overall electrical and thermal resistivities*

$$\rho = \rho_{\text{imp}} + \rho_{e-e}(T) + \rho_{e-p}(T), \qquad (137)$$

$$W = W_{\text{imp}}(T) + W_{e-e}(T) + W_{e-p}(T), \qquad (138)$$

one must be able to resolve the electron–electron terms against the contributions arising from impurity scattering and from the electron–phonon interaction. Let us

---

* Depending on what impurities are present in a given metal, there might be additional terms in Eq. (137) representing Kondo-like resistivity and possibly a T-dependent electron-impurity term, which, unlike in Eq. (85), arises due to inelastic scattering.

Sec. 4 · SCATTERING PROCESSES

illustrate how challenging the task is for the easier of the two cases—determination of the coefficient A_{ee}.

In copper, for example, the impurity resistivity in the purest samples of copper is on the order of 10^{-12} Ω m. Electrical resistivity due to electron–electron processes is on the order of 10^{-16} Ω m at 1 K. Even if one measures the electrical resistivity at very low temperatures ($T < 1$ K) in order to eliminate a contribution from the electron–phonon scattering, the ratio $\rho_{e-e}(1\text{ K})/\rho_{imp} \sim 10^{-4}$. To resolve this with an accuracy of 1%, one is called upon to make precision measurements at the level of one part per million. Such precision was impossible to achieve prior to the development of high-sensitivity SQUID-based voltage detectors (e.g., Ref. 120). However, once these supersensitive devices became available and were incorporated in dilution refrigerators or helium-3 cryostats, a wealth of data appeared in the literature ready to be analyzed.

Even in the purest of specimens, one eventually enters a temperature regime where ρ_{imp} becomes the dominant resistive process. Because of its presumed temperature independence, one can eliminate the impurity contribution by simply taking a derivative of Eq. (137) with respect to temperature and isolate the electron–electron term by writing

$$\frac{1}{2T}\frac{d\rho}{dT} = \frac{1}{2T}\frac{d\rho_{imp}}{dT} + \frac{1}{2T}\frac{d\rho_{e-e}(T)}{dT} + \frac{1}{2T}\frac{d\rho_{e-p}(T)}{dT} = A_{ee} + \frac{1}{2T}\frac{d\rho_{e-p}(T)}{dT}. \qquad (139)$$

If one now extends measurements to sufficiently low temperatures to suppress the electron–phonon term, one obtains A_{ee}. This scheme works well regardless of whether the electron–phonon term is a power law of temperature or, as in the case of alkali metals, an exponentially decreasing contribution on account of a phonon drag effect.[121]

The values of A_{ee} for transition metals (Table 6) are by far the largest, and thus the data have been collected with relative ease for most transition elements. Interpretation of the data, however, is often complicated by the presence of other interaction mechanisms, most notably electron-magnon scattering which also manifests its presence by a T^2 temperature dependence. It is interesting to compare the coefficients A_{ee} (experimental and theoretical values) for alkali metals with those for noble metals. The data in Table 6 indicate that there is little difference between the two, at least if one focuses on the lighter alkalis. This is a rather fortuitous result. A small value of Δ for alkalis (a few opportunities for U-processes) is compensated by their quite large value of $(\tau_{ee})^{-1}/n$, resulting in A_{ee}'s comparable to those of noble metals. One should note that, in some cases, specifically that of potassium and copper, the experimental data are strongly sample dependent. This is indicated by a wide range of values entered in Table 6. As one goes toward polyvalent metals, A_{ee} becomes rather large. Polyvalent metals are often superconductors, and this fact seriously curtails the temperature range where the electron–electron interaction has a chance to be clearly manifested. Frequently, the data are an admixture of electron–electron and electron–phonon interactions, and one must try to separate the two by modeling the behavior of the electron–phonon term. We include in Table 6 a T^2 term for the resistivity of Nb$_3$Sn, one of the highest T_c conventional superconductors that, close to its transition temperature $T_c \sim 18$ K, displays a robust quadratic T-dependence of resistivity that extends to about 40 K. While other scattering mechanisms may be at play, it is nevertheless intriguing to consider a proposal by Kaveh and Wiser[88] that such a giant T^2 term is not inconsistent with the electron–electron interaction, especially when one takes into account a strong enhancement in A_{ee} due to phonon mediation. Surely, in this high-transition-temperature conventional superconductor, such an enhancement would be particularly large even though it must disappear at high-temperatures where the resistivity becomes sublinear. There are other features in the transport behavior of these so-called A15 compounds that make Kaveh and Wiser argue that the electron–electron interaction is a reasonable prospect in spite of the controversy such a proposal causes.

For completeness we also provide in Table 6 the data on two archetypal semimetals, bismuth and graphite. Low and ultralow-temperature measurements show a quadratic variation of resistivity with temperature. Whether this is a signature of intervalley electron–electron—or, more precisely, electron–hole—scattering[93] or a footprint of highly elongated (cigarlike) carrier pockets of electrons on the carrier–phonon scattering[116] has been a puzzle and a source of controversy for some time. If we take a position that electron–hole scattering is important, it cannot be the usual intrapocket variety but rather the interpocket (or intervalley) mechanism. The reason is that the carrier pockets in Bi and graphite are very small (carrier densities are four orders of magnitude lower than in typical metals) and take up only a small fraction of space in the Brillouin zone. The U-processes, the only processes in the case of intravalley scattering that can give rise to resistivity, are simply not accessible in the case of Bi and graphite. On the other hand, interpocket scattering, with or without umklapp processes, is in principle a resistive mechanism, just as the case of *sd* scattering in transition metals. In compensated semimetals, i.e., systems with an equal number of electrons and holes, even with just N-processes, the interpocket scattering is a highly resistive process.[122] This should be evident when one realizes that, in response to an electric field, electrons and holes move in opposite directions, and thus an interchange of momenta slows down both kinds of carriers. If a semimetal is perfectly compensated, the T^2 dependence of resistivity should be present at all temperatures. In the case of incomplete compensation, the initial T^2 dependence at low temperatures gives way to saturation at high-temperatures.

It is important to remember that, within the spirit of the relaxation time approximation, the relaxation time τ_{e-e} in Eq. (131) represents an average of the time between electron–electron collisions over the entire Fermi surface. However, not all collisions are equally effective in hindering the transport, i.e., contributing to the resistivity. We mentioned that N-processes cannot give rise to resistivity. One must therefore somehow capture this notion of ineffectiveness of some of the collisions, and this is accomplished by introducing a parameter Δ that measures the fraction of U-processes among all electron–electron processes.[82] The parameter Δ plays a similar role as the parameter $(1 - \cos\theta)$—it measures the effectiveness of scattering processes in *degrading* the electrical current.

In alkali metals normal electron–electron processes are far more frequent than the U-processes, and thus Δ is very small. In noble metals, and especially in polyvalent metals, the fraction of U-processes drastically rises and rivals that of N-processes. This, of course, reflects the complicated, multisheet Fermi surface providing many more opportunities for U-processes. Hitherto one has to be careful when calculating coefficients A_{ee}. One cannot take a position that all A_{ee} should be roughly the same because all metals have comparable carrier densities and the strength of the electron–electron interaction is approximately the same. The structure and the shape of the Fermi surface really matters, and through the parameter Δ it has a considerable influence on the transport properties. Moreover, it depends

Sec. 4 • SCATTERING PROCESSES

on whether a metal contains isotropic or anisotropic scatterers. The latter situation might lead to a considerable enhancement in A_{ee} at low temperatures if N-processes dominate because the anisotropic electron scattering makes the N-processes resistive.[88]

4.3.2. Effect of e-e Processes on Thermal Resistivity

In the preceding paragraphs we have described a successful approach to determine the electron–electron contribution to electrical resistivity. Of course, one might hope to use the same strategy to evaluate the coefficient B_{ee} of the electron–electron term in the thermal resistivity, Eqs. (136) and (138). Unfortunately, this is not possible, at least not by focusing on the low-temperature transport. The difficulty arises not just because now all three terms in Eq. (138) are temperature dependent,

$$W = B_{\text{imp}}T^{-1} + B_{ee}T + B_{e-p}T^2, \qquad (140)$$

but, primarily, because the electron–electron contribution is really small and thermal transport measurements do not have a prayer of achieving the desired precision of a few parts per million. To measure thermal gradients (or temperature differences) is a far more challenging task than to measure voltage differences, and the precision one can achieve at low temperatures is at best only about 0.1–1%. So, experimental attempts to isolate and measure B_{ee} at low temperatures are simply futile.

However, as shown by Laubitz,[123] there is some prospect of accessing and determining B_{ee} via measurements at high-temperatures. Although it is not easy and the utmost care must be exercised to carry the experiments through to their successful conclusion, there is, nevertheless, a chance. The approach is based on an observation that for noble metals the experimentally derived Wiedemann–Franz ratio, $L_{\text{exp}} = \rho_{\text{exp}}(T)/TW_{\text{exp}}(T)$ never quite reaches the theoretical Lorenz number $L_o = (\pi^2/3)(k_B/e)^2$, even at temperatures several times the Debye temperature. The essential point here is that L_{exp} is lower than L_o by a couple of percent as determined by the *electronic thermal resistivity* only; i.e., after any, however small, lattice thermal conductivity contribution is subtracted and does not enter consideration. Then, because at high-temperatures (well above the Debye temperature) both impurities and phonons scatter electrons purely elastically, these two scattering processes yield the Sommerfeld value $L_o = 2.44 \times 10^{-8}$ V^2/K^2 for the Lorenz number; i.e.,

$$W_{imp}(T) + W_{e-p}(T) = \frac{\rho_{imp} + \rho_{e-p}(T)}{TL_o}. \qquad (141)$$

To isolate a small term W_{e-e} against the background of a very large W_{e-p}, one writes, as Laubitz did, the total measured electronic thermal resistivity W^{exp} and electrical resistivity ρ^{exp} as

$$\frac{W^{\text{exp}}}{T} - \frac{\rho^{\text{exp}}}{L_o T^2} = \frac{W_{e-e}^{\text{exp}}(T)}{T} - \frac{\rho_{e-e}^{\text{exp}}(T)}{L_o T^2} = B_{ee} - \frac{A_{ee}}{L_o}. \qquad (142)$$

In principle, Eq. (142) allows for a determination of B_{ee} via high-temperature measurements. In reality, because one deals with a very subtle effect, great care is needed to account for any small contribution that we otherwise would have completely neglected. For instance, we take for granted that the Lorenz ratio is equal to $L_o = \rho\kappa/T$, while actually, taking into account second-order effects in the thermal conductivity [see Eq. (73)], it should be $L_o - S^2$, where S is the thermopower. Under normal circumstances we would completely neglect S^2 because in noble metals at ambient temperature it is on the order of 10^{-11}–10^{-12} V^2/K^2 (three to four orders of magnitude less than L_o). Likewise, because the experiment is run at high but finite temperatures and not at $T \to \infty$, there is a remaining "tail" of phonons that scatter electrons inelastically and one should account for it. This is accomplished by introducing a term C/T^2. Thus, in its practical form, Eq. (142) is written as

$$\Delta W^{\text{exp}}(T) \equiv W^{\text{exp}} - \frac{\rho^{\text{exp}}}{(L_o - S^2)T} = \frac{C}{T^2} + \left(B_{ee} - \frac{A_{ee}}{L_o}\right)T. \tag{143}$$

Using this approach, Laubitz and his colleagues were able to extract the values of B_{ee} for several alkali, noble, and polyvalent metals. Table 7 lists the experimental results together with a few calculated values. The unusually large probable errors included in the entries in Table 7 should not be surprising, as it is truly a proverbial needle in the haystack type of measurement.

The data in Table 7 indicate good agreement between theoretical and experimental values for the alkali metals, whereas calculated values are roughly a factor of 4 larger than the experimental values for noble metals. That A_{ee} and B_{ee} share a common factor $(\tau_{ee})^{-1}/n$,

$$A_{ee} \propto \frac{\tau_{ee}^{-1}}{n}\Delta$$

and

$$B_{ee} \propto \frac{\tau_{ee}^{-1}}{n} \tag{144}$$

(the effectiveness parameter Δ enters only in A_{ee} because for the thermal transport all scattering processes hinder heat flow and thus no Δ is needed) suggests that

TABLE 7 Values of the coefficient B_{ee} (exp) Given in Units of 10^{-6} mW^{-1} Obtained from High-Temperature Measurements of the Thermal Conductivity and Electrical Resistivity of Several metals.[a]

Metal	B_{ee} (calc)	B_{ee} (exp)	Reference
Na	1.1	1.1±0.6	34
K	3.5	2.8±0.5	30
Rb	4.9	3.5±0.5	41
Cs	9.0	12.4±2	125
Cu	0.22	0.05±0.02	123
Ag	0.23	0.05±0.02	123
Au	0.29	0.09±0.03	123
Pb		∼0.5	126
Al		<0.04	126

[a] Calculated values of B_{ee} (calc) are from Refs. 115 and 121. The entries for Pb and Al are completely unreliable because they are only an estimate and upper bound, respectively. References relate to the experimental data.

Sec. 4 • SCATTERING PROCESSES

there is a relationship between A_{ee} and B_{ee}. Indeed, Lawrence and Wilkins[82] derived the following relation:

$$A_{ee} = B_{ee} L_o \frac{5\Delta}{8 + 3\Delta}, \quad (145)$$

where $L_o = (\pi^2/3)(k_B/e)^2$. This relation in principle provides a consistency check between A_{ee} and B_{ee}. However, there is a major complication that stems from the fact that A_{ee} and B_{ee} are actually temperature dependent, and if a comparison is being made it must be done in the same temperature regime—either at very high-temperatures or at very low temperatures (but remember, no experimental data are available for B_{ee} at low temperatures). The temperature dependence enters because of two phenomena: phonon mediation in the electron–electron processes, and the influence of anisotropic scattering centers such as dislocations.

Phonon mediation was introduced by MacDonald[118] and by MacDonald and Geldart[124] to account for the fact that the overall electron–electron interaction consists of two mutually opposing contributions—the repulsive Coulomb interaction and the attractive phonon mediated interaction. While the influence of phonon mediation differs in different classes of metals (a very small, ~10%, enhancement in noble metals; a factor of 2 decrease in alkali metals; and a huge, more than an order of magnitude, enhancement in polyvalent metals), it is effective only at low temperatures and does not influence data at high-temperatures, $T > \theta_D$. Thus, following Kaveh and Wiser[88], we make a sketch of the behavior of $B_{ee}(T)$ as a function of the reduced temperature, T/θ_D (Fig. 6).

The electrical resistivity term A_{ee} is subject to the same phonon mediation influence. However, because it also depends on the collision effectiveness parameter Δ,

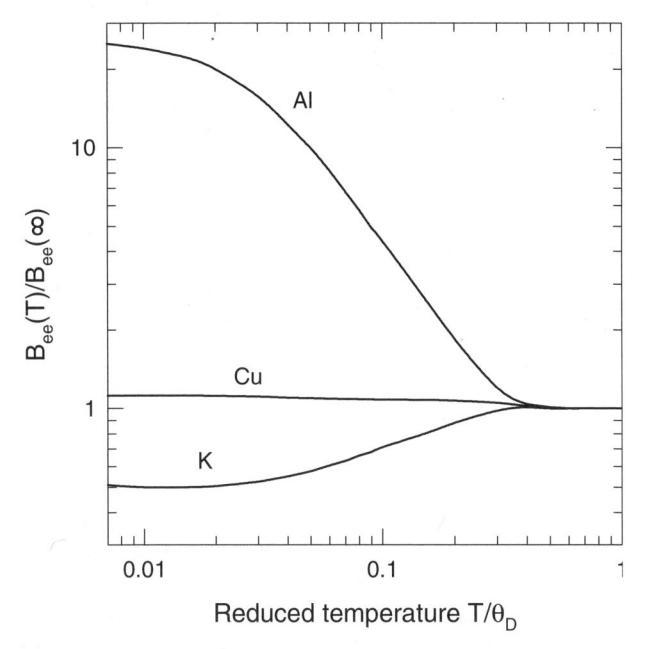

FIGURE 6 A plot of the temperature dependence of the coefficient B_{ee} [electron–electron contribution to the electronic thermal resistivity defined in Eq. (136)] normalized to its high-temperature value $B_{ee}(\infty)$. The curves represent schematic behavior for a typical alkali metal (K), a typical noble metal (Cu), and a polyvalent metal (Al). [After Kaveh and Wiser in Ref. 88]

the situation is a bit more complicated. It is well known, and we have indicated so in Table 6, that for some metals (potassium and copper in particular) the literature data on A_{ee} cover a wide range of values. This has nothing to do with the sensitivity or precision of the experimental technique but is due to samples with different concentration of defects, especially defects such as dislocations that scatter electrons anisotropically. We have pointed out that N-processes are not an impediment to electric current. This is strictly true only if electrons do not encounter anisotropic scattering centers. Since the samples used in the experiments are often annealed at different temperatures and for different durations of time, they contain dislocations of usually unspecified, but certainly unequal, concentrations. A problem with dislocations is that they tend to scatter electrons anisotropically. So, how does this give rise to an additional temperature dependence of A_{ee}?

At low temperatures where impurity scattering dominates, one is *de facto* in the anisotropic regime if the sample contains a high density of anisotropic scatterers such as dislocations. On the other hand, high-temperature transport is dominated by electron–phonon interactions, and at $T \geq \theta_D$ the scattering is essentially isotropic. This qualitatively different nature of scattering at low and high-temperatures may give rise to an additional (beyond phonon mediation) temperature dependence of A_{ee}. Whether this actually happens depends on a particular class of metals and on the relative importance of normal and umklapp scattering in each specific situation. For instance, in the case of alkali metals, the Fermi surface is essentially spherical and is contained well within the first Brillouin zone. In such a case, the scattering is dominated by N-processes, the effectiveness parameter Δ is very small, and if only isotropic scattering centers were present the coefficient A_{ee} would vanish or be exceedingly small. However, if alkali metals contain high densities of dislocations, N-processes may turn resistive at low temperatures because scattering of electrons on dislocations is anisotropic. This can lead to a dramatic enhancement in $(A_{ee})^{LT}$ by a few orders of magnitude. On the other hand, if U-processes dominate the scattering events, the effectiveness parameter Δ is large (close to unity), and whether the electrons do or do not encounter dislocations is not going to have an additional substantial influence on the already large A_{ee}. Thus, what gives the anisotropic scattering enhancement is the ratio $\Delta^{anis}(0)/\Delta^{iso}(0)$ taken at low temperatures (marked here as 0 K). The upper bound to the increase in Δ on account of electron-dislocation scattering is in essence given by $1/\Delta_U$, where Δ_U is the umklapp electron–electron value of the effectiveness parameter Δ. Large Δ_U, i.e., very frequent U-processes, cannot do much for the enhancement, but small Δ_U, i.e., the dominance of N-processes, has a great potential for an enhancement if anisotropic scattering centers are present. Kaveh and Wiser[127] related the ratio $\Delta^{anis}(0)/\Delta^{iso}(0)$ to the anisotropic $\rho_o = \rho_{imp} + \rho_{disloc} \cong \rho_{disloc}$ and isotropic $\rho_o = \rho_{imp} + \rho_{disloc} \cong \rho_{imp}$ limits of scattering. If the preponderance of scattering events involves U-processes, there is no anisotropy enhancement. If, however, scattering is dominated by N-processes, then as the electrons scatter on dislocations they make N-processes resistive and an enhancement is realized. Since the parameter Δ contains information on both umklapp electron–electron and electron-dislocation scattering, it follows that if the former one is weak then the electron-dislocation scattering can have a great effect on A_{ee} (the case of alkali metals). In the opposite case when U-processes dominate, essentially no enhancement arising from the electron-dislocation scattering is possible. Kaveh and Wiser provide extrapolations for various regimes of dislocation density in different classes of metals. We conclude this section by sketching in Fig. 7 a possible trend in the temperature

Sec. 5 • LATTICE THERMAL CONDUCTIVITY

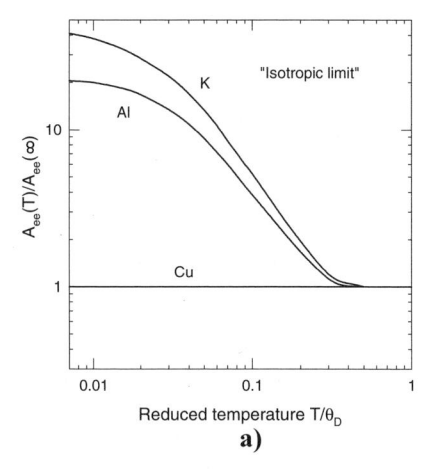

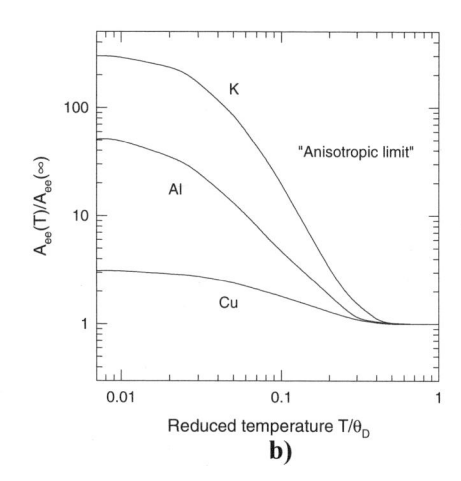

FIGURE 7 Temperature dependence of the coefficient A_{ee} [(electron–electron contribution to the electrical resistivity defined in Eq. (135)] normalized to its value at high-temperatures, $A_{ee}(\infty)$. (a) The coefficient A_{ee} includes the effect of phonon mediation for a sample in the "isotropic limit", i.e., when the residual resistivity is dominated by isotropic scattering centers. (b) The coefficient A_{ee} in the "anisotropic limit" refers to a situation where scattering is dominated by anisotropic scattering centers. [After Kaveh and Wiser in Ref. 88].

dependence of $A_{ee}(T)$ for representatives of the three main classes of metals we discussed.

5. LATTICE THERMAL CONDUCTIVITY

Lattice vibrations are an essential feature of all crystalline solids, regardless of whether they are metals or insulators, and therefore lattice (or phonon) thermal conductivity is always present in any solid material. The question is, how large a fraction of the total thermal conductivity does it represent; in other words, does it dominate the heat conduction process or is it a very small contribution that can be neglected?

In Chapter 1.1, heat conduction via lattice vibrations was discussed, and it would seem that we could apply the results found there to the case of metals and have the answer ready. While it is true that phonons will be governed by the relations developed in Chapter 1.1, the problem is that the underlying principles refer primarily to dielectric crystals void of any free charge carriers. In contrast, an environment of metals presents a unique situation—on the order of 10^{23} electrons per cubic centimeter—something that phonons do not encounter in dielectric solids. It is this interaction of phonons with electrons and its influence on the phonon thermal conductivity that motivates interest in the study of lattice thermal conductivity in metals.

5.1. Phonon Thermal Resistivity Limited by Electrons

The ideal thermal resistivity of metals W_{e-p} was discussed in Sect. 4.2. In Eq. (106) we considered the effect of the electron–phonon interaction on the distribution function of electrons. We are now interested in its influence on the phonon distribution function N_q. The problem was dealt with first by Bethe and

Sommerfeld.[128] In the process of changing its state from **k** to **k+q** and vice versa, an electron emits or absorbs a phonon of energy $\hbar\omega_q$ with momentum $\hbar\mathbf{q}$. The collision term for this process is

$$\left(\frac{\partial N_q}{\partial t}\right)_{coll} =$$
$$-\frac{2\pi}{\hbar}|C_q|^2 \sum_{k,\sigma} \left[(1 - f_{k+q})f_k N_q - (1 - f_k)f_{k+q}(N_q + 1)\right]\delta(E_{k+q} - E_k - \hbar\omega_q). \quad (146)$$

The summation now is done over all possible electron states that fulfill the energy and wave vector conditions of Eqs. (94) and (95). Retracing the steps taken in deriving Eq. (110), we linearize the collision term using Eqs. (107) and (108) to obtain

$$\left(\frac{\partial N_q}{\partial t}\right)_{coll} =$$
$$-\frac{2\pi}{\hbar}|C_q|^2 \sum_{k,\sigma} f^o(E_k)\left[1 - f^o(E_{k+q})\right] N^o(\omega_q)\delta(E_{k+q} - E_k - \hbar\omega_q)(\psi_k - \psi_{k+q} + \phi_q). \quad (147)$$

In deriving the relaxation time τ_k for the ideal electronic thermal conductivity, Eq. (110), we made an assumption that the phonon system is at equilibrium. In the present situation of phonons scattering on electrons, we apply a similar condition, only now we demand that the electron population be at equilibrium, i.e., we set $\psi=0$. With this simplification and introduction of a relaxation time for this process, τ_{p-e}, Eq. (147) becomes

$$\frac{1}{\tau_{p-e}} = \frac{2\pi}{\hbar}|C_q|^2 \sum_{k,\sigma} f^o(E_k)\left[1 - f^o(E_{k+q})\right]\delta(E_{k+q} - E_k - \hbar\omega_q). \quad (148)$$

Substituting for $|C_q|^2$ from Eq. (101), integrating over k, and writing the phonon thermal resistivity due to scattering on electrons as $W_{p-e} = (\kappa_{p-e})^{-1} = 3/(C_l v_s^2 \tau_{p-e})$, where C_l is the lattice specific heat and v_s is the longitudinal sound velocity, we obtain

$$\kappa_{p-e}(T) = G\left(\frac{T}{\theta_D}\right)^2 J_3\left(\frac{\theta}{T}\right), \quad (149)$$

where the coefficient G is given by

$$G = \frac{k_B^3 h \theta_D^2 M}{2\pi^2 m_e^2 a^3} \sum_{j=1}^{3} \frac{1}{C_j^2}. \quad (150)$$

Here M is the mass of an ion, C_j^2 is a coupling constant that specifies the interaction between electrons and phonons, and the index j indicates the three possible branches, one longitudinal and two transverse. Equation (149) was originally derived by Makinson.[129] Although cast in a different form, Klemens' coefficient[130] G is identical to Eq. (150). Wilson[131] uses a coefficient that is a factor of 3 smaller. Asymptotic forms of Eq. (149) for high and low temperatures are

Sec. 5 · LATTICE THERMAL CONDUCTIVITY

$$\kappa_{p-e}(T) \cong \frac{G}{2}, \qquad T \geq \theta_D, \tag{151}$$

$$\cong 7.18 G \left(\frac{T}{\theta_D}\right)^2, \qquad T \ll \theta_D. \tag{152}$$

Thus, while at high-temperatures the phonon thermal conductivity due to scattering on electrons is temperature independent, it decreases rapidly at low temperatures; i.e., the electrons are very effective in limiting the heat current due to phonons. At high-temperatures the phonon–phonon umklapp processes are far more effective than electrons in scattering phonons, and the total lattice thermal conductivity never reaches the values indicated in Eq. (151). On the other hand, phonon–electron scattering is an important limiting process for the lattice thermal conductivity of metals at low temperatures, and we shall return to this topic when we discuss thermal conductivity of metals in Sec. 6.

Equations (149)–(152) capture the essence of the phonon–electron interaction even though they were developed for a highly simplified model. We can inquire how the thermal conductivity κ_{p-e} (or thermal resistivity W_{p-e}) compares to the ideal thermal conductivity of metals and thus gain some feel for the magnitude of the lattice thermal conductivity and in what temperature range it might have its greatest presence.

Using Eqs. (124), (125), (151), and (152), we form a ratio $\kappa_{p-e}/\kappa_{e-p}$ for the high- and low-temperature range:

$$\frac{\kappa_{p-e}}{\kappa_{e-p}} = \frac{G/2}{4L_0 \theta_D / A} = \frac{9C^2}{8\pi^2 N_a^2} \sum_j \frac{1}{C_j^2} = \frac{81}{8\pi^2 N_a^2} = \frac{1.03}{N_a^2} \qquad \text{for } T \geq \theta_D \tag{153}$$

$$= \frac{7.18 \times 37.81 \times 81}{\pi^2 N_a^2} \left(\frac{N_a}{2}\right)^{2/3} \left(\frac{T}{\theta_D}\right)^4 = 1404 \, N_a^{-4/3} \left(\frac{T}{\theta_D}\right)^4 \qquad \text{for } T \ll \theta_D. \tag{154}$$

In Eqs. (153) and (154) we used $[\sum C_j^2]^{-1} = [\sum C_j^{-2}]/9$, assuming that all three polarization branches (one longitudinal and two transverse) interact with the electrons on an equal footing; i.e., $C_{\text{long}}^2 = C_{\text{trans}}^2 = C^2/3$. The same assumption was used in Ref. 129, except that Makinson apparently took $[\sum C_j^2]^{-1} = [\sum C_j^{-2}]/3$. The high-temperature result, Eq. (153), might imply comparable thermal conductivity contributions provided no other scattering processes are present. This, however, is an erroneous assumption since the dominant scattering process for phonons at high-temperatures is not the phonon–electron interaction but U-processes. Their strong presence makes the lattice thermal conductivity in pure metals quite negligible. We also assumed free electrons; i.e., $(q_D/k_F)^2 = (2/N_a)^{2/3}$, where N_a is the number of electrons per metal ion.

Turning to the low-temperature ratio, in order for the asymptotic behavior to be valid the ratio θ_D/T should be at least 10. From Eq. (154) it follows that the thermal conductivity κ_{p-e} is much smaller than κ_{e-p} and progressively so as the temperature decreases. Thus, the lattice thermal conductivity will be an insignificant fraction of the total thermal conductivity of pure metals at low temperatures. In fact, in pure metals, the lattice thermal conductivity can safely be neglected except at intermediate temperatures (between the high and low asymptotic regimes), and even there it amounts to no more than about 2%.

As pointed out, numerical factors in the formulas of transport coefficients reflect approximations and simplifications necessary to keep the problem tractable. In Eqs. (153) and (154) the situation is compounded by the fact that one is never fully certain exactly how large the contribution of transverse phonons is. The difference between the Bloch model ($C^2_{\text{long}} = C^2$, $C^2_{\text{trans}} = 0$) and the Makinson model ($C^2_{\text{long}} = C^2_{\text{trans}} = C^2/3$) is a factor of 3 (actually 9 if properly averaged), and this does not even address the issue of the actual intrinsic strength of the electron–phonon interaction. Thus, one should not be surprised to find very different numerical factors used by different authors when making comparisons such as in Eq. (154). We shall return to this issue in Sec. 6.

Although in a cross comparison of the phonon and ideal thermal conductivities we explored a possibility of equal interactions for different polarizations, the equation that governs the lattice thermal conductivity [Eq. (149)] was derived in the spirit of the Bloch model, assuming that only longitudinal phonons participate in the phonon-electron interaction. If this were so, a contribution of transverse phonons to the low-temperature lattice thermal conductivity of metals would be unchecked and κ_{p-e} would be increasing with decreasing T in a fashion similar to that in dielectric solids. In reality, transverse phonons do contribute to the heat transport. Because of frequent N-processes at low temperatures, there is no reason to assume that a redistribution of the momentum among phonons via three-phonon N-processes somehow discriminates between transverse and longitudinal modes. While normal three-phonon processes do not contribute to the thermal resistance directly, they play an important role in establishing thermal equilibrium because they tend to equalize the mean-free paths of phonons of the same frequency but different polarization. Thus, in practical situations, transverse phonons are an integral part of the heat transport process and must be considered on equal footing with that of the longitudinal phonons.

When metals contain a significant number of impurities or, for that matter, form alloys, we are presented with a somewhat different situation than in pure metals. Even though the electronic thermal conductivity remains the dominant contribution, the lattice thermal conductivity may represent a significant fraction of the overall thermal conductivity and, again, its presence should be most noticeable at intermediate temperatures. Because the defect and impurity scatterings have a much stronger effect on the mean free path of electrons than on phonons, it is primarily the electronic thermal conductivity that decreases. As a result, the lattice thermal conductivity is less affected and attains a proportionally higher influence, although rarely exceeding 50% of the electronic contribution. It is only in the case of semimetals (carrier densities of a couple of orders of magnitude lower than in good metals) where the lattice thermal conductivity might dominate at all except at sub-Kelvin temperatures.

Experimentally, one has a couple of options for extracting and assessing the lattice thermal conductivity. Because the lattice thermal conductivity contribution in pure metals is very small, it is much easier to work with dilute alloys and extrapolate back to the environment of a pure metal. This is certainly the most accurate, but also the most laborious, approach, necessitating preparation of a series of dilute alloys. The most reliable results are obtained when measuring at low temperatures, say at a liquid helium temperature range. This is usually a low-enough temperature for the defect/impurity scattering to dominate, and the scattering processes are therefore elastic. One can then use the measured electrical resistivity and apply with confidence the Wiedemann–Franz law to obtain the electronic

Sec. 5 · LATTICE THERMAL CONDUCTIVITY

thermal conductivity. The difference between the measured (total) thermal conductivity and the electronic thermal conductivity is the lattice contribution. Performing a series of such measurements on progressively more dilute alloys, one can readily extrapolate the data to zero-impurity content, i.e., to the phonon thermal conductivity of a pure metal. This approach, of course, assumes that there are no dramatic band structure changes within the range of solute concentrations used in the study. Numerous such investigations were done in the late 1950s and throughout the 1960s, a time of very active studies of the transport properties of metals (e.g., Refs. 132–135).

In principle, one should also be able to use the same approach at high-temperatures where the scattering of electrons is expected to be elastic. This might work for metals with a rather low Debye temperature, but most of the metals have Debye temperatures (or temperatures Θ_s) above room temperature, and the usual longitudinal steady-state technique of measuring thermal conductivity is not well suited to temperatures much in excess of 300 K. Thus, one might not be measuring safely within the high-temperature regime ($T \geq \theta_D$), and the scattering processes might not be strictly elastic. Consequently, such measurements would not be as reliable as those performed at low temperatures. What constitutes a safe margin in the context of the high-temperature limit can be gleaned from Fig. 2 in Chapter 1, where the Lorenz ratio, normalized to the Sommerfeld value, is plotted as a function of reduced temperature for different impurity concentrations in a monovalent metal.

One can also estimate (with the same uncertainty as discussed in the preceding paragraph) the magnitude of the lattice thermal conductivity by inspecting how much the Lorenz ratio exceeds the Sommerfeld value L_o. But this is nothing qualitatively new because the Lorenz ratio is not a parameter measured directly but is calculated from the data of the thermal conductivity and electrical resistivity, as described.

Also note that electronic and phonon thermal conductivities may be separated with the aid of a transverse magnetic field. In the context of semiconductors, the phenomenon is often referred to as the Maggi–Righi–Leduc effect. This technique requires very high carrier mobilities, so with a magnetic field readily available in a laboratory (5–10 T) one can saturate the magnetothermal conductivity and isolate (or more frequently extrapolate toward) the lattice thermal conductivity, presuming it is field independent. In ordinary metals this approach would not work because the carrier mobility is not high enough and the lattice thermal conductivity is too small. However, in semimetals such as Bi and its isoelectronic alloys the technique works quite well.[136]

5.2. Other Processes Limiting Phonon Thermal Conductivity in Metals

The scattering processes that limit phonon thermal conductivity in metals are primarily U-processes at high-temperatures and phonon–electron scattering at low temperatures. The influence of the latter is seen clearly from Eq. (152), which implies that as the temperature decreases the lattice thermal resistivity rises as an inverse square power of temperature. The influence of the U-processes can be gleaned from a formula derived by Slack:[137]

$$\kappa_p^U = \frac{3 \times 10^{-5} M_a a \theta_o^3}{\gamma^2 n_c^{2/3}} \frac{1}{T}, \qquad (155)$$

where M_a is the average atomic mass, a is the lattice constant, n_c is the number of atoms in a primitive cell, θ_o is the temperature obtained from low-temperature heat capacity, and γ is the high-temperature Grüneisen constant. The characteristic $1/T$ dependence of the lattice thermal conductivity of metals, as well as its magnitude, is quite similar to that of dielectric solids possessing comparable density and shear modulus.

Boundary scattering and impurity scattering are considered less effective in the phonon thermal resistivity of metals than in dielectric solids because other processes, notably the phonon–electron and U-processes, are the mean free path-limiting processes. Moreover, point defects do not scatter low-frequency phonons (Rayleigh scattering that governs phonon interaction with point defects has an ω^4-frequency dependence), and thus the effect of alloying on the phonon thermal conductivity of metals is always much smaller (by a factor of 2–5 for solute concentrations greater than 2 at.%) than the effect of the same solute concentration on the electrical resistivity. Of course, solutes may differ regarding their valence state as well as in terms of their mass and size difference *vis-à-vis* the solvent atoms. Appropriate treatments for these attributes of impurities are available in the literature.[138] In general, the environment of metals is not the best choice for the study of the effect of point defects or boundary scattering on the phonon thermal conductivity; dielectric crystals are far more appropriate for this task.

By adding more and more impurity, and especially when the impurity is a strong scatterer of electrons, one may arrive at a situation where the electron mean free path is very short. What constitutes a very short mean free path is conveniently judged by a parameter $l_e q$, the product of the electron mean free path with the phonon wave vector. Specifically, electron–phonon interaction as used in transport theory assumes $l_e q > 1$, which is often interpreted as a condition for an electron to be able to sample all phases of phonons with which it interacts. For very short electron mean free paths, this condition may no longer be satisfied. In that case we enter a troublesome regime addressed in the theory of Pippard[139] and described in Chapter 1.1. Under these circumstances, the lattice thermal conductivity of metals deviates from T^2 dependence and becomes proportional to a linear function of temperature.[140]

Of the extended defects, dislocations have been the topic of several investigations (see, e.g. Refs. 141–143). For obvious reasons their influence on the lattice thermal conductivity of metals can only be studied in alloys where the lattice thermal conductivity is large enough and one has a chance to resolve its various contributions. However, even in this case the task is challenging because the temperature dependence of phonon-dislocation scattering is proportional to T^{-2}; i.e., it has the same functional dependence as phonon-electron scattering at low temperatures.[144] One thus has to unambiguously separate the two contributions in order to get a meaningful result. This calls for measurements on a mechanically strained sample that is annealed in stages, taking care that the only effect of annealing is a reduction in the number of dislocations. Of course, one has to somehow independently establish the density of dislocations in order to make the study quantitative. The upshot of such measurements is a seemingly larger number of dislocations (a factor of about 10) required to produce the observed scattering rate,[145]

$$\frac{1}{\tau_{\text{str}}(\omega)} \propto N_d \frac{\gamma^2 B^2 \omega}{2\pi}, \tag{156}$$

than what any reasonable and independent estimate of their numbers suggests. In Eq. (156), N_d is the dislocation density and B is the Burgers vector of the dislocation. One can express the strength of dislocation scattering in terms of the increase in the phonon thermal resistivity per dislocation, i.e., $W_{p-d}T^2/N_d$. Typical experimental values for Cu [143] are 5×10^{-8} cm^3 K^3/W, while the improved theory[72] yields $\sim 5\times 10^{-9}$ cm^{3}K^3/W. Thus, it appears that the scattering power of dislocations is about a factor of 10 more than what the theory predicts. However, one must be cautious here because, as pointed out by Ackerman,[146] it is not always clear what kind of dislocations are present and what their orientation is. Different averaging factors may strongly alter the outcome of the analysis. It is interesting, however, that the discrepancy between the actual and implied dislocation densities in metal alloys is significantly smaller than the discrepancy observed in defected dielectric crystals, where it can reach on the order of 10^2–10^3. Finally, we mention that conventional superconductors are perhaps the best system in which to study the influence of dislocations in a metallic matrix. The draw here is the fact that well below T_c essentially all electrons have condensed into Cooper pairs, and thus the expected T^{-2} phonon-dislocation scattering term has no competition from the phonon–electron scattering, because there is none. In this case even pure metals are accessible for the study, and the data exist for Nb, Al, Ta, and Pb,[147–149] among others. Invariably, the data indicate exceptionally strong phonon-dislocation scattering that cannot be due to just sessile dislocations but requires a resonant scattering due to vibrating dislocations.

6. THERMAL CONDUCTIVITY OF REAL METALS

Having described the relevant interaction processes, we now consider how they influence the thermal conductivity of real metals. We should remember that metals present a wide spectrum of band structures that challenge both experimentalists and theorists. They include metals with relatively simple, near-spherically-symmetric Fermi surfaces found in alkali metals and, more or less, noble metals, and extend to a bewildering array of multipiece carrier pockets in polyvalent metals for which spherical symmetry represents a big stretch of the imagination. Thus, it would be very unreasonable to expect that the theoretical predictions developed based on a highly simplified and "fit for all"–type physical picture could capture the nuances of the physical environment of various metals. We should expect to obtain a pretty good description of transport phenomena in alkalis and noble metals but, for transition metals, rare-earth metals, and polyvalent metals in general, our expectations should be tempered since we might establish no more than general trends.

6.1. Pure Metals

Potential problems arise in electrical transport and thermal transport, although in the latter case they seem to be more severe. We mentioned that the Bloch–Grüneisen T^5 temperature dependence is rarely seen. Even more puzzling is the low-temperature behavior of electrical resistivity of alkali metals where an exponential term is observed at temperatures below 2 K.[143] This is caused by a phonon drag contribution arising from an exponentially decaying umklapp scattering term. U-processes no doubt leave their fingerprints on alkali metals also in the form of a rather large value of the electron–electron coefficient A_{ee} (Fig. 7). The neglect of umklapp

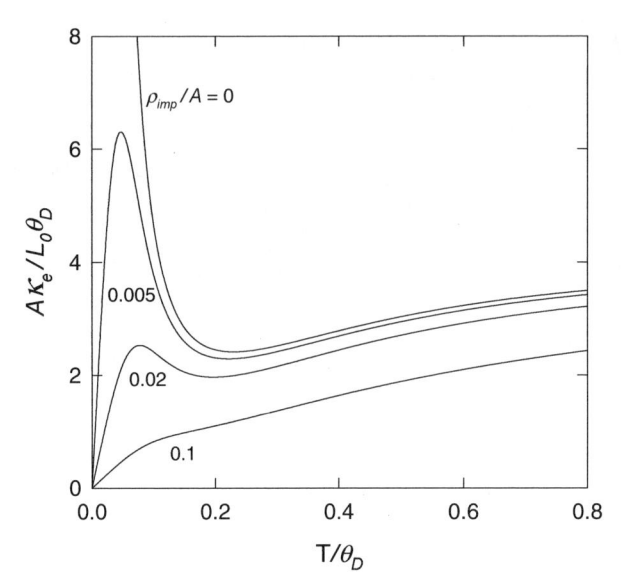

FIGURE 8 A plot of the theoretical electronic thermal conductivity versus reduced temperature using the ideal thermal resistivity given in Eq. (123). We assume a monovalent metal, and the parameter ρ_{imp}/A determines the influence of impurity scattering. A pronounced minimum near $T/\theta_D = 0.2$ has never been observed in real metals.

scattering indeed seems to be the most serious shortcoming of the theoretical treatment of transport. An excellent example of how the omission of U-processes affects electronic thermal conductivity is provided in Fig. 8, where the theoretical κ_{e-p} [the inverse of W_{e-p} from Eq. (123), which was derived assuming no participation of U-processes] is plotted against the reduced temperature. The curve shows a minimum near $\theta_D/5$ and seems to "hang" at high-temperatures, where it is nearly 60% above the limiting high-temperature ideal thermal conductivity. The minimum is particularly notable when the theory assumes a monovalent metal. Although the minimum is not as pronounced in higher-order calculations of ideal thermal conductivity,[150] it takes the participation of U-processes to obliterate it in the theoretical curves. An experimental fact is that such a minimum has never been seen in the thermal conductivity data of monovalent metals. With U-processes properly accounted for, the minimum is a nonissue also in the theoretical description of transport.

As an illustration of the trend among the different classes of metals, we plot in Fig. 9 thermal conductivities for an alkali metal (K), a noble metal (Cu), a transition metal (Ni), and a rare-earth metal (Gd). In each case, the respective curve represents the behavior of a pure metal. If impurities were added, we would see a considerably diminished peak and a gradual shift of its position toward higher temperatures. The curves generally conform to the predictions for thermal conductivity of metals. We note an essentially constant thermal conductivity at high-temperatures giving way to a rapidly rising thermal conductivity at lower temperatures; a peak developing at low temperatures near $T \sim \theta_D/15$ as a result of the competing influence of electron–phonon and electron-impurity scattering; and, finally, an approximately linear decrease of thermal conductivity with decreasing temperature as impurity scattering dominates the transport. As discussed in Sec. 4.3.2, electron–electron processes are too weak to be detected in the data of thermal conductivity.

The different magnitudes of thermal conductivity in Fig. 9 are striking, particularly in the case of the transition and rare-earth metals. These two classes of metals

Sec. 6 · THERMAL CONDUCTIVITY OF REAL METALS

are clearly poor conductors of heat. The low values of thermal conductivity in transition metals and even lower values in rare earths are associated with magnetic interactions and spin disorder. Although not shown here, transition metals and rare earths usually have much more anisotropic thermal properties, due primarily to their anisotropic Fermi surfaces.

Because the numerical values of the theoretical transport coefficients serve more as a guide than as hard and reliable numbers, it is often good practice to cross-compare asymptotic forms of the transport coefficients for consistency. With ideal electrical resistivity and ideal electronic thermal resistivity given in Eqs. (120), (121), (124), and (125), and by selecting two temperatures—one in the low-temperature domain $T_{\text{L.T.}}$ and one in the high-temperature regime $T_{\text{H.T.}}$ (some authors designate this temperature as $T = \infty$)—we can derive relationships that are expected to hold between various transport coefficients. They are

$$\rho_{\text{e-p}}(T_{\text{L.T.}}) = \frac{497.6}{\theta_R^4} \frac{T_{\text{L.T.}}^5}{T_{\text{H.T.}}} \rho_{\text{e-p}}(T_{\text{H.T.}}), \tag{157}$$

$$W_{\text{e-p}}(T_{\text{L.T.}}) = \frac{95.28}{\theta_D^2} N_a^{2/3} T_{\text{L.T.}}^2 W_{\text{e-p}}(T_{\text{H.T.}}), \tag{158}$$

$$W_{\text{e-p}}(T_{\text{L.T.}}) = \frac{95.28}{497.6} \frac{\theta_D^2}{L_o T_{\text{L.T.}}^3} N_a^{2/3} \rho_{\text{e-p}}(T_{\text{L.T.}}). \tag{159}$$

We have already noted that the domains where the asymptotic forms of the electrical and thermal resistivities (or conductivities) are valid depend on a particular cutoff temperature. As first pointed out by Blackman,[151] Bloch theory considers electrons interacting with longitudinal phonons only, and thus the appropriate cutoff temperature should be the one related to the longitudinal low-frequency phonons, Θ_s. On the other hand, most texts use the Debye temperature, θ_D. Since θ_D is related to a mean of $1/v^3$ over all polarizations,

$$\frac{1}{\theta_D^3} = \left(\frac{k_B}{\hbar \omega_D}\right)^3 = \frac{4\pi k_B^3}{9Nh^3}\left(\frac{1}{v_{\text{long}}^3} + \frac{2}{v_{\text{trans}}^3}\right), \tag{160}$$

Θ_s is considerably larger than θ_D (typically $\Theta_s \sim 1.5\theta_D$). The Debye temperature θ_D is normally obtained from the low-temperature specific heat data, where transverse and longitudinal modes, are present. Therefore, the use of θ_D implies the participation of transverse modes in the electron–phonon interaction, and θ_D is the lowest temperature one can justify as a cutoff. One also has an option to define the Debye-like temperature θ_R from a fit of electrical resistivity by using Eq. (119). In fact, an estimate of this temperature may also be obtained from Eq. (157). The best fits are usually obtained with a cutoff temperature that is close to the Debye temperature. This further attests to the importance of transverse phonons to heat transport. There is indeed no reason to assume that transverse phonons would somehow be inactive as far as transport properties are concerned.

In principle, the materials that should best conform to a theoretical description are alkali metals. They are monovalent with very nearly spherical Fermi surfaces and, thus, best fit the assumptions used in the derivation of Bloch's transport theory. Consequently, alkali metals had been the center of attention of experimental studies in spite of their high chemical reactivity that complicates sample pre-

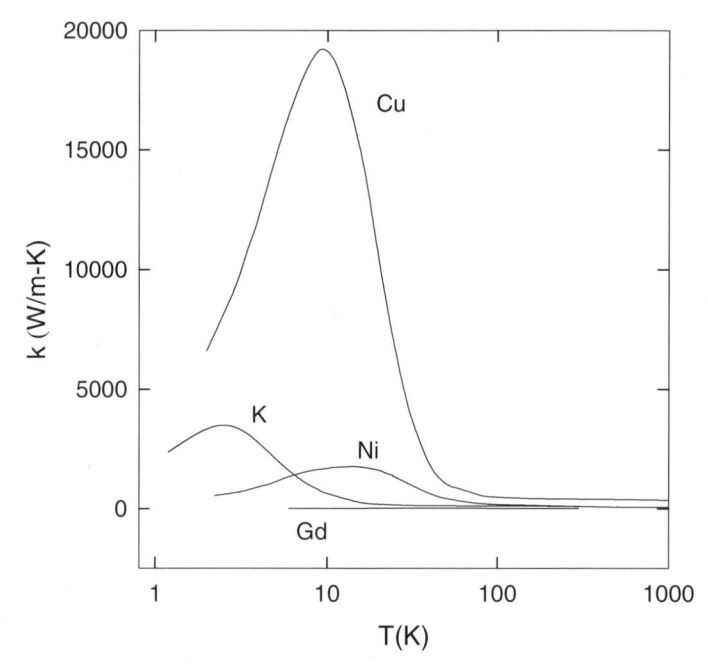

FIGURE 9 Thermal conductivity plotted as a function of log T (in order to have a better perspective of the position of the maxima) for different classes of metals, including a noble metal (Cu), an alkali metal (K), a transition metal (Ni), and a rare-earth metal (Gd). The curves are constructed from the data in Refs. 10, 30, 152, 135, 153, 154, respectively. The position of the peaks is at about $\theta_D/15$. The same vertical scale is used for all four metals to emphasize the relative magnitudes of thermal conductivity at low temperatures. Gadolinium has a perfectly developed peak near 25 K, but its thermal conductivity is so low (∼20 W/m-K at its peak) that on the scale of the figure it looks completely flat.

paration and mounting. Next in the order of complexity are the noble metals that likewise are monovalent but with a Fermi surface that crosses the zone boundary. In these two classes of metals it is reasonable to take the number of electrons per atom, N_a, as equal to unity.

In gathering experimental data to be used in cross comparisons of transport parameters, it is essential that all transport measurements be made on the same sample and, preferably, using the same contacts. This requires a dedicated effort, and although numerous transport studies have been made during the past 50 years, only a limited set of them conforms to this requirement. Indeed, the most interesting are those related to alkali metals and noble metals.

Relevant experimental data for alkali metals and noble metals together with the theoretical predictions based on Eqs. (157)–(159) are presented in Table 8. We note that even for these simplest of metals, significant disagreements arise between the theory and measurements—high-temperature (room temperature) experimental data are generally much larger in relation to the low-temperature data than what the theory predicts. Discrepancies seem to be larger for thermal resistivities than electrical resistivities. A part of the problem is no doubt the complexity of calculating ideal thermal resistivity, which, even to lowest order (thermal conductivity is a second-order effect), represents quite an involved computation. Sondheimer[150] undertook the challenging task of calculating thermal conductivity to third order and found that the numerical factor in Eq. (158) should be somewhat smaller. Subsequently, Klemens[155] solved the Boltzmann integral equation numerically in the limit

Sec. 6 · THERMAL CONDUCTIVITY OF REAL METALS

$T \to 0$ and obtained a further small reduction in the numerical factor. It turns out that the (Bloch–Wilson) coefficient in Eq. (158), the Sondheimer higher-order solution, and the Klemens numerical solution are in the ratio 1.49 : 1.11 : 1.00; i.e., the Klemens solution would yield a factor of 63.95 in place of 95.6 in Eq. (158), and, likewise, the same factor of 95.6 in Eq. (159) should be replaced by 63.95. But such changes are more or less "cosmetic" and do not solve the key problem—the high-temperature theoretical resistivities being too low. Following Klemens,[156] the discrepancies are conveniently assessed by using the parameters X and Y which are obtained from Eqs. (158) and (159) by dividing the equations by $W_{e-p}(T_{L.T.})$, substituting the Klemens coefficient 63.95 in place of Bloch's 95.28, and taking $N_a = 1$:

$$X = 63.95 \left(\frac{T_{L.T.}^2}{W_{e-p}(T_{L.T.})} \right) \left(\frac{W_{e-p}(T_{H.T.})}{\theta_D^2} \right), \qquad (161)$$

$$Y = \frac{63.95 \, \theta_D^2}{497.6 \, L_o} \left(\frac{T_{L.T.}^2}{W_{e-p}(T_{L.T.})} \right) \left(\frac{\rho_{e-p}(T_{L.T.})}{T_{L.T.}^5} \right). \qquad (162)$$

Obviously, theoretical values of X and Y in these equations are equal to unity and as such will serve as a benchmark against which we compare the "experimental" values of X and Y when the respective transport parameters are entered in Eqs. (161) and (162).

Before we proceed, it is useful to draw attention to Eqs. (158) and (159) [or to Eqs. (161) and (162)] and point out what is being compared in each case. Equations (158) and (161) relate ideal thermal resistivities at the low- and high-temperature regimes–the same transport parameter but in two very different transport domains, one dominated by high-frequency phonons, the other by low-frequency phonons. On the other hand, Eqs. (159) or (162) explore a relationship between low-temperature ideal electrical and thermal resistivities, i.e., different transport coefficients subjected to the influence of basically the same low-frequency phonons.

The data in Table 8 indicate that for the monovalent metals studied, the values of X and Y are significantly (in some cases more than an order of magnitude) greater than unity. As far as the parameter X is concerned, this implies that the theoretical ratio $W_{e-p}(T_{H.T.})/W_{e-p}(T_{L.T.})$ is much smaller than its experimental counterpart. The question is, what is the cause of such a discrepancy? Since electrons in metals

TABLE 8 Transport Parameters of Monovalent Metals.

Metal	$[W_{e-p}(T)/T^2]_{L.T.}$ $(W^{-1}mK^{-1})$	$W_{e-p}(T_{H.T.})$ $(W^{-1}mK)$	$[\rho_{e-p}(T)/T^5]_{L.T.}$ (ΩmK^{-5})	θ_D (K)	X	Y	Ref.
Na	3.8×10^{-6}	7.3×10^{-3}	5.4×10^{-17}	150	5.3	1.8	157
K	1.2×10^{-5}	7.0×10^{-3}	3.5×10^{-15}	100	3.7	16	158
Rb	9.2×10^{-5}	1.7×10^{-2}	4.5×10^{-14}	60	3.3	9	158
Cs	2.2×10^{-4}	1.7×10^{-2}	6.5×10^{-13}	25	8	10	158
Cu	2.6×10^{-7}	2.6×10^{-3}	2.6×10^{-18}	315	6.2	5.4	159
Ag	6.4×10^{-7}	2.4×10^{-3}	1.1×10^{-17}	215	5.2	4.2	160
Au	1.3×10^{-7}	6.4×10^{-3}	3.9×10^{-17}	170	5.8	4.5	161

are highly degenerate, apart from a slight "smearing" of the Fermi level at high-temperatures, the electron system itself is substantially temperature independent. This, however, cannot be said about phonons. Because the dominant phonon frequency is proportional to temperature, phonon transport at low temperatures is dominated by long-wavelength phonons (small wave vectors) that reach only a short distance toward the zone boundary. At high temperatures, on the other hand, plenty of phonons with large wave vectors reach very close to the Brillouin zone boundary. This has two important consequences for the high-temperature phonon spectrum that have no counterpart in the low-temperature domain, nor, for that matter, have they been considered in the Bloch–Wilson theory on which Eqs. (158) and (159) are based:

a. Phonon dispersion must be taken into account
b. U-processes have high probability of occurrence.

Phonon dispersion decreases the frequency of large wave vector phonons and increases the high-temperature thermal resistivity. Klemens[138] estimates that dispersion could account for up to a factor of 2.3 in the underestimate of the theoretical high-temperature ideal thermal resistivity. Likewise, the participation of U-processes in the high-temperature transport could lead to a comparably large factor. Thus, the bulk of the discrepancy in the parameter X could be eliminated by including the foregoing two modifications to the Bloch–Wilson theory. From the experimental point of view, one should also adjust the data to account for the fact that measurements are carried out under constant pressure whereas the theory assumes conditions of constant volume; i.e., the dimensions of the sample [and therefore all parameters entering Eqs. (158) or (161)] may change slightly due to thermal expansion.

One can avoid the difficulties with comparing the ideal thermal resistivity across a wide temperature range by using Eq. (159) instead, i.e., focusing on the parameter Y, which relates ideal electrical and thermal resistivities at the same low-temperature regime. As is clear from Table 8, here too one observes serious discrepancies between experiment and theory, with the latter markedly underestimating reality. In this case there is no question of influence of dispersion on the phonon spectrum nor of U-processes on the thermal resistivity which is given by vertical processes that are more efficient in impeding heat flow in this temperature range than horizontal processes. Rather, one must focus on the electrical resistivity and inquire why horizontal scattering processes are so much more effective than the theory would predict. One possible reason might be the participation of transverse; as well as longitudinal, phonons in the electron–phonon interaction. But we have already taken this into account by using the Debye temperature θ_D as the cutoff temperature. Klemens[162] proposed an explanation based on the assumption that the Fermi surfaces are not really spherical even in the case of alkali metals. While departures from sphericity have no effect on thermal resistivity because, as mentioned earlier, scattering is governed by the vertical processes that change the energy of electrons but do not substantially alter their direction, nonsphericity may have a strong influence on the electrical resistivity. In Sec. 4.2 we pointed out that the relaxation times for electrical and thermal processes at low temperatures are not the same because what matters in relaxing the electron distribution perturbed by an electric field is the change of the electron momentum (electron wave vector **k**) away from the direction of the electric field. At low temperatures only phonons with small

Sec. 6 · THERMAL CONDUCTIVITY OF REAL METALS

wave vectors **q** are available to scatter electrons, and it takes many collisions to change the electron momentum by a significant amount—like taking an electron and moving it across the Fermi surface. Thus, an electron can be viewed as diffusing across the Fermi surface (making many small steps) as it encounters low-q phonons and it interacts with them via N-processes. This is the picture underscoring the Bloch theory. Departures from sphericity, and especially if some segments of the Fermi surface come very close and touch the zone boundary, are going to have a profound effect on the diffusion process. This creates a great opportunity for U-processes to partake in the conduction process at temperatures where they would otherwise have been frozen out had the Fermi surface been spherical. Since such U-processes very effectively short out the diffusion motion of an electron on the Fermi surface, they increase the electrical resistivity. Klemens estimates that the increase could be as much as four times that of the normal diffusion process, and thus the discrepancies between the theoretical and experimental values of the parameter Y could be explained, at least in the case of noble metals.

Of critical importance in the thermal transport of metals is the behavior of the Lorenz ratio. Its magnitude and temperature dependence shed light not only on the nature of scattering of charge carriers, but it also offers a simple and convenient way of assessing how well a given metal conducts heat. Moreover, because thermal conductivity measurements are far more challenging, considerably more expensive to carry out, and much less precise than measurements of electrical resistivity, being able to make use of the resistivity data in conjunction with the knowledge of how the Lorenz ratio behaves also has an economic impact.

In Sec. 4.2 we presented a form of the Lorenz ratio that follows from the electron–phonon interaction, Eq. (126). Including a contribution due to impurity scattering, we can write the overall Lorenz ratio as

$$L = \left(\frac{\rho}{WT}\right)_e = L_o \frac{\rho_{imp}/A + (T/\theta_D)^5 J_5(\theta_D/T)}{\frac{\rho_{imp}}{A} + \left(\frac{T}{\theta_D}\right)^5 J_5\left(\frac{\theta_D}{T}\right)\left[1 + \frac{3}{\pi^2}\left(\frac{k_F}{q_D}\right)^2\left(\frac{\theta_D}{T}\right)^2 - \frac{1}{2\pi^2}\frac{J_5(\theta/T)}{J_7(\theta/T)}\right]}. \quad (163)$$

At high-temperatures the second and third terms in brackets of Eq. (163) vanish, and the Lorenz ratio equals L_o. At low temperatures the second term representing vertical scattering processes becomes very important and will drive the Lorenz ratio well below L_o (perhaps as much as $L_o/2$) before the impurity scattering will tend to restore its value back to the Sommerfeld value at the lowest temperatures. Thus, if we have a pure metal, it is quite legitimate to estimate its thermal conductivity at high-temperatures using the high-temperature value of its electrical resistivity in conjunction with the Wiedemann–Franz law and $L = L_o$. In the case of noble metals the thermal conductivity may be marginally overestimated but by no more than 2%. A similar situation is in the case of alkali metals, except that here the error could approach 10%. The worst-case scenario is the Group IIA metals, Be in particular, and alkaline earths Ba and Sr, where the experimental Lorenz ratio appears anomalously low. Using L_o in place of L might lead to an overestimate of the electronic thermal conductivity by up to 50% for Be and 25%–30% for Sr and Ba, respectively. However, lattice thermal conductivity is a large fraction of heat conduction in Be and alkaline earths, and by neglecting this contribution one effectively compensates for the overestimate of the electronic term. In the case of transition metals, except for Fe and perhaps Os, the experimental Lorenz ratio is actually larger than unity, especially for Pd and Pt (~25% and 15%, respectively). Yet it is Fe rather than Pd or Pt where the lattice thermal contribution is significant. Thus, in either case, the calculated value of the thermal conductivity will be underestimated. Large values of L in the case of Pd and Pt were explained to be due to Fermi surface smearing,[38] i.e., less than full degeneracy of electrons at high-temperatures. For other pure metals the application of the Wiedemann–Franz law at high-temperatures will not lead to errors larger than about 5%–7%. Compilations of experimental data on the Lorenz ratio of pure metals can be found in Klemens and Williams[163] and in Ref. 10.

If one is interested in the thermal conductivity of pure metals at temperatures below ambient, one moves away from a comfort zone of essentially elastic electron scattering and one has to deal with the fact that the Lorenz ratio will be governed by Eq. (163). It is instructive to plot the temperature dependence of the Lorenz ratio and note how large reductions in L are likely to arise and at what temperatures they arise (see Fig. 10). Generally, the purer the metal the more dramatic will be the reduction in L. Nevertheless, there are no documented cases where the Lorenz ratio would fall much below about $L_o/2$ before the impurity scattering at very low temperatures would take over and return the ratio to its Sommerfeld value. Thus, the use of the Wiedemann–Franz law, even in this regime where it is patently invalid, will yield a thermal conductivity that will not be more than a factor of 2 larger than its real value, and such an estimate may often be sufficient for technological applications.

6.2. Alloys

Depending on what one counts as a metal, there are some 80 metals in the periodic table from which one can make an essentially unlimited number of alloys. It turns out that for industrial applications one virtually always uses alloys because of their mechanical and other advantages over the pure metals. Because it is impossible and impractical to measure thermal conductivity of every one of the alloys one can think of, it would be an advantage to have a simple prescription to assess their thermal conductivities even though the estimates will have a margin of error.

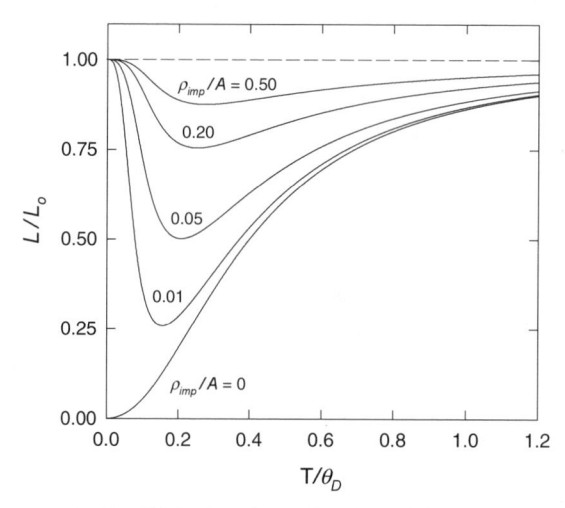

FIGURE 10 The Lorenz ratio, Eq. (163), plotted as a function of the reduced temperature for progressively higher level of impurity (larger ρ_{imp}/A).

In metals with a large concentration of impurities and in alloys, we encounter a situation where electrons are strongly scattered by solute atoms. The immediate consequence is a reduced ability of electrons to carry current and heat, and thus the lattice contribution will have a proportionally larger influence on the overall thermal conductivity. Although this complicates the analysis, we have discussed in Sec. 5.1 how one can isolate and estimate the lattice thermal conductivity as well as what the dominant scattering mechanisms are that limit the flow of phonons. However, large concentrations of impurities and the presence of solute atoms simplify the problem in some sense, at least as far as the electronic thermal conductivity is concerned—they dramatically expand the domain of temperatures where scattering is elastic and thus the Wiedemann–Franz law is valid. Therefore, in numerous situations, a knowledge of the lattice thermal conductivity and the Wiedemann–Franz law is all that one needs in order to assess the heat carrying ability of alloys. In particular, this is the case of high-thermal-conductivity alloys based on Cu, Ag, Au, and Al.

Often, thermal conductivity of alloys is described with the aid of the so-called Smith-Palmer equation[164] introduced in 1935. It is an empirically based relation between thermal conductivity and the parameter T/ρ (ρ being electrical resistivity) of the general form

$$\kappa = C \frac{L_o T}{\rho} + D, \qquad (164)$$

where C and D are constants. It assumes a temperature-independent phonon thermal conductivity via the constant D, and thus is suited to situations where point-defect and/or phonon–electron processes are dominant. As pointed out in Ref. 163, the Smith–Palmer formula seems to fit the thermal conductivity data better at high-temperatures (above 500 K) than at lower temperatures. Klemens and Williams[163] discuss the relevance of the Smith–Palmer equation to different industrially useful families of alloys and provide an extensive list of references to the original literature.

7. CONCLUSION

Heat conduction in metals is a topic of interest for its intrinsic scientific merit as well as for its relevance to a wide range of technological applications. In this chapter we reviewed the basic physical principles that form the pillars of the transport theory of metals. We tried to present a historical perspective by noting the important developments that have helped to form the understanding of the structure of metals, the energy content of the electron gas, and the key empirical findings relevant to the transport of charge and heat in metallic systems. We then discussed how the Boltzmann equation addresses the nonequilibrium nature of the electron distribution function created by an electric field or a thermal gradient. We inquired how the electrons, colliding with the crystal lattice imperfections, establish a steady-state distribution, and how the nature of the conduction process differs in the case of electrical and thermal transport. We considered the three main interaction processes the electrons engage in: scattering on static defects, electron–phonon interaction, and interactions with other electrons. In each case we outlined the main steps leading to a formula for the relaxation time from which we derived expressions for the electrical and thermal resistivities or their inverses—electrical and thermal conductivities. From the expressions for the electrical and thermal conductivity we derived formulas for one of the most important transport parameters of metals—the Lorenz ratio—and we discussed under what conditions the Wiedemann–Franz law is valid. We stressed that the Wiedemann–Franz law is valid provided that electrons undergo elastic scattering, and we noted that this is the case at high-temperatures $(T \geq \theta_D)$ where electrons scatter through large angles, and at very low temperatures where impurity scattering dominates the transport behavior. At intermediate temperatures, phonons more effectively impede the flow of heat than the flow of electric charge, and the relaxation time for thermal conductivity is considerably shorter than the relaxation time for electrical conductivity. This divergence in the relaxation times invalidates the Wiedemann–Franz law. Although the flow of electrons represents the dominant heat conduction mechanism, we considered the contribution of lattice vibrations, and we noted that in most of the pure metals this contribution may be neglected, at least at ambient and very low temperatures. In very impure metals and in alloys this is not the case, and the lattice (phonon) contribution represents a significant fraction of the overall heat conductivity. We discussed how one can isolate the electronic and lattice thermal conductivity contributions and how one can benefit from using the Wiedemann–Franz law in analyzing and assessing heat transport in metals. We supported our text with numerous references to the original research work, and we noted several monographs and review articles that readers might find useful in pursuing the subject matter in greater detail. We hope that this chapter will serve well to provide a basic understanding about the fascinating and important topic of heat conduction in metals.

8. REFERENCES

1. R. BERMAN, *Thermal Conduction in Solids* (Oxford University Press, 1976).
2. N. W. ASHCROFT AND N. D. MERMIN, *Solid State Physics* (Saunders, Philadelphia, 1976).
3. J. M. ZIMAN, *Electrons and Phonons* (Clarendon Press, Oxford, 1960).
4. H. SMITH AND H. H. JENSEN, *Transport Phenomena* (Clarendon Press, Oxford, 1989).
5. G. D. MAHAN, *Many-Particle Physics* (Plenum Press, New York, 1990).
6. G. WIEDEMANN AND R. FRANZ, Ann. Phys. **89**, 497–531 (1853).
7. L. LORENZ, Ann. Phys. **13**, 422–447 (1881).
8. P. DRUDE Ann. Phys. **1**, 566–613 (1900).
9. J. J. THOMSON, Phil. Mag. **44**, 293–316 (1897).
10. LANDOLT-BÖRNSTEIN, New Series, Vol. **15**, edited by O. Madelung (Springer-Verlag, Berlin, 1991).
11. T. MATSUMURA AND M. J. LAUBITZ, Can. J. Phys. **48**, 1499–1503 (1970).
12. J. G. COOK, J. P. MOORE, T. MATSUMURA, AND M. P. VAN DER MEER, in *Thermal Conductivity* 14, edited by P. G. Klemens and T. K. Chu (Plenum Press, New York, 1976), p. 65.
13. M. J. LAUBITZ AND D. L. MCELROY, Metrologia **7**, 1–15 (1971).
14. J. G. COOK AND M. J. LAUBITZ, Can. J. Phys. **56**, 161–174 (1978).
15. G. W. C. KAYE AND T. H. LABY, *Table of Physical and Chemical Constants* (Longmans Green, London, 1966).
16. J. G. COOK AND M. J. LAUBITZ AND M. P. VAN DER MEER, Can. J. Phys. **53**, 486–497 (1975).
17. L. J. WITTENBERG in Proc. 9th Rare Earth Conf., edited by P. E. Field (Virginia Polytechnic Institute, Blacksburg, VA p. 386 (1971).
18. M. J. LAUBITZ AND T. MATSUMURA, Can. J. Phys. **51**, 1247–1256 (1973).
19. J. P. MOORE, R. K. WILLIAMS AND D. L. MCELROY, in *Thermal Conductivity* 7, edited by D. R. Flynn and B. A. Peavy (NBS Spec. Pub. No. 302, Washington, 1968), p. 297.
20. A. W. LEMMON, H. W. DEEM, E. A. ELDRIDGE, E. H. HALL, J. MATOLICH AND J. F. WALLING, BMI/NASA Report BATT-4673-T7 (1964).

Sec. 8 · REFERENCES

21. J. G. Hust and A. B. Lankford, (NBS Int. Rep. 84–3007, U.S. Dept. Commerce, 1984).
22. B. W. Jolliffe, R. P. Tye and R. W. Powell, J. Less-Common Met. **11**, 388–394 (1966).
23. J. P. Moore, D. L. McElroy and M. Barisoni, *Thermal Conductivity* 6, edited by M. L. Minges and G. L. Denman (AFML, Dayton, Ohio, 1966), p. 737.
24. R. W. Powell, M. J. Woodman and R. P. Tye, B. J. Appl. Phys. **14**, 432–435 (1963).
25. J. E. Campbell, H. B. Goodwin, H. J. Wagner, R. W. Douglas and B. C. Allen, DMIC Rep. 160, (1961) p. 1.
26. H. Reddemann, Ann. Phys. (Leipzig), (5) **14**, 139–163 (1932).
27. L. Binkele, High Temp.-High Pressures **21**, 131–137 (1989).
28. M. Barisoni, R. K. Williams and D. L. McElroy, *Thermal Conductivity* 7, edited by D. R. Flynn and B. A. Peavy (NBS Spec. Pub. No.302, Washington, 1968), pp. 279–292.
29. R. W. Powell and R. P. Tye, Int. J. Heat Mass Transfer **10**, 581–596 (1967).
30. J. G. Cook, Can. J. Phys. **57**, 1216–1223 (1979).
31. C. C. Bidwell, Phys. Rev. **28**, 584–597 (1926).
32. H. Masumoto, Sci. Rep. Tohoku Univ. **13**, 229–242 (1925).
33. J. P. Moore, R. K. Williams and R. S. Graves, Rev. Sci. Instrum. **45**, 87–95 (1974).
34. J. G. Cook, M. P. van der Meer and M. J. Laubitz, Can. J. Phys. **50**, 1386–1401 (1972).
35. R. K. Williams, W. H. Butler, R. S. Graves and J. P. Moore, Phys. Rev. **B28**, 6316–6324 (1983).
36. N. G. Bäcklund, *Thermal Conductivity* 8, edited by C. Y. Ho, and R. E. Taylor (Plenum Press, New York, 1969), p. 355.
37. W. Hemminger, Int. J. Thermophys. **10**, 765–777 (1989).
38. M. J. Laubitz, and T. Matsumara, Can. J. Phys. **50**, 196–205 (1972).
39. E. D. Devyatkova, V. P. Zhuze, A. V. Golubov, V. M. Sergeeva and I. A. Smirnov, Sov. Phys.-Solid State **6**, 343–346 (1964).
40. J. F. Andrew and P. G. Klemens, in *Thermal Conductivity* 17, edited by J. G. Hust (Plenum Press, New York, 1983), p. 209.
41. J. G. Cook, Can. J. Phys. **57**, 871–883 (1979).
42. R. W. Powell, R. P. Tye and M. J. Woodman, J. Less-Common Met. **5**, 49–56 (1963).
43. R. W. Powell, R. P. Tye and M. J. Woodman, J. Less-Common Met. **12**, 1–10 (1967).
44. S. Arajs and G. R. Dunmyre, Physica **31**, 1466–1472 (1965).
45. R. K. Williams, R. S. Graves, T. L. Hebble, D. L. McElroy., and J. P. Moore, Phys. Rev. **B26**, 2932–2942 (1982).
46. D. E. Baker, J. Less-Common Met. **8**, 435–436 (1965).
47. R. L. Anderson, D. H. Grotzky and W. E. Kienzle, in *Thermal Conductivity* 9, edited by H. R. Shanks (Plenum Press, New York, 1970), p. 326.
48. V. E. Peletskii, High Temp.-High Pressures **17**, 111–115 (1985).
49. A. Eucken and K. Dittrich, Z. Phys. Chem. (Leipzig) **125**, 211–228 (1927).
50. W. W. Tyler, A. C. Wilson and G. J. Wolga, Trans. Metall. Soc. AIME **197**, 1238–1239 (1953).
51. W. D. Jung, F. A. Schmidt and G. C. Danielson, Phys. Rev. **B15**, 659–665 (1977).
52. K. E. Wilkes, R. W. Powell and D. P. de Witt, in *Thermal Conductivity* 8, edited by C. Y. Ho and R. E. Taylor (Plenum Press, New York, 1969), p. 301.
53. H. W. Deem, USAEC Rep. BMI-849 (1953).
54. M. Ohyama, J. Phys. Soc. Jpn. **23**, 522–525 (1967).
55. D. T. Morelli and C. Uher, Phys. Rev. **B28**, 4242–4246 (1983).
56. Kh. I. Amirkhanov, A. Z. Daibov and V. P. Zhuze, Dokl. Akad. Nauk SSSR **98**, 557–560 (1954).
57. C. F. Gallo, B. S. Chandrasekharand P. H. Sutter, J. Appl. Phys. **34**, 144–152 (1963).
58. A. Eucken and O. Neumann, Z. Phys. Chem. (Leipzig) **111**, 431–446 (1924).
59. H. J. Goldsmid, *Electronic Refrigeration* (Pion, London, 1986).
60. F. J. Blatt, P. A. Schroeder, C. L. Foiles and D. Greig, *Thermoelectric Power of Metals* (Plenum Press, New York, 1976).
61. T. M. Tritt, *Semiconductors and Semimetals*, Vol. **69** (Academic Press, San Diego, 2001).
62. M. Köhler, Z. Phys., **124**, 772–789 (1948).
63. F. J. Blatt, *Physics of Electronic Conduction in Solids* (McGraw-Hill, New York, 1968).
64. R. Kubo, J. Phys. Soc. Jpn. **12**, 570–586 (1957).
65. D. A. Greenwood, Proc. Phys. Soc. **71**, 585–596 (1958)
66. F. S. Edwards, Proc. Phys. Soc. **86**, 977–988 (1965).
67. W. Kohn and J. M. Luttinger, Phys. Rev. **108**, 590–611 (1957).
68. L. S. Rodberg and R. M. Thaler, *Introduction to the Quantum Theory of Scattering* (Academic Press, San Diego, 1967).

69. A. MATTHIESSEN, Rep. Br. Assoc. **32**, 144–150 (1862).
70. J. BASS, Adv. Phys. **21**, 431–604 (1972).
71. J. A. ROWLANDS AND S. B. WOODS, J. Phys. F **8**, 1929–1939 (1978).
72. P. G. KLEMENS, in *Solid State Physics*, Vol. **7**, edited by F. Seitz and D. Turnbull (Academic Press, New York, 1958), p. 1.
73. A. HAUG, *Theoretical Solid State Physics*, Vol. **2** (Pergamon Press, Oxford, 1972).
74. J. DE LAUNAY, in *Solid State Physics*, Vol. **2**, edited by F. Seitz and D. Turnbull, (Academic Press, New York, 1956).
75. F. J. BLATT, in *Solid State Physics*, Vol. **4**, edited by F. Seitz and D. Turnbull (Academic Press, New York, 1957), pp. 199–366.
76. F. BLOCH, Z. Phys. **52**, 555–600 (1928).
77. E. GRÜNEISEN, Ann. Phys. Leipzig **16**, 530–540 (1933).
78. O. DREIRACH, J. Phys. F: Metal Phys. **3**, 577–584 (1973).
79. S. N. KHANNA AND A. JAIN, J. Phys. F: Metal Phys. **4**, 1982–1986 (1974).
80. G. K. WHITE AND S. B. WOODS, Can. J. Phys. **33**, 58–73 (1955).
81. D. K. WAGNER AND R. BOWERS, Adv. Phys. **27**, 651–746 (1978).
82. W. E. LAWRENCE AND J. W. WILKINS, Phys. Rev. B**7**, 2317–2332 (1973).
83. W. J. DE HAAS AND J. H. DE BOER, Physica **1**, 609–616 (1934).
84. L. LANDAU AND I. POMERANCHUK, Zh. Eksp. Teor. Fiz. **7**, 379–385 (1937).
85. W. G. BABER, Proc. R. Soc. A, **158**, 383–396 (1937).
86. N. F. MOTT, Proc. Phys. Soc. (London) **47**, 571–588 (1935).
87. C. POTTER AND G. J. MORGAN, J. Phys. F **9**, 493–503 (1979).
88. M. KAVEH AND N. WISER, Adv. Phys. **33**, 257–372 (1984).
89. A. SCHMID, Z. Phys. **271**, 251–256 (1974).
90. B. L. ALTSHULER, A. G. ARONOV AND P. A. LEE, Phys. Rev. Lett. **44**, 1288–1291 (1980).
91. A. H. MACDONALD, R. TAYLOR AND D. J. W. GELDART, Phys. Rev. B **23**, 2718–2730 (1981).
92. J. E. BLACK, Can. J. Phys. **56**, 708–714 (1978).
93. C. A. KUKKONEN AND P. F. MALDAGUE, J. Phys. F: Metal Phys. **6**, L301–302 (1976).
94. M. SINVANI, A. J. GREENFIELD, M. DANINO, M. KAVEH AND N. WISER, J. Phys. F:Metal Phys. **11**, L73–78 (1981).
95. B. LEVY, M. SINVANI, AND A. J. GREENFIELD, Phys. Rev. Lett. **43**, 1822–1825 (1979).
96. H. VAN KEMPEN, J. S. LASS, J. H. RIBOT AND P. WYDER, Proc. 14th Int. Conf. on Low Temperature Physics, Vol. **3**, Amsterdam; (North-Holland, 1976), pp. 94–97.
97. J. A. ROWLANDS, C. DUVVURY AND S. B. WOODS, Phys. Rev. Lett. **40**, 1201–1204 (1978).
98. C. W. LEE, M. L. HEARLE, V. HEINEN, J. BASS, W. P. PRATT, J. A. ROWLANDS AND P. SCHROEDER, Phys. Rev. B **25**, 1411–1414 (1982).
99. B. R. BARNARD AND A. D. CAPLIN, Commun. Phys. **2**, 223–227 (1977).
100. M. KHOSHNEVISAN, W. P. JR. PRATT, P. A. SCHROEDER S. STEENWYK AND C. UHER, J. Phys. F **9**, L1–L5 (1979).
101. E. BORCHI AND S. DE GENNARO, J. Phys. F **10**, L271–L274 (1980).
102. A. BERGMANN, M. KAVEH AND N. WISER, J. Phys. F **12**, 2985–3008 (1982).
103. M. KHOSHNEVISAN, W. P. JR. PRATT, P. A. SCHROEDER, S. STEENWYK, Phys. Rev. B **19**, 3873–3878 (1979).
104. J. H. RIBOT, J. BASS, F. VAN KEMPEN, R. J. M. VAN VUCHT AND P. WYDER, J. Phys. F **9**, L117–122 (1979).
105. J. G. BEICHTMAN, C. W. TRUSSEL AND R. V. COLEMAN, Phys. Rev. Lett. **25**, 1291–1294 (1970).
106. D. RADHAKRISHMA AND M. NIELSEN, Phys. Stat. Solidi **11**, 111–115 (1965).
107. G. K. WHITE AND R. J. TAINSH, Phys. Rev. Lett. **19**, 165–166 (1967).
108. G. W. WEBB, Phys. Rev. **181**, 1127–1135 (1969).
109. T. L. RUTHRUFF, C. G. GRENIER AND R. G. GOODRICH, Phys. Rev. B **17**, 3070–3073 (1973).
110. J. T. SCHRIEMPF AND W. M. MACINNES, Phys. Lett. A **33**, 511–512 (1970).
111. C. UHER AND P. A. SCHROEDER, J. Phys. F: Metal Phys. **8**, 865–871 (1978).
112. C. UHER M. KHOSHNEVISAN, W. P. PRATT, AND J. BASS, J. Low. Temp. Phys. **36**, 539–566 (1979).
113. J. T. SCHRIEMPF, Solid State Commun. **6**, 873–876 (1968).
114. C. UHER, C. W. LEE AND J. BASS,, Phys. Lett. A**61**, 344–346 (1977).
115. G. W. WEBB, Z. FISK, J. J. ENGELHARDT AND S. D. BADER, Phys. Rev. B **15**, 2624–2629 (1977).
116. C. UHER AND W. P. PRATT, Phys. Rev. Lett. **39**, 491–494 (1977).
117. D. T. MORELLI AND C. UHER Phys. Rev. B **30**, 1080–1082 (1984).
118. A. H. MACDONALD, Phys. Rev. Lett. **44**, 489–493 (1980).

Sec. 8 · REFERENCES

119. C. Herring, Phys. Rev. Lett. **19**, 167–168 (1967).
120. W. P. Pratt, C. Uher, P. A. Schroeder and J. Bass, in *Thermoelectricity in Metallic Conductors*, edited by F. J. Blatt and P. A. Schroeder (Plenum Press, New York, 1978), pp. 265–280.
121. M. Kaveh and N. Wiser, Phys. Rev. Lett. **26**, 635–636 (1971).
122. P. F. Maldague and C. A. Kukkonen, Phys. Rev. B **19**, 6172–6185 (1979).
123. M. J. Laubitz, Phys. Rev. B **2**, 2252–2254 (1970).
124. A. H. MacDonald and D. J. W. Geldart, J. Phys. F **10**, 677–692 (1980).
125. J. G. Cook, Can. J. Phys. **60**, 1759–1769 (1982).
126. M. J. Laubitz and J. G. Cook, Phys. Rev. B **7**, 2867–2869 (1973).
127. M. Kaveh and N. Wiser, Phys. Rev. B **21**, 2291–2308 (1980).
128. A. Sommerfeld and H. Bethe, in *Handbuch der Physik*, Vol. **24**, Part 2 (Springer, Berlin, 1933), p. 33.
129. R. E. B. Makinson, Proc. Camb. Phil. Soc. **34**, 474–497 (1938).
130. P. G. Klemens, Aust. J. Phys. **7**, 57–63 (1954).
131. A. H. Wilson, *Theory of Metals* (Cambridge University Press, 1958).
132. W. R. Kemp, P. G. Klemens, R. J. Tainsh and G. K. White, Acta Metall. **5**, 303–309 (1957).
133. G. K. White and R. J. Tainsh, Phys. Rev. **119**, 1869–1871 (1960).
134. D. Greig and J. P. Harrison, Phil. Mag. **12**, 71–79 (1965).
135. T. Farrell and D. Greig, J. Phys. C **2**, 1465–1473 (1969).
136. C. Uher and H. J. Goldsmid, Phys. Stat. Solidi (b) **65**, 765–772 (1974).
137. G. A. Slack, in *Solid State Physics*, Vol. **34**, edited by H. Ehrenreich, F. Seitz and D. Turnbull (Academic Press, New York, 1979), pp. 1–71.
138. P. G. Klemens, in *Encyclopedia of Physics*, Vol. **14**, edited by S. Flügge (Springer-Verlag, Berlin, 1956), pp. 198–218.
139. A. B. Pippard, Phil. Mag. **46**, 1104–1114 (1955).
140. P. Lindenfeld and W. B. Pennebaker, Phys. Rev. **127**, 1881–1889 (1962).
141. J. N. Lomer and N. H. Rosenberg, Phil. Mag. **4**, 467–483 (1959).
142. P. Charsley, J. A. M. Salter and D. W. Leaver, Phys. Stat. Solidi **25**, 531–540 (1968).
143. W. R. G. Kemp, P. G. Klemens and R. J. Tainsh, Phil. Mag. **4**, 845–857 (1959).
144. W. R. G. Kemp and P. Klemens, Aust. J. Phys. **13**, 247–254 (1960).
145. F. R. N. Nabarro, Proc. R. Soc. **A209**, 278–290 (1951).
146. M. W. Ackerman, Phys. Rev. **B5**, 2751–2754 (1972).
147. A. C. Anderson and S. C. Smith, J. Phys. Chem. Solids, **34**, 111–122 (1973).
148. S. G. O'Hara and A. C. Anderson, Phys. Rev. **B9**, 3730–3734 (1974).
149. S. G. O'Hara and A. C. Anderson, Phys. Rev. **B10**, 574–579 (1974).
150. E. H. Sondheimer, Proc. Roy. Soc. **A203**, 75–98 (1950).
151. M. Blackman, Proc. Phys. Soc. London **A64**, 681–690 (1951).
152. R. S. Newrock and B. W. Maxfield, Phys. Rev. **B7**, 1283–1295 (1973).
153. L. Binkele, High Temp.-High Pressures **18**, 599–607 (1986).
154. W. J. Nellis and S. Legvold, Phys. Rev. **180**, 581–590 (1969).
155. P. G. Klemens, Aust. J. Phys. **7**, 64–69 (1954).
156. P. G. Klemens, Aust. J. Phys. **7**, 70–76 (1954).
157. R. Berman and D. K. C. MacDonald, Proc. R. Soc. London **A209**, 368–375 (1951).
158. D. K. C. MacDonald, G. K. White and S. B. Woods, Proc. R. Soc. London **A235**, 358–374 (1956).
159. R. Berman and D. K. C. MacDonald, Proc. R. Soc. London, **A211**, 122–128 (1952).
160. K. Mendelssohn and M. H. Rosenberg, Proc. Phys. Soc. London **A65**, 385–388 (1952).
161. G. K. White, Proc. Phys. Soc. London **A66**, 559–564 (1953).
162. P. G. Klemens, Proc. Phys. Soc. London **A67**, 194–196 (1954).
163. P. G. Klemens and R. K. Williams, Int. Metals Rev. **31**, 197–215 (1986)
164. C. S. Smith and E. W. Palmer, Trans. AIME **117**, 225–243 (1935).

Chapter 1.3

THERMAL CONDUCTIVITY OF INSULATORS AND GLASSES

Vladimir Murashov and Mary Anne White

*Department of Chemistry and Institute for Research in Materials
Dalhousie University, Halifax, Nova Scotia, Canada*

1. INTRODUCTION

Heat transport in a system is governed by the motion of "free particles" which try to restore thermodynamic equilibrium in the system subjected to a temperature gradient. For insulating materials we can generally ignore contributions of mobile electrons to thermal conduction processes, which dominate the thermal conductivity of metals, and concentrate instead on propagation of heat through acoustic phonons.[1] In general, we begin with the concept of a perfect harmonic solid (i.e., one in which all interactions are well represented by harmonic oscillators) being a perfect carrier to heat (i.e. with an infinite thermal conductivity, κ). Since this is not the case for any insulating material under any circumstances, we seek to understand thermal conductivity in terms of thermal resistance mechanisms.

In this chapter we present theories of thermal conductivity of insulating materials and glasses, and, where useful to illustrate the theories, some experimental findings. Experimental data for some 400 insulating solids have been compiled, and they were published in 1970.[2] Although the thermal conductivities of many more substances have been determined since then, no thorough updated compilation has been published. Several reviews of thermal conductivity of nonmetallic solids and of glasses have been published previously. [3-6]

Our discussion starts with the most ordered simple solids, adding degrees of disorder until we finish with a discussion of the thermal conductivity of glasses. All the while, we should keep in mind that, generally, thermal conductivity is an anisotropic property and that, furthermore, most theories apply to isochoric conditions, whereas most experiments are carried out isobarically. (Ross and co-workers[4,5,7] have shown the importance of this correction, especially for soft or disordered solids; for example, it changes κ by 30% for adamantane at room temperature, although the effect is much smaller at low temperatures.[5])

2. PHONONIC THERMAL CONDUCTIVITY IN SIMPLE, CRYSTALLINE INSULATORS

2.1. Acoustic Phonons Carry Heat

The common approach to understanding thermal conductivity of simple, crystalline dielectric solids is based on Debye's equation for heat transfer in gases, treating the lattice vibrations as a "gas" of phonons:[8,9]

$$\kappa = \frac{1}{3} C v \lambda, \qquad (1)$$

where C is the heat capacity per unit volume, v is the average phonon group velocity, and λ is the mean free path of phonons between collisions. For a more general anisotropic case, elements of the thermal conductivity tensor can be expressed as a sum over all wave vectors, k, of the first Brillouin zone for each polarization branch, m:[10]

$$\kappa_{ij} = \sum_{k,m} v_i(k)\, v_j(k)\, \tau_j(k)\, C_m(k), \qquad (2)$$

where τ is the phonon relaxation time.

The contribution of the optic branches of the dispersion curve to the heat capacity at constant volume, C_v, is approximated by the Einstein model of isolated atomic vibrations.[11] (See Chapter 1.1.) Phonons from the optic branches usually are ineffective carriers of thermal energy, but, as we shall show, they can attenuate heat flux by the acoustic modes in certain circumstances, especially for more complex insulators. However, most of the thermal energy in insulators is carried by acoustic phonons.[12] The Debye approximation of lattice dynamics as collective vibrations of atoms gives a good estimate of the acoustic contribution to the heat capacity (see also Chapter 1.1):[8,9]

$$C_v = 9 N k_B x^{-3} \int_0^x \frac{x^4 e^x}{(e^x - 1)^2} dx, \qquad (3)$$

where $x = \theta_D/T$, θ_D ($= h v_D/k_B$) is the Debye (characteristic) temperature, and v_D is the Debye cutoff frequency. The heat capacity of a solid with n atoms per unit cell is best represented by three Debye (acoustic) modes and $(3n - 3)$ Einstein (optic) modes. Although the heat capacity of an insulating solid can be modeled to within a few percent if the mode frequencies are well known from vibrational experiments, the models for thermal conductivity are much less quantitatively advanced.

Resistance to heat flow in dielectric solids (or, in other words, interference to phonon motion) arises from scattering of phonons by defects in the crystal structure (lattice defects, grain boundaries, isotopes, impurities, etc.) and from collisions of phonons with each other, resulting in an alteration of phonon frequencies and momenta. In the (ideal) harmonic solid, such phonon–phonon interactions are not

possible (the phonons do not couple), so the thermal conductivity is infinite. In solids with (real) anharmonic interparticle interactions, there are two kinds of three-phonon processes, which can be generally described in terms of their energy (represented by the frequency, ω) and momenta (represented by the wave vector, k):
(1) An $A - event$ involves annihilation of two phonons and creation of a third phonon:

$$\omega(k_1) + \omega(k_2) = \omega(k_3), \tag{4}$$

$$k_1 + k_2 = k_3 + G. \tag{5}$$

(2) A $B - event$ represents annihilation of one phonon with creation of two phonons:

$$\omega(k_1) = \omega(k_2) + \omega(k_3), \tag{6}$$

$$k_1 + G = k_2 + k_3. \tag{7}$$

G is a reciprocal lattice vector.

Depending on the extent of involvement of the crystal as a whole in the scattering, there are so-called normal processes (N-process, $G = 0$) and Umklapp processes (U-process, $G \neq 0$). N-processes do not interfere with the phonon stream, but influence heat transfer indirectly through a change of the phonon frequency distribution.[13] U-processes provide the dominant thermal resistance mechanism in insulating solids. In this process the sum of wave vectors of colliding phonons falls outside of the first Brillouin zone (see also Chapter 1.1), and thus the resultant phonon wave vector opposes the phonon stream, effectively giving rise to thermal resistivity.

Callaway[14] assumed that different scattering mechanisms act independently and introduced a total phonon relaxation rate, $1/\tau_{tot}$, as the sum of scattering rates due to diverse elastic scattering mechanisms ($1/\tau_S$) and phonon–phonon scattering ($1/\tau_N + 1/\tau_U$), for N- and U-processes, respectively. Consideration of the Boltzmann equation for phonon distribution;[15]

$$\frac{\partial N}{\partial t} = -(v \bullet \nabla T)\frac{\partial N}{\partial T}, \tag{8}$$

where N is the phonon concentration and t is time, in the Debye regime of heat transfer [considering heat transfer via acoustic phonons given by Eq. (3)], yields the thermal conductivity coefficient, κ, as a sum of two parts κ_1 and κ_2:[15]

$$\kappa_1 = \frac{k_B}{2\pi^2 v}\left(\frac{2\pi k_B T}{h}\right)^3 \int_0^{\theta_D/T} \tau_{tot} \frac{x^4 e^x}{(e^x - 1)^2} dx \tag{9}$$

and

$$\kappa_2 = \frac{k_B}{2\pi^2 v}\left(\frac{2\pi k_B T}{h}\right)^3 \frac{\left(\int_0^{\theta_D/T} \frac{\tau_{tot}}{\tau_N} x^4 e^x (e^x - 1)^{-2} dx\right)^2}{\int_0^{\theta_D/T} \frac{\tau_{tot}}{\tau_N \tau_U} x^4 e^x (e^x - 1)^{-2} dx}. \tag{10}$$

When N-processes are dominant, the relative relaxation processes are $\tau_U \gg \tau_N, \tau_{tot}$

≈ τ_N, so κ_2 is the main term. Alternatively, if resistive processes dominate, $\tau_N \gg \tau_U$, $\tau_{tot} \approx \tau_U$, and κ_1 contributes more to the overall thermal conductivity. Complete exclusion of the U-processes results in an infinite conductivity, as the denominator in Eq. (10) goes to zero. This is in agreement with the concept of zero thermal resistivity in ideal (harmonic) crystals. The N-processes are important only at very low temperatures and in nearly perfect, low-anharmonicity crystals.

At present there are several expressions for the relaxation rate due to the U-processes, $1/\tau_U$,[16–19] but for dielectric solids, the most widely applied formula is as follows:[16]

$$\frac{1}{\tau_U} = A\omega^2 T e^{-B\Theta_D/T}, \tag{11}$$

where A and B are constants.

At very low temperatures ($T \leq 10$ K for a single crystal with linear dimensions ≤ 0.01 m), boundary scattering becomes important. The boundary scattering rate, $1/\tau_b$, can be expressed as[20]

$$\frac{1}{\tau_b} = \frac{1.12v}{d}, \tag{12}$$

where d is the dimension of the single crystal.

Doped crystals can have large contributions to the resistivity mechanism from impurity scattering. The rate of this scattering, $1/\tau_i$, was shown to be proportional to the fourth power of phonon frequency:[21]

$$\frac{1}{\tau_i} = A'\omega^4 \tag{13}$$

with a temperature-independent parameter, A', given by[10]

$$A' = \frac{V_0}{4\pi v^3}\left(\frac{\Delta M}{M}\right)^2, \tag{14}$$

where V_0 is the effective volume of the defects, M is the mass of the regular elementary unit of the substance, and ΔM is the difference in mass between the defect and the regular unit.

2.2. Temperature Dependence of κ

At quite high-temperatures ($T \gg \theta_D$), the thermal resistivity is dominated by umklapp processes, and the relaxation, given by Eq. (11), leads to a reduction in the mean free phonon path (λ) as the temperature increases. Correspondingly, $\kappa \sim T^{-1}$ for well-ordered solids at $T > \theta_D$.[5] As the temperature decreases, λ grows longer until it approaches its limit, which is governed by imperfections or by boundary scattering. Since the velocity of sound varies little with temperature, the other major factor to consider for κ is the heat capacity [see Eq. (1)], which is approximately constant at $T \gg \theta_D$, and gradually decreases as $T \to 0$ K. A schematic view of the thermal conductivity and the corresponding mean free phonon path in a simple, insulating solid is shown in Fig. 1.

Sec. 3 · MORE COMPLEX INSULATORS: THE ROLE OF OPTIC MODES

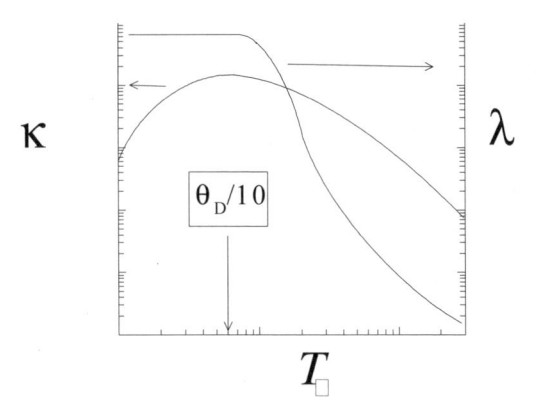

FIGURE 1 Schematic view of the thermal conductivity, κ, and the phonon mean free path, λ, as functions of temperature for a simple, insulating solid. Note that the peak in κ occurs at a temperature about 10% of the Debye characteristic temperature, θ_D.

2.3. Impurities

Impurities predominantly lower the phonon mean free path by adding additional scattering centers, while having little effect on the heat capacity or velocity of sound of a solid. Therefore, the impurities have the largest effect on κ at low temperatures where the phonon mean free path would otherwise be limited by boundaries (for a pure single insulating crystal).

The effect of impurities has been most effectively studied in the case of the thermal conductivity of diamond. Owing to the strong carbon–carbon interaction in diamond, the Debye temperature of diamond is very high, about 2200 K,[22] leading to an exceptionally high velocity of sound (ca. 1800 m s^{-1}). Hence, the thermal conductivity of diamond is extraordinarily large, ca. 2000 W m^{-1} K^{-1} at room temperature.[23] However, isotopically purified diamond (99% ^{12}C) has an even higher thermal conductivity, about 4000 W m^{-1} K^{-1} at room temperature, and increasing to 41,000 W m^{-1} K^{-1} at $T = 100$ K.[24] The increase in thermal conductivity on removal of isotopic impurities has been explained in terms of the N-processes.[25]

Although the thermal conductivity usually is lowered with impurities, because the impurities gives rise to phonon scattering, the thermal conductivity of a semiconductor can be *increased* on addition of n- or p-type impurities, because they provide increased electronic heat conduction which can more than compensate for the reduction in phononic heat conduction.

3. MORE COMPLEX INSULATORS: THE ROLE OF OPTIC MODES

3.1. Molecular and Other Complex Systems

Although the acoustic phonons are the predominant carriers of thermal energy in simple insulators, and acoustic modes are usually considered to be well separated from optic modes, this is not always the case, especially in materials with large numbers of atoms per unit cell. As mentioned, for n atoms per unit cell, 3 modes would be acoustic, and the remaining $(3n - 3)$ degrees of freedom would be associated with optical modes. If n is large, and some of the optic modes are close to the frequency range of the acoustic modes, in principle, this could give rise to interactions between the acoustic modes and the optic modes, making understanding of the thermal conductivity much more difficult in most cases. Furthermore, since many of the optic modes are at very high frequencies, the Dulong–Petit-type considerations of the heat capacity are not appropriate for molecular systems (although they are often used[5]), so Eq. (1) cannot be applied without experimental information concerning heat capacities.

The number of optical modes is especially important for molecular solids because of the potential for large numbers of atoms per unit cell. Furthermore, molecules can have additional degrees of freedom which can interfere with heat conduction. For example, whereas the thermal conductivities of solid Ar, Kr and Xe can be expressed as a universal function of reduced thermal resistivity, $\kappa^{-1}/((r_0/k_B)(M/\epsilon)^{1/2}$, where r_0 is the nearest-neighbor distance, M is the molecular mass and ϵ is the static binding energy, as a function of reduced temperature, $T/(\epsilon/k_B)$, for $T > 0.25\theta_D$,[26] the thermal conductivity of solid N_2, in both the α-phase and the β-phase, is less than that of the solidified inert gases because of additional phonon scattering mechanisms associated with phonon–libron interactions. Furthermore, the thermal conductivity of N_2 in the orientationally disordered β-phase is about 20% less than in the ordered α-phase.[26, 27] This is attributed to the interaction of heat-carrying phonons with fluctuations associated with orientational disorder of the N_2 molecules.[26] Similarly, measurements of κ of CH_4 from 22 to 80 K (in the high-temperature orientationally disordered solid phase) show[28] κ to be small (~ 0.4 W m^{-1} K^{-1}) and with only a slight maximum at about 50 K.

Another interesting case is C_{60}. In its high-temperature orientationally disordered phase, κ is small (~ 0.4 W m^{-1} K^{-1}) and essentially independent of temperature.[29] This has been attributed to efficient phonon scattering due to orientation disorder, with a calculated mean free path of only a few lattice spacings. On cooling below the ordering transition at 260 K, κ of C_{60} increases abruptly by about 25%.[29] Within the low-temperature simple cubic phase of C_{60}, the approximate temperature dependence is $\kappa \sim T^{-1}$, as one would expect from phonon–phonon Umklapp scattering processes, but a more accurate fit to the temperature dependence is achieved by inclusion of scattering from misoriented molecules[29] or from point defects.[30]

The crystalline phases of ice present many fascinating findings with regard to thermal conductivity. For the normal phase of ice, Ih, $\kappa \sim T^{-1}$ for $T > 40$ K, consistent with acoustic phonons carrying the heat and being primarily limited by three-phonon Umklapp processes.[31] At 'high' temperatures, κ of ice Ih is close to a value calculated based on phononic thermal conductivity in a simple lattice with vibrating point masses of average molecular mass 18, with O–H rotations and vibrations making relatively little contribution.[31] Furthermore, κ for ice Ih is abnormally high at the melting point compared with liquid water; the phonon mean free path for ice Ih at 273 K is about 30 lattice constants, which shows that κ is greater than its minimum value.[31] When a small amount of NaOH is added to ice, this orders the protons and produces ice XI; κ increases ~ 15% increase at the Ih→XI transformation, consistent with Umklapp processes and point defects as the main thermal resistance mechanisms in Ih and XI, with the XI phase closer to a harmonic lattice due to proton ordering.[32] In general, within the various ice phases, κ, expressed as $\kappa \sim T^{-n}$, falls into two groups, either with $n \geq 1$ or $n \sim 0.7$.

Sec. 3 • MORE COMPLEX INSULATORS: THE ROLE OF OPTIC MODES

The phases for which $n \geq 1$ are either proton disordered (phase Ih) or essentially completely antiferroelectrically ordered (phases II, VIII, and IX). The phases with $n < 1$ (III, V, VI, VI', and VII) are paraelectric.[33]

Thermal conductivity of n-alkanes near room temperature is of considerable interest because of the use of these materials as latent–energy storage materials. At 0°C, the thermal conductivity of n-C_mH_{2m+2} (m = 14 to 20) was found to show two distinct trends (one for m = even, one for m = odd; the reason for the differentiation is that packing is different for the two types of zigzag chains) with κ increasing linearly with m in both cases.[34] The thermal conductivity of the odd-carbon members was about 30% lower than for even carbon numbers. (It should be noted that the apparent thermal conductivity of n-$C_{20}H_{42}$ was found to depend on the conditions under which the sample was solidified.[35]) In contrast with globular-shaped molecules where κ increases by only about 5% on solidification, κ for C_mH_{2m+2} (m = 9,11,13,...,19) increases by about 35% on solidification.[36,37] Further substantial increase in κ was observed on cooling from the high-temperature disordered solid phase to the low-temperature solid phase, with the thermal conductivity increasing with carbon chain length.[36,37] The latter finding was attributed to strong intrachain bonds giving rise to more efficient heat conduction.[37] In the low-temperature phase under isobaric conditions, κ was found to depend more strongly on temperature than T^{-1}, due to thermal expansion effects.[37]

Although there is no strict theoretical basis, as there is for simpler, insulating solids, we can conclude that molecular solids in which there is dynamic disorder showed lowered κ, commensurate with shorter mean free paths because of additional coupling mechanics (including possibly coupling of the acoustic modes with low-lying optical modes) and a temperature dependence for $T > \theta_D$ of the form $\kappa \sim T^{-n}$, where $n > 1$.

3.2. Optic–Acoustic Coupling

Although much of the heat flux is carried by acoustic modes, and optic modes are usually considered to be far higher in energy than acoustic modes, in some systems, the optic modes can be close enough in energy to the acoustic modes to allow optic–acoustic coupling. We now examine a few such cases.

In inclusion compounds — that is, systems in which there is a host lattice in which other atoms or molecules reside as loosely associated guests — it is possible to have resonant scattering due to the coupling of the lattice acoustic modes with the localized low-frequency optical modes of the guest species. In some cases this can be the dominant phonon scattering mechanism,[38,39] and it can be represented by a phenomenological expression for the resonant scattering rate, $1/\tau_R$:[40,41]

$$\frac{1}{\tau_R} = N_0 D \frac{\omega_0^2 \omega^2}{(\omega_0^2 - \omega^2)^2}, \qquad (15)$$

where N_0 is the concentration of guest, D is a coefficient depicting the strength of host–guest coupling, and ω_0 is the characteristic resonant frequency. If this mechanism sufficiently contributes to the overall thermal resistance, this can lead to low values of κ and positive values of $d\kappa/dT$ at temperatures above about 10 K, similar to the behavior of κ for a glass (*vide infra*).

A similar mechanism has been used to describe the interaction of side-chain motions in a polymer with acoustic phonons.[42] Furthermore, other molecular systems exhibit similar contributions to thermal resistance.[43–45] This mechanism has been suggested[46] and proven[47] to reduce thermal conductivity for applications in thermoelectrics. In general, this mechanism could be important in any systems with low-lying optical energy levels, e.g., due to dynamic disorder of ions or molecular units.

4. THERMAL CONDUCTIVITY OF GLASSES

4.1. Comparison with Crystals

To a first approximation, the major distinctions between thermal conductivities of glasses and crystalline solids are that glasses have lower thermal conductivities and $d\kappa/dT$ is positive for glasses and negative for simple crystalline solids.[48] [This is somewhat of an oversimplification, as we also know (see Sec. 3.2) that crystalline materials can have low thermal conductivities with positive temperature coefficients of κ, when optic–acoustic resonance scattering is important.] For example, the thermal conductivity of crystalline SiO_2 is about 10 W m^{-1} K^{-1} at room temperature, whereas that of amorphous SiO_2 is about 1 W m^{-1} K^{-1}.[48] Furthermore, the thermal conductivities of glasses of a wide range of composition are similar both in magnitude and in temperature dependence.[6] The higher thermal conductivity of crystalline cellulose, compared with amorphous cellulose, is the secret to successful popping corn.[49]

Based on the Debye model of the thermal conductivity of a solid [eq. (1)], Kittel[50] explained the behavior of glasses in terms of an approximately constant mean free path for the lattice phonons, so the thermal conductivity closely follows the heat capacity. He found that the value of the phonon mean free path in a glass at room temperature is on the order of magnitude of the scale of disorder in the structure of the glass, viz. about 7 Å.

4.2. More Detailed Models

Many glasses have been shown to follow an almost universal temperature dependence of κ, including a very-low-temperature ($T < 1$ K) region with a steep positive slope, followed by a plateau between about 1 K and 20 K, and then a region of positive $d\kappa/dT$ above about 20 K.[6]

The thermal conductivity of glasses below the plateau varies as T^n, where $n \sim 2$; it has been attributed to the scattering of phonons from low-lying energy states,[6] which can be interpreted in a two-level system model.[51,52] A Boson peak at a few meV in the Raman spectrum has been associated with these levels.

In the region of the plateau, there are many different interpretations of the origins of the flat thermal conductivity, ranging from scattering from structural disorder,[53] to tunneling interactions,[54] to resonance scattering from localized vibrational modes[55] and a soft-potential model in which the tunneling states and the localized resonant modes have a common origin.[56]

Kittel's explanation of the thermal conductivity of a glass above the plateau showed that the phonon mean free path is of the order of the interatomic spacing, which is random in a glass, so the concept of phonons is not as apt a description as a random walk of localized (Einstein) oscillations.[6] Cahill and Pohl considered a model of oscillators that are so strongly damped that they pass on their energy within half a period of oscillation; the thermal conductivity can be described as

$$\kappa_{\text{Eins}} = 2\,k_B{}^2 l^{-1} h^{-1} \theta_E x^2 e^x \,(e^x - 1)^{-2} \qquad (16)$$

where l is the interatomic spacing, θ_E is the Einstein temperature, and $x = \theta_E/T$.[6] Their more detailed approach considers a distribution of oscillators, representing the thermal conductivity in terms of an integration over frequencies, ω:[6]

$$\kappa = \tfrac{1}{3} \int_0^\infty \tfrac{dC}{d\omega} v(\omega)\, l(\omega)\, d\omega . \qquad (17)$$

Orbach and co-workers have devised a different model, in which the localized phonons are thermally activated to hop among inequivalent localized sites.[57] The effect of localized modes on thermal and other properties has been discussed by Buchenau.[58]

4.3. The Exception: Recent Amorphous Ice Results

Although most amorphous phases are well described as above, recent studies of the low-density form of amorphous ice show its thermal conductivity to be more like that of a simple, crystalline solid, although the thermal conductivity of high-density amorphous ice is typical for a glass.[59] It has been concluded that the thermal resistance in low-density amorphous ice is dominated by rather weak phonon–phonon scattering, in sharp contract with results for other glasses.[59]

5. MINIMUM THERMAL CONDUCTIVITY

As mentioned, a perfect harmonic crystal has no thermal resistance mechanism; hence, it has infinite thermal conductivity. Therefore it is worth considering the lower limit to the thermal conductivity. This point was addressed by Slack,[3] from the perspective of insulators in which the phonons are maximally coupled, and as a function of the number of atoms in the unit cell. The experimental thermal conductivity of amorphous SiO_2 above the plateau is close to the calculated minimum thermal conductivity, and for some crystalline systems the experimental thermal conductivity approaches the minimum as the temperature approaches the melting point.[3] Further support for the concept of a minimal thermal conductivity comes from experimental studies of mixed crystals with controlled disorder.[60] In the context of a random walk between Einstein oscillators of varying sizes, the minimum thermal conductivity can be expressed as[6]

$$\kappa_{\min} = \tfrac{2}{2.48} k_B n^{2/3}\, v \left(\tfrac{T}{\theta_c}\right) \int_0^{2\theta_c/T} \tfrac{x^3 e^x}{(e^x - 1)^2}\, dx \qquad (18)$$

where θ_c is the cutoff frequency for that mode.

The concept of minimal thermal conductivity can be a useful guide in attempts to develop materials with exceptionally low thermal conductivities (e.g., for thermoelectrics). The thermal conductivity of a dielectric can be reduced toward its theoretical minimum by increasing the size of the unit cell, the presence of heavy atoms, amorphization, random atomic substitution, increased optic–acoustic coupling and increasing the lattice symmetry.[3]

6. RADIATION

We have concentrated our discussion on heat transfer by thermal conductivity through the sample, but wish to conclude with a reminder that in materials at very high-temperatures, especially above 700 K, radiative heat transfer must also be included. The measured thermal conductivity would be the sum of the intrinsic phononic thermal conductivity and the radiative contribution, where the latter is given by

$$\kappa_{\text{radiation}} = \frac{16}{3}\eta^2 \frac{\sigma}{k_r} T^3 \tag{19}$$

where η is the refractive index, σ is the Stefan–Boltzmann constant, and k_r is the Rosseland mean absorption coefficient.[61] The form of the radiative term shows that it becomes increasingly important as the temperature is increased.

7. REFERENCES

1. R. L. SPROULL *The Conduction of Heat in Solids* Sci. Am. December 92–102 (1962).
2. Y. S. TOULOUKIAN, R. W. POWELL, C. Y. HO AND P. G. KLEMENS, eds. *Thermophysical Properties of Matter* Vol. 2, *Thermal Conductivity, Nonmetallic Solids* (Plenum, New York, 1970).
3. G. A. SLACK *The Thermal Conductivity of Nonmetallic Crystals* Solid State Phys. **34**, 1–71 (1979).
4. R. G. ROSS, P. ANDERSSON, B. SUNDQVIST AND G. BÄCKSTRÖM *Thermal Conductivity of Solids and Liquids Under Pressure* Rep. Progr. Phys. **47**, 1347–1402 (1984).
5. R. G. ROSS *Thermal Conductivity and Disorder in Nonmetallic Materials* Phys. Chem. Liq. **23**, 189–210 (1991).
6. D. G. CAHILL AND R. O. POHL *Lattice Vibrations and Heat Transport in Crystals and Glasses* Ann. Rev. Phys. Chem. **39**, 93–121 (1988).
7. R. G. ROSS *Thermal Conductivity of Solids under Pressure* High Temp.–High Press. **21**, 261–266 (1989).
8. C. KITTEL *Introduction to Solid State Physics*, 7th ed. (Wiley, New York, 1996).
9. M. A. WHITE *Properties of Materials* (Oxford University Press, New York, 1999).
10. P. G. KLEMENS *Thermal Conductivity and Lattice Vibrational Modes*, in *Solid State Physics*, Vol. 7, edited by F. Seitz and D. Turnbull, (Academic Press, 1958), pp. 1–98.
11. P. BRÜESCH *Phonons: Theory and Experiments* I (Springer-Verlag, Berlin, 1982).
12. D. G. CAHILL AND R. O. POHL *Lattice Vibrations and Heat Transport in Crystals and Glasses* Ann. Rev. Phys. Chem. **39**, 93–121 (1988).
13. J. CALLAWAY *Quantum Theory of the Solid State*, 2nd ed. (Academic Press, San Diego, 1991).
14. J. CALLAWAY *Model for Lattice Thermal Conductivity at Low Temperatures* Phys. Rev. **113**, 1046–1051 (1959).
15. R. BERMAN *Thermal Conductivity in Solids* (Clarendon Press, Oxford, 1976).
16. P. G. KLEMENS in *Thermal Encyclopedia of Physics*, Vol. **14**, edited by S. Flugge (Springer-Verlag, Berlin, 1956), p. 198.
17. R. BERMAN AND J. C. F. BROCK *The Effect of Isotopes on Lattice Heat Conduction. I. Lithium Fluoride.* Proc. R A **298**, 46–65 (1965).
18. J. CALLAWAY AND H. C. VON BAEYER *Effect of Point Imperfections on Lattice Thermal Conductivity* Phys. Rev. **120**, 1149–1154 (1960).
19. J. E. PARROT AND A. D. STUCKES *Thermal Conductivity of Solids* (Pion, London, 1975).
20. H. B. G. CASIMIR *Note on the Conduction of Heat in Crystals* Physica 5, 495–500 (1938).
21. P. G. KLEMENS *Thermal conductivity of dielectric solids at low temperatures* Proc. R. Soc., A **208**, 108–133 (1951).
22. J. E. DESNOYERS AND J. A. MORRISON *The Heat Capacity of Diamond between 12.8 and 277 K* Phil. Mag. **8**, 42–48 (1952).
23. G. A. SLACK *Nonmetallic Crystals with High Thermal Conductivity* J. Phys. Chem. Solids **34**, 321–335 (1973).

24. L. WEI, P. K. KUO, R. L. THOMAS, T. R ANTHONY, AND W. F. BANHOLZER *Thermal Conductivity of Isotopically Modified Single Crystal Diamond* Phys. Rev. Lett. **70**, 3764–3767 (1993).
25. J. R. OLSON, R. O. POHL, J. W. VANDERSANDE, A. ZOLTAN, T. R. ANTHONY, AND W. F. BANHOLZER *Thermal Conductivity of Diamond between 170 and 1200 K and the isotope effect* Phys. Rev. B **47**, 14850–14856 (1993).
26. L. A. KOLOSKOVA, I. N. KRUPSKII, V. G. MANZHELII, AND B. YA. GORODILOV *Thermal Conductivity of Solid Carbon Monoxide and Nitrogen* Fizika Tverdogo Tela (Sankt-Peterburg) **15**, 1913–1915 (1973).
27. P. STACHOWIAK, V. V. SUMAROKOV, J. MUCHA, AND A. JEŻOWSKI *Thermal Conductivity of Solid Nitrogen* Phys. Rev. B **50**, 543–546 (1994).
28. V. G. MANZHELII AND I .N. KRUPSKII *Heat Conductivity of Solid Methane*, Fizika Tverdogo Tela (Sankt-Peterburg) **10**, 284–286 (1968).
29. R. C. YU, N. TEA, M. B. SALAMON, D. LORENTS, AND R. MALHOTRA *Thermal Conductivity of Single-Crystal C_{60} Fullerene* Phys. Rev. Lett. **68**, 2050–2053 (1992).
30. B. SUNDQVIST *Point Defects and Thermal Conductivity of C_{60}* Phys Rev. B **48**, 14712–14713 (1993).
31. G. A. SLACK *Thermal Conductivity of Ice* Phys. Rev. B **22**, 3065–3071 (1980).
32. O. ANDERSSON AND H. SUGA *Thermal Conductivity of the Ih and XI Phases of Ice* Phys. Rev. B **50**, 6583–6588 (1994).
33. R. G. ROSS, P. ANDERSSON, AND G. BÄCKSTRÖM *Effects of H and D Order on the Thermal Conductivity of Ice Phases* J. Chem. Phys. **68**, 3967–3972 (1978).
34. D. W. YARBOROUGH AND C. N. KUAN *The Thermal Conductivity of Solid n-Eicosane, n-Octadecane, n-Heptadecane, n-Pentadecane, and n-Tetradecane* Thermal Conductivity **17**, 265–274 (1981).
35. P. C. STRYKER AND E. M. SPARROW *Application of a Spherical Thermal Conductivity Cell to Solid n-Eicosane Paraffin* Int. J. Heat Mass Transfer **33**, 1781–1793 (1990).
36. H. FORSMAN AND P. ANDERSSON *Effects of Temperature and Pressure on the Thermal Conductivity of Solid n-Undecane* Ber. Bunsenges. Phys. Chem. **87**, 490–495 (1983).
37. H. FORSMAN AND P. ANDERSSON *Thermal Conductivity at High Pressure of Solid Odd-Numbered n-Alkanes Ranging from Nonane to Nonadecane* J. Chem. Phys. **80**, 2804–2807 (1984).
38. J. S. TSE AND M. A. WHITE *Origin of Glassy Crystalline Behaviour in the Thermal Properties of Clathrate Hydrates: A Thermal Conductivity Study of Tetrahydrofuran Hydrate* J. Phys. Chem, **92**, 5006–5011 (1988).
39. J. S. TSE, V. P. SHPAKOV, V. R. BELOSLUDOV, F. TROUW, Y. P. HANDA, AND W. PRESS *Coupling of Localized Guest Vibrations with the Lattice Modes in Clathrate Hydrates* Europhys. Lett. **54**, 354–360 (2001).
40. C. T. WALKER AND R. O. POHL *Phonon Scattering by Point Defects* Phys. Rev. **131**, 1433–1442 (1963)
41. V. NARAYANAMURTI AND R. O. POHL *Tunneling States of Defects in Solids* Rev. Mod. Phys. **42**, 201–236 (1970).
42. M. N. WYBOURNE, B. J. KIFF AND D. N. BATCHELDER *Anomalous Thermal Conduction in Polydiacetylene Single Crystals* Phys. Rev. Lett. **53**, 580–583 (1984).
43. M. ZAKRZEWSKI AND M. A. WHITE *Thermal Conductivities of a Clathrate with and without Guest Molecules* Phys. Rev. B **45**, 2809–2817 (1992).
44. D. MICHALSKI AND M.A. WHITE *Thermal Conductivity of a Clathrate with Restrained Guests: The CCl_4 Clathrate of Dianin's Compound* J. Phys. Chem. **99**, 3774–3780 (1995).
45. D. MICHALSKI AND M. A. WHITE *Thermal Conductivity of an Organic Clathrate: Possible Generality of Glass-like Thermal Conductivity in Crystalline Molecular Solids* J. Chem. Phys. **106**, 6202–6203 (1997).
46. G. A. SLACK *New Materials and Performance Limits for Thermoelectric Cooling*, in CRC Handbook of Thermoelectrics (1995), pp. 407–440.
47. B. C. SALES, D. MANDRUS AND R. K. WILLIAMS. *Filled Skutterudite Antimonides: A New Class of Thermoelectric Materials.* Science **272**, 1325–1328 (1996).
48. A. EUCKEN *The Change in Heat Conductivity of Solid Metalloids with Temperature* Ann. Phys. **34**, 185–222 (1911).
49. W. J. DA SILVA, B. C. VIDAL, M. E. Q. MARTINS, H. VARGAS, A. C. PEREIRA, M. ZERBETTO AND L. C. M. MIRANDA *What Makes Popcorn Pop* Nature **362**, 417 (1993).
50. C. KITTEL *Interpretation of Thermal Conductivity of Glasses* Phys. Rev. **75**, 972–974 (1949)
51. P. W. ANDERSON, B. I. HALPERIN, AND C. M. VARMA *Anomalous Low-Temperature Thermal Properties of Glasses and Spin Glasses* Philos. Mag. **25**, 1–9 (1972).
52. W. A. PHILLIPS *Tunneling States in Amorphous Solids* J. Low Temp. Phys. **7**, 351–360 (1972).

53. R. Orbach *Dynamics of Fractal Networks* Science **231**, 814–819 (1986).
54. M. W. Klein *Dielectric Susceptibility and Thermal Conductivity of Tunneling Electric Dipoles in Alkali Halides* Phys. Rev. B 40, 1918–1925 (1989)
55. E. R. Grannan, M. Randeria and J.P. Sethna *Low-temperature Properties of a Model Glass. I. Elastic Dipole Model* Phys. Rev. B 41, 7784–7798 (1990)
56. M. A. Ramos and U. Buchenau *Low-temperature Thermal Conductivity of Glasses within the Soft-Potential Model* Phys. Rev. B 55, 5749–5754 (1997)
57. S. Alexander, C. Laermans, R. Orbach and H. M. Rosenberg *Fracton Interpretation of Vibrational Properties of Cross-linked Polymers, Glasses, and Irradiated Quartz* Phys Rev. B 28, 4615–4619 (1983)
58. U. Buchenau *Dynamics of Glasses* J. Phys. Condens. Matter. 13 7827-7846 (2001)
59. O. Andersson and H Suga *Thermal Conductivity of Amorphous Ices* Phys. Rev. B 65, 140201 (4 pages) (2002)
60. D. G. Cahill, S. K. Watson and R. O. Pohl *Lower Limit to the Thermal Conductivity of Disordered Crystals* Phys. Rev. B 46, 6131–6139 (1992)
61. U. Fotheringham *Thermal Properties of Glass* Properties of Optical Glass, 203–230 (Springer-Verlag, 1995)

Chapter 1.4

THERMAL CONDUCTIVITY OF SEMICONDUCTORS

G. S. Nolas

Department of Physics, University of South Florida, Tampa, FL, USA

H. J. Goldsmid

School of Physics, University of New South Wales, Sydney, NSW, Australia

1. INTRODUCTION

The conduction of heat in semiconductors has been the subject of intensive study during the past 50 years. From the practical point of view, thermal conductivity is an important parameter in determining the maximum power under which a semiconductor device may be operated. Moreover, thermal conductivity is one of the most important parameters determining the efficiency of those semiconductors used in thermoelectric energy conversion.[1] However, a number of factors have also made the study of semiconductors particularly important in advancing our knowledge of the mechanisms of heat conduction in solids. For example, the need for single crystals of exceptional perfection and purity has made available samples for measurement that display effects that cannot be observed in less perfect specimens.

Certain semiconductors have rather high electrical restivities, so heat conduction is then, in effect, due solely to lattice vibrations. On the other hand, in some materials the electronic component of thermal conductivity is large enough to be important, and, indeed, this is always the case for semiconductors in thermoelectric applications. The separation of the lattice and electronic contributions to the thermal conductivity is often necessary. It becomes particularly interesting when the semiconductor contains both electrons and positive holes since there can then be a large bipolar heat conduction effect. In a semiconductor with only one type of charge carrier, the ratio of the electronic thermal conductivity to the electrical conductivity has more or less the same value that it would have in a metal. In other words, the ratio satisfies the Wiedemann–Franz law or, rather, the version of this law that is appropriate for a nondegenerate or partly degenerate conductor. However, when bipolar conduction takes place, the ratio of electronic to lattice conductivity can become much larger.

It is often useful to be able to preselect those semiconductors in which the value of the lattice conductivity will be very high or very low. It is also helpful if one can predict the types of treatment of a given material that will lead to a substantial reduction of the thermal conductivity. Thus, in this chapter, we shall give due attention to both these matters. Certain crystal structures allow one to modify the thermal conductivity in subtle ways. We are often especially interested in processes that reduce lattice conductivity but have relatively little effect on electronic properties. These processes include, for example, the formation of solid solutions, addition of impurities, and reduction of grain size.

The range of thermal conductivity in semiconductors is exceedingly large. A semiconducting diamond, for example, has a thermal conductivity higher than that of any metal, while some clathrates have values comparable with those of glasses. Table 1 gives values for the thermal conductivity at room temperature for various semiconductors. In many cases the total thermal conductivity will be virtually the same as the lattice contribution, but in others the electronic component will be substantial and will vary according to the concentration of charge carriers. The lattice conductivity, itself, may be significantly smaller for less than perfect crystals.

2. ELECTRONIC THERMAL CONDUCTIVITY IN SEMICONDUCTORS

2.1. Transport Coefficients for a Single Band

One success of the classical theory of metals was its explanation of the Wiedemann–Franz law. This law states that the ratio of thermal conductivity to electrical conductivity is the same for all metals at a given temperature. The actual value of the ratio was given only approximately, but this shortcoming was made good when Sommerfeld's quantum theory of metals was applied.[2] Other properties of metals required the additional refinement of band theory, and this, of course, is an essential feature of any discussion of semiconductors. We shall first discuss thermal conductivity when the charge carriers reside in a single band.

It will be useful to obtain expressions for the electrical conductivity and the Seebeck coefficient as well as for the electronic thermal conductivity. The electrical conductivity is needed to calculate the so-called Lorenz number, and the Seebeck coefficient is required when we consider bipolar thermal conduction. We write equations for electric current and heat flux when an electric field and a temperature gradient are applied. For our purposes we assume that the material is isotropic and that all flows are in the x-direction. We suppose that the charge carriers are scattered in such a way that their relaxation time, τ, may be expressed in terms of the energy, E, by the relation $\tau = \tau_0 E^r$, where τ_0 and r are constants. Then by solving the Boltzmann equation one finds that the electric current, i, is given by

$$i = \mp \int_0^\infty euf(E)g(E)dE, \qquad (1)$$

Sec. 2 · ELECTRONIC THERMAL CONDUCTIVITY IN SEMICONDUCTORS

where $-e$ is the electronic charge, u is the velocity of the charge carriers in the x-direction, $f(E)$ is the Fermi distribution function, and $g(E)$ is the carrier distribution function, g being proportional to $E^{1/2}$. Here and in subsequent equations the upper sign refers to electrons and the lower sign to holes. The rate of flow of heat per unit cross-sectional area is

$$w = u(E - \zeta)f(E)g(E)dE, \tag{2}$$

where ζ is the Fermi energy and $E - \zeta$ represents the total energy transported by a carrier.

Since there can be no transport when $f = f_0$, we may replace f by $f - f_0$ in the preceding equations. Also, in general, the thermal velocity of the charge carriers will always be much greater than any drift velocity, and we may set

$$u^2 = \frac{2E}{3m^*}, \tag{3}$$

where m^* is the effective mass of the carriers. From Eqs. (1–3) we find

$$i = \mp \frac{2e}{3m^*} \int_0^\infty g(E)\tau_e E \frac{\partial f_0(E)}{\partial E} \left(\frac{\partial \zeta}{\partial x} + \frac{E - \zeta}{T} \frac{\partial T}{\partial x} \right) dE, \tag{4}$$

and

$$w = \pm \frac{\zeta}{e} i + \frac{2}{3m*} \int_0^\infty g(E)\tau_e E^2 \frac{\partial f_0(E)}{\partial E} \left(\frac{\partial \zeta}{\partial x} + \frac{E - \zeta}{T} \frac{\partial T}{\partial x} \right) dE. \tag{5}$$

The electrical conductivity, σ, may be found by setting the temperature gradient, $\partial T/\partial x$, equal to zero. The electric field, E, is given by $\pm(\partial \zeta/\partial x)e^{-1}$ so that

$$\sigma = \frac{i}{E} = -\frac{2e}{3m^*} \int_0^\infty g(E)\tau_e E \frac{\partial f_0(E)}{\partial E} dE. \tag{6}$$

On the other hand, when the electric current is zero, Eq. (4) shows that

$$\frac{\partial \zeta}{\partial x} \int_0^\infty g(E)\tau_e E \frac{\partial f_0(E)}{\partial E} dE + \frac{1}{T}\frac{\partial T}{\partial x} \int_0^\infty g(E)\tau_e E(E - \zeta) \frac{\partial f_e(E)}{\partial E} dE = 0. \tag{7}$$

The condition $i = 0$ is that for the definition of the Seebeck coefficient and the thermal conductivity. The Seebeck coefficient, α, is equal to $(\partial \zeta / \partial x)[e(\partial T/\partial x)]^{-1}$ and is given by

$$\alpha = \pm \frac{1}{eT} \left[\zeta - \int_0^\infty g(E)\tau_e E^2 \frac{\partial f_0(E)}{\partial E} dE \bigg/ \int_0^\infty g(E)\tau_e E \frac{\partial f_0(E)}{\partial E} dE \right]. \tag{8}$$

The Seebeck coefficient is negative if the carriers are electrons and positive if they are holes.

The electronic thermal conductivity, κ_e, is equal to $-w(\partial T/\partial x)^{-1}$ and is obtained from Eqs. (5) and (7), thus

$$\kappa_e = \frac{2}{3m^*T} \left\langle \left\{ \left[\int_0^\infty g(E)\tau_e E^2 \frac{\partial f_0(E)}{\partial E} dE \right]^2 \bigg/ \right. \right.$$

$$\left. \left. \int_0^\infty g(E)\tau_e E \frac{\partial f_0(E)}{\partial E} dE \right\} - \int_0^\infty g(E)\tau_e E^3 \frac{\partial f_0(E)}{\partial E} dE \right\rangle. \tag{9}$$

The integrals in Eqs. (6), (8), and (9) may be expressed as

$$K_s = -\frac{8\pi}{3}\left(\frac{2}{h^2}\right)^{3/2}(m^*)^{1/2}T\tau_0\int_0^\infty E^{s+r+3/2}\frac{\partial f_0(E)}{\partial E}dE, \quad (10)$$

where g and τ_e have been eliminated in favor of m^*, r, and τ_0. One then finds that

$$\int_0^\infty E^{s+r+3/2}\frac{\partial f_0(E)}{\partial E}dE = -\left(s+r+\frac{3}{2}\right)\int_0^\infty E^{s+r+1/2}f_0(E)dE \quad (11)$$

and

$$K_s = \frac{8\pi}{3}\left(\frac{2}{h^2}\right)^{3/2}(m^*)^{1/2}T\tau_0\left(s+r+\frac{3}{2}\right)(k_BT)^{s+r+3/2}F_{s+r+1/2}, \quad (12)$$

where

$$F_n(\xi) = \int_0^\infty \xi^n f_0(\xi)d\xi, \quad (13)$$

$\xi = E/k_BT$. Numerical values of the functions F_n, which are known as the Fermi–Dirac integrals, may be found elsewhere.

The expressions for the transport coefficients in terms of the integrals K_s are

$$\sigma = \frac{e^2}{T}K_0, \quad (14)$$

$$\alpha = \pm\frac{1}{eT}\left(\zeta - \frac{K_1}{K_0}\right), \quad (15)$$

and

$$\kappa_e = \frac{1}{T^2}\left(K_2 - \frac{K_1^2}{K_0}\right). \quad (16)$$

TABLE 1 Thermal Conductivity of Typical Samples of Some Elemental and Binary Compound Semiconductors at 300 K.[a]

Semiconductor	Thermal conductivity (W m^{-1}K^{-1})
Diamond	200
Silicon	124
Germanium	64
ZnS	14
CdTe	5.5
BN	20
AlSb	60
GaAs	37
InAs	29
InSb	16
SnTe	9.1
PbS 2.3	2.3
PbTe	2.3
Bi$_2$Te$_3$	2.0

[a.] From *CRC Handbook of Chemistry and Physics*.

Sec. 2 · ELECTRONIC THERMAL CONDUCTIVITY IN SEMICONDUCTORS

It may be necessary to use numerical methods to find the transport properties from Eqs. (15) and (16). However, if the Fermi energy is either much greater than or much less than zero, one can use simple approximations to the Fermi distribution function.

2.2. Nondegenerate and Degenerate Approximations

The nondegenerate approximation is applicable when $\zeta/k_B T \ll 0$; that is, when the Fermi level lies within the forbidden gap and far from the edge of the band in which the carriers reside. The approximation is very good when $\zeta \ll -4k_B T$, and is often used when $\zeta \ll -2k_B T$. It is assumed that the number of minority carriers is negligibly small.

When $\zeta/k_B T \ll 0$, the Fermi–Dirac integrals are given by

$$F_n(\eta) = \exp(\eta) \int_0^\infty \xi^n \exp(-\xi) d\xi = \exp(\eta)\Gamma(n+1), \tag{17}$$

where $\eta g = g\zeta v < /i > k_B T$ and the gamma function is such that

$$\Gamma(n+1) = n\Gamma(n). \tag{18}$$

When n is integral, $\Gamma(n+1) = n!$, while for half-integral values of n the gamma function can be found from the relation $\Gamma(1/2) = \pi^{1/2}$.

With the gamma function, the transport integrals become

$$K_s = \frac{8\pi}{3}\left(\frac{2}{h^2}\right)^{3/2}(m^*)^{1/2}T\tau_0(k_B T)^{s+r+3/2}\Gamma\left(s+r+\frac{5}{2}\right)\exp(\eta). \tag{19}$$

Then the electrical conductivity of a nondegenerate semiconductor is

$$\sigma = \frac{8\pi}{3}\left(\frac{2}{h^2}\right)^{3/2}e^2(m^*)^{1/2}\tau_0(k_B T)^{r+3/2}\Gamma\left(r+\frac{5}{2}\right)\exp(\eta). \tag{20}$$

It is often useful to express the electrical conductivity as

$$\sigma = ne\mu, \tag{21}$$

where n is the concentration of the charge carriers,

$$n = \int_0^\infty f(E)g(E)dE = 2\left(\frac{2\pi m^* k_B T}{h^2}\right)^{3/2}\exp(\eta), \tag{22}$$

and their mobility μ is

$$\mu = \frac{4}{3\pi^{1/2}}\Gamma\left(r+\frac{5}{2}\right)\frac{e\tau_0(k_B T)^r}{m^*}. \tag{23}$$

The carrier concentration has the value that would result if there were $2(2\pi m^* k_B T/h)^{3/2}$ states located at the band edge. Thus, this quantity is known as the effective density of states.

The Seebeck coefficient of a nondegenerate semiconductor is

$$\alpha = \pm\frac{k_B}{e}\left[\eta - \left(r+\frac{5}{2}\right)\right]. \tag{24}$$

It is usual to express the electronic thermal conductivity in terms of the Lorenz number, L, defined as $\kappa_e/\sigma T$. Then, from Eqs. (14) and (16),

$$L = \frac{1}{e^2 T^2}\left(\frac{K_2}{K_0} - \frac{K_1^2}{K_0^2}\right). \tag{25}$$

Using the nondegenerate approximation, we obtain

$$L = \left(\frac{k_B}{e}\right)^2 \left(r + \frac{5}{2}\right). \tag{26}$$

The Lorenz number does not depend on the Fermi energy if r is constant.

We now discuss the degenerate condition $\zeta/k_B T \gg 0$. This means that the Fermi level lies well inside the band that holds the charge carriers and the conductor is a metal. The Fermi–Dirac integrals take the form of the rapidly converging series

$$F_n(\eta) = \frac{\eta^{n+1}}{n+1} + n\eta^{n-1}\frac{\pi^2}{6} + n(n-1)(n-2)\eta^{n-3}\frac{7\pi^4}{360} + \ldots \tag{27}$$

One uses as many terms in the series as are necessary to yield a finite or nonzero value for the appropriate parameter.

The electrical conductivity of a degenerate conductor needs only the first term in the series

$$\sigma = \frac{8\pi}{3}\left(\frac{2}{h^2}\right)^{3/2} e^2 (m^*)^{1/2} \tau_0 \zeta^{r+3/2}. \tag{28}$$

On the other hand, if only the first term in Eq. (27) were included, the Seebeck coefficient would be zero. To obtain a nonzero value, the first two terms are used, whence

$$\alpha = \mp \frac{\pi^2}{3}\frac{k_B}{e}\frac{\left(r + \frac{3}{2}\right)}{\eta}. \tag{29}$$

The first two terms are also needed for the Lorenz number, which is given by

$$L = \frac{\pi^2}{3}\left(\frac{k_B}{e}\right)^2. \tag{30}$$

This shows that the Lorenz number should be the same for all metals and, in particular, it should not depend on the scattering law for the charge carriers. These features agree with the well-established Wiedemann–Franz–Lorenz law, which states that all metals have the same ratio of thermal to electrical conductivity and that this ratio is proportional to the absolute temperature. Figure 1 shows the Lorenz number plotted against the reduced Fermi energy, η, for different values of the scattering parameter, r.

2.3. Bipolar Conduction

So far it has been assumed that the carriers in only one band contribute to the transport processes. Now we consider what happens when there is more than one type of carrier. The most important example is that of mixed and intrinsic semiconductors in which there are significant contributions from electrons in the con-

Sec. 2 · ELECTRONIC THERMAL CONDUCTIVITY IN SEMICONDUCTORS

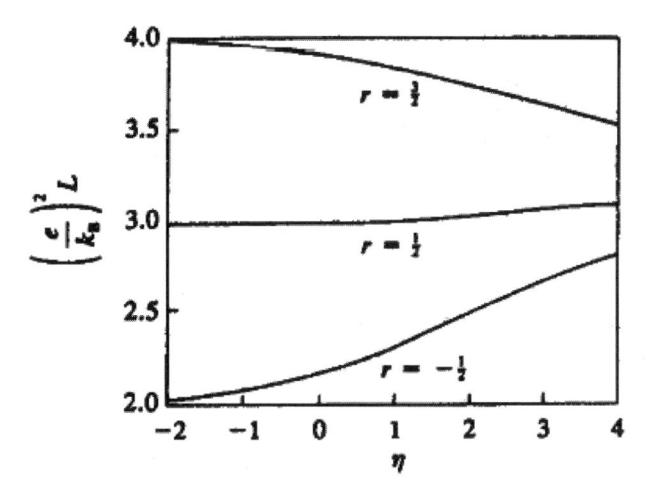

FIGURE 1 Dimensionless Lorenz number plotted against reduced Fermi energy for different values of the scattering parameter r.

duction band and holes in the valence band. The problem will be discussed for two types of carrier (represented by subscripts 1 and 2).

There will be contributions to the electric current density, i, from both types of carriers. These contributions may be expressed in terms of partial electrical conductivities and partial Seebeck coefficients. Thus,

$$i_1 = \sigma_1 \left(E - \alpha_1 \frac{\partial T}{\partial x} \right), \quad i_2 = \sigma_2 \left(E - \alpha_2 \frac{\partial T}{\partial x} \right). \tag{31}$$

where E is the electric field. The partial coefficients may be found by using the expressions that have already been obtained for single bands. When the electric field and the temperature gradient are in the same direction, the thermoelectric emf opposes E for a positive Seebeck coefficient and assists E when the Seebeck coefficient is negative. If the temperature gradient is zero,

$$i = i_1 + i_2 = (\sigma_1 + \sigma_2)E, \tag{32}$$

and the electrical conductivity is simply

$$\sigma = \sigma_1 + \sigma_2. \tag{33}$$

If the electric current is equal to zero, as it is when i_1 and i_2 are equal and opposite,

$$(\sigma_1 + \sigma_2)E = (\alpha_1\sigma_1 + \alpha_2\sigma_2)\frac{\partial T}{\partial x} \tag{34}$$

and the Seebeck coefficient is

$$\alpha = \frac{\alpha_1\sigma_1 + \alpha_2\sigma_2}{\sigma_1 + \sigma_2}. \tag{35}$$

If we now examine the thermal flow, the heat flux densities due to the two carrier types are

$$w_1 = \alpha_1 T i_1 - \kappa_{e,1}\frac{\partial T}{\partial x}, \quad w_2 = \alpha_2 T i_2 - \kappa_{e,2}\frac{\partial T}{\partial x}, \tag{36}$$

where the contributions from the Peltier effect have been expressed in terms of the partial Seebeck coefficients by use of Kelvin's relation. Since the thermal conductivity is defined in the absence of an electric current, i_1 and i_2 must again be equal and opposite. We find that

$$i_1 = -i_2 = \frac{\sigma_1\sigma_2}{\sigma_1+\sigma_2}(\alpha_1-\alpha_2)\frac{\partial T}{\partial x}. \tag{37}$$

By substitution into Eq. (36), the total heat flux density is

$$w = w_1 + w_2 = -\left[\kappa_{e,1}+\kappa_{e,2}+\frac{\sigma_1\sigma_2}{\sigma_1+\sigma_2}(\alpha_2-\alpha_1)^2 T\right]\frac{\partial T}{\partial x}, \tag{38}$$

and

$$\kappa_e = \kappa_{e,1}+\kappa_{e,2}+\frac{\sigma_1\sigma_2}{\sigma_1+\sigma_2}(\alpha_2-\alpha_1)^2 T. \tag{39}$$

The remarkable feature of Eq. (39) is that the total electronic thermal conductivity is not just the sum of the partial conductivities, $\kappa_{e,1}$ and $\kappa_{e,2}$. The presence of the third term arises from the Peltier heat flows that can occur when there is more than one type of carrier, even when the total electric current is zero. The difference, $\alpha_2 - \alpha_1$, between the partial Seebeck coefficients is, of course, greatest when the two carriers are holes and electrons, respectively. Thus, we expect κ_e to be substantially greater than $\kappa_{e,1}+\kappa_{e,2}$ for an intrinsic semiconductor. The additional contribution, known as the bipolar thermodiffusion effect[6] is observed most easily in semiconductors that have a small energy gap. Although $\alpha_2-\alpha_1$ is then smaller than it would be for a wide-gap semiconductor, intrinsic conduction takes place with reasonably large values for the partial electrical conductivities, σ_1 and σ_2. If σ_1 and σ_2 are too small, the electronic component of the thermal conductivity, even including bipolar thermodiffusion, is masked by the lattice component.

One of the first materials to display the bipolar effect was the semiconductor bismuth telluride.[7] Figure 2 shows the thermal conductivity plotted against temperature for lightly and heavily doped samples. It will be seen that, as the lightly doped sample becomes intrinsic at the higher temperatures, its thermal conductivity becomes much greater than that of the extrinsic heavily doped sample, even though it has a much smaller electrical conductivity. In fact, the Lorenz number is an order of magnitude greater for intrinsic bismuth telluride than it is for an extrinsic sample.

2.4. Separation of Electronic and Lattice Thermal Conductivities

The total thermal conductivity of a semiconductor is usually simply the sum of the lattice contribution, κ_L, and the electronic component, κ_e. Thus, the knowledge of the electronic contribution allows the lattice component to be determined. Although the calculation of the electronic thermal conductivity using the relationships derived from Eq. (16) often gives reasonable results, it sometimes leads to unacceptable conclusions. For example, Sharp et al.[8] found that certain polycrystalline samples of a Bi–Sb alloy would have negative lattice conductivity if theoretical

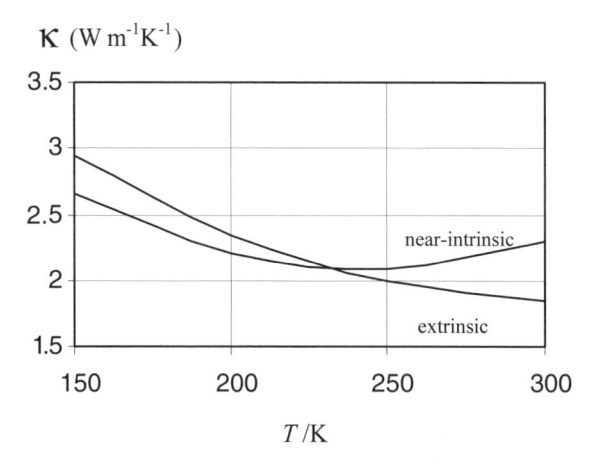

FIGURE 2 Plot of thermal conductivity against temperature for lightly and heavily doped samples of bismuth telluride.

values for the Lorenz number were assumed. It is, therefore, important to be able to determine the electronic thermal conductivity by an experimental method.

One possibility is to measure the thermal conductivity for differently doped samples of a given semiconductor. These samples will then have different values of the electrical conductivity and of the electronic thermal conductivity. One can then plot the thermal conductivity against the electrical conductivity; extrapolation of the plot to zero electrical conductivity should then give a value for the lattice component. As will be seen from Figure 3, in which the procedure has been carried out for bismuth telluride at 300 K, there can be some difficulties.[9] The onset of intrinsic conduction at low values of the electrical conductivity causes the Lorenz number to become exceptionally large, as already mentioned. Also, there is the possibility that the doping agents used to alter the carrier concentration may themselves scatter the phonons and, thereby, change the lattice conductivity. Nevertheless, measurements on differently doped samples offer one of the best methods of separating the two components of thermal conductivity.

When the mobility of the charge carriers is sufficiently high, another method of determining the electronic thermal conductivity becomes possible. The thermal resistivity, like the electrical resistivity, is increased by the application of a transverse magnetic field. In most cases the lattice conductivity remains unchanged. In certain cases it is possible to apply a magnetic field that is large enough for the electronic thermal conductivity to become negligible in comparison with the lattice contribution. However, even when the mobility of the charge carriers is not large enough for this to occur in the available magnetic field, the magnetothermal resistance effect is still useful. For example, the thermal conductivity can be plotted against the electrical conductivity as the magnetic field changes. Extrapolation to zero electrical conductivity allows the Lorenz number to be found.

In one situation, however, the electronic thermal conductivity does not become zero, no matter how large the magnetic field strength is, namely when the Nernst–Ettingshausen figure of merit is large. The residual electronic thermal conductivity in this case is due to the transverse thermoelectromagnetic effects.[10] The effect is

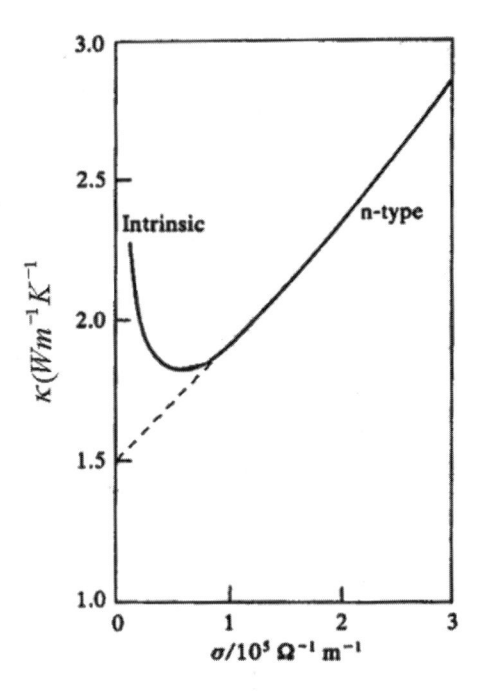

FIGURE 3 Plot of thermal conductivity against electrical conductivity for bismuth telluride at 300 K.

noticeable in Bi and Bi–Sb alloys, the latter being narrow-gap semiconductors for a range of Bi:Sb ratios. Uher and Goldsmid[11] have shown how the electronic thermal conductivity can be determined in single crystals of Bi and its alloys by measuring the magnetothermal resistance for two orientations of the sample. It is rather difficult to predict the magnitude of the Nernst–Ettingshausen figure of merit and, thence, the size of that part of the thermal conductivity in a high magnetic field due to the transverse effects. However, it is quite simple to calculate the ratio of the residual electronic thermal conductivities for two orientations. It is then a straightforward matter to separate out the lattice conductivity.

3. PHONON SCATTERING IN IMPURE AND IMPERFECT CRYSTALS

3.1. Pure Crystals

Phonons can be scattered by impurities and by crystal defects. Thus, when good heat conduction from a semiconductor device is required, it is an advantage for the material to have a rather high thermal conductivity. It is, in fact, fortunate that pure silicon has such a thermal conductivity and that most semiconductor devices require rather pure and perfect material. However, for those semiconductors that are used in thermoelectric energy conversion, a low thermal conductivity is needed. It is, therefore, of interest to consider the means by which the lattice thermal conductivity can be made smaller.

The lattice conductivity, κ_L, can be expressed in terms of the mean free path, l_t,

Sec. 3 • PHONON SCATTERING IN IMPURE AND IMPERFECT CRYSTALS

of the phonons by the equation

$$\kappa_L = \frac{1}{3} c_v v l_t. \qquad (40)$$

Here c_v is the specific heat per unit volume due to the lattice vibrations, and v is the speed of sound. For many purposes it is a reasonable approximation to use Debye theory[12] for the specific heat and to assume that the speed of the phonons is independent of frequency. This is particularly true for imperfect crystals, since the higher-frequency phonons, for which Debye theory is most likely to break down, are then rather strongly scattered and do not make much of a contribution to heat conduction.

If the thermal vibrations were perfectly harmonic, the free path length of the phonons would be infinite in a perfect unbounded crystal. However, as the temperature rises above absolute zero, the vibrations become more and more anharmonic. This causes the mean free path to vary inversely with temperature at high-temperatures. It was shown by Peierls[13] that the phonon scattering events are of two kinds. The normal (or N-)processes conserve momentum and do not lead directly to any thermal resistance, though they do redistribute the momentum within the phonon system. Then there are the umklapp (or U-) processes, in which momentum is not conserved, and these events are responsible for the observed finite thermal conductivity.

At high-temperatures the specific heat and the sound velocity may be regarded as constant. As the temperature is raised, the increasing probability of U-processes means that κ_L varies as $1/T$. This is consistent with the observations of Eucken[14] and later workers. At very low temperatures U-processes become rather improbable, and it was predicted by Peierls that there should be an exponential increase in lattice conductivity when $T \ll \theta_D$, θ_D being the Debye temperature. In fact, such an exponential variation is rarely observed because other factors invariably take over in real crystals. For example, the free path length of the phonons will certainly be limited by the crystal boundaries unless there is completely specular reflection. Bearing in mind that the specific heat itself tends to zero as $T \to 0$, the lattice conductivity also tends to zero. There is, then, some temperature at which the thermal conductivity has a maximum value. The inverse variation of κ_L with temperature is often observed when T is as small as θ_D or even less, despite the original prediction that this would only occur for $T \gg \theta_D$.

3.2. Scattering of Phonons by Impurities

Impurities scatter phonons because they produce local variations of the sound velocity through change in density or elastic constants. However, point-defect scattering presents us with a theoretical problem. The relaxation time for such scattering should vary as $1/\omega^4$, where ω is the angular frequency of the phonons. We expect, then, that low-frequency phonons will be little affected by point-defect scattering. We also know that U-processes are ineffective for low-frequency phonons, so we would expect exceedingly large values for lattice conductivity at low temperatures. Such values are not, in fact, observed. This dilemma can be resolved by taking into account the N-processes. Although the N-processes do not change the momentum in the phonon system, they do redistribute this quantity between different phonons. This redistribution can pass on the momentum to higher-frequency phonons that are more strongly influenced by U-processes and impurity scattering. How, then, does one take into account the effect of the N-processes?

The problem has been solved by Callaway,[15] who suggested that the N-processes cause relaxation toward a phonon distribution that carries momentum, whereas the U-processes and impurity scattering cause relaxation toward a distribution that does not carry momentum. This means that the relaxation time, τ_R, for the non-momentum-conserving processes, has to be replaced by a relaxation time, τ_{eff}, which is related to τ_R and the relaxation time, τ_N, for the N-processes by

$$\frac{1}{\tau_{\text{eff}}} = \left(\frac{1}{\tau_R} + \frac{1}{\tau_N}\right)\left(1 + \frac{\beta}{\tau_N}\right)^{-1} = \frac{1}{\tau_c}\left(1 + \frac{\beta}{\tau_N}\right)^{-1}, \quad (41)$$

where τ_c is the relaxation time that would have been expected if the N-processes did not conserve momentum, and β is a constant that must be selected so that the N-processes are indeed momentum conserving. If it is assumed that the Debye model for the phonon distribution is valid, as it should be for the low-frequency modes with which we are most concerned, then

$$\beta = \int_0^{\theta_D/T} \frac{\tau_c}{\tau_N} x^4 \frac{\exp x}{(\exp x - 1)^2} dx \bigg/ \int_0^{\theta_D/T} \frac{1}{\tau_N}\left(1 - \frac{\tau_c}{\tau_N}\right) x^4 \frac{\exp x}{(\exp x - 1)^2} dx, \quad (42)$$

where $x = \hbar\omega/2\pi k_B T$.

In certain situations this complex expression may be simplified. For example, if the scattering by imperfections is very strong, with a relaxation time $\tau_I \ll \tau_N$, one can use the approximation $1/\tau_{\text{eff}} \approx 1/\tau_I + 1/\tau_N$. Parrott[16] has used a high-temperature approximation that should be valid for most thermoelectric materials. He supposed that the relaxation times for the U- and N-processes are proportional to ω^{-2}, while for point defects τ_I is proportional to ω^{-4}. We use the following expressions for the relaxation times: $1/\tau_I = A\omega^4$, $1/\tau_U = B\omega^2$, $1/\tau_N = C\omega^2$, where A, B, and C are constants for a given sample. Also, at high-temperatures $x \ll 1$ for the whole phonon spectrum, so $x^2 \exp x/(\exp x - 1)^2 \approx 1$. This allows us to obtain the relatively simple equation

$$\frac{\kappa_L}{\kappa_0} = \left(1 + \frac{5k_0}{9}\right)\left[\frac{\tan^{-1} y}{y} + \left(1 - y\frac{\tan^{-1} y}{y}\right)^2 \left(\frac{y^4(1 + k_0)}{5k_0} - \frac{y^2}{3} - \frac{\tan^{-1} y}{y}\right)^{-1}\right], \quad (43)$$

where $k_0 = C/B$ is the strength of the N-processes relative to the U-processes, and y is defined by the equation

$$y^2 = \frac{\omega_D^2}{\omega_0}\left(1 + \frac{5k_0}{9}\right)^{-1}, \quad (44)$$

with ω_0 given by

$$\left(\frac{\omega_0}{\omega_D}\right)^2 = \frac{k_B}{2\pi^2 v\kappa_0 \omega_D A}. \quad (45)$$

The value of k_0 can be found experimentally by measuring the thermal conductivity of two crystals, one pure and one containing impurities. It is, of course, necessary that one know the value of the parameter which determines the relaxation time for the defect scattering. This parameter may be calculated most easily when the scattering by density fluctuations outweighs that due to variations in the elastic constants. The reduction of thermal conductivity due to mass–defect scattering was certainly observed for germanium crystals with different isotopic concentrations.[17] Mass–defect scattering has also been dominant in certain semiconductor solid solutions, though in others it must be presumed that variations in the elastic constants have also been an important factor. When mass–defect scattering occurs, then

$$A = \frac{\pi}{2v^3 N} \sum_i \frac{x_i (M_i - \overline{M})^2}{\overline{M}^2}, \quad (46)$$

where x_i is the concentration of unit cells of mass M_i, $\overline{M}$ is the average mass per unit cell, and N is the number of unit cells per unit volume.

At high-temperatures it is quite a reasonable approximation to ignore the N-processes in many systems. If we assume that the U-processes dominate the N-processes, $k_0 \to 0$, and Eq. (43) becomes

$$\frac{\kappa_L}{\kappa_0} = \frac{\omega_0}{\omega_D} \tan^{-1}\left(\frac{\omega_D}{\omega_0}\right). \quad (47)$$

Note that new materials exist for which these simple ideas about scattering are inapplicable. For example, there are semiconductors, such as those with clathrate structures,[18] in which the crystals contain open cages in which foreign atoms can reside. These atoms are loosely bound and are known as rattlers. They are very effective in scattering phonons, sometimes reducing the thermal conductivity to a value that is close to that in an amorphous substance.[19,20] This type of material will be discussed elsewhere.

3.3. Boundary Scattering

The scattering of phonons on the crystal boundaries has been known since the work of Casimir,[21] but for many years it was thought to be essentially a low-temperature phenomenon. However, it is now thought that boundary scattering may occur at larger grain sizes and higher temperatures than was previously thought possible. It is indeed possible that boundary scattering may sometimes have a greater effect on lattice conductivity than on carrier mobility, even when the mean free path is larger for electrons or holes than it is for phonons.

An enhanced boundary scattering effect was proposed by Goldsmid and Penn[22] on the basis that, although the number of low–frequency phonon modes is small, they make a substantial contribution to the thermal conductivity because they have a large free path length. This is illustrated by Figure 4, in which the contribution to the lattice conductivity from phonons of angular frequency ω is plotted against ω. The area under the curve up to the Debye frequency, ω_D, indicates the contribution from all the phonons. The upper curve represents schematically the behavior of a large pure crystal. The double-hatched region represents the reduction in thermal conductivity due to boundary scattering when the grain size is still substantially larger than the mean free path. The effect of boundary scattering becomes relatively larger for a solid solution in which the high–frequency phonons are strongly scattered, thus removing the contribution shown by the single-hatched region. Since solid solutions are the most frequently employed material for thermoelectric applications, it is possible that boundary scattering of phonons may improve the figure of merit.

If one uses the Debye model for the phonon distribution, one finds that the lattice conductivity of a sample of a solid solution having an effective grain size L is given by

$$\kappa_L = \kappa_S - \frac{2}{3}\kappa_0\sqrt{\frac{l_t}{3L}}, \qquad (48)$$

where κ_S is the lattice conductivity of a large crystal of the solid solution and l_t is the phonon mean free path.[23] The parameter κ_0, the thermal conductivity in the absence of point–defect scattering, can be estimated from the values for the different components of the solid solution.

4. PREDICTION OF THE LATTICE THERMAL CONDUCTIVITY

It would be useful if we could predict the types of semiconductors, that would be likely to have a very high or a very low thermal conductivity. We outline here the kind of approach that has been used for this purpose in the high–temperature region. Bearing in mind the approximate nature of the predictive methods, we assume the Debye model.

One of the earliest approaches to the problem was that of Leibfried and Schlömann,[24] who used the variational method to show that

$$\kappa_L = 3.5\left(\frac{k_B}{h}\right)^3 \frac{MV^{1/3}\theta_D^3}{\gamma^2 T}, \qquad (49)$$

where M is the average atomic mass, V is the average atomic volume, and γ is the Grüneisen parameter.

An alternative approach that leads to more or less the same result was based on the proposal by Dugdale and MacDonald[25] that the lattice conductivity should be related to the thermal expansion coefficient α_T. The anharmonicity of the thermal

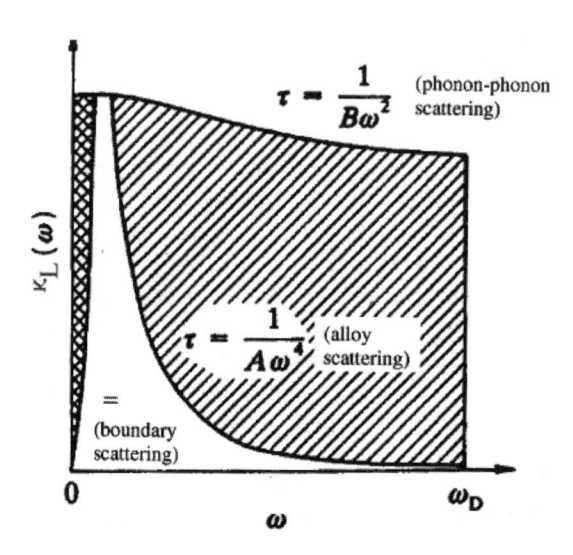

FIGURE 4 Schematic plot of the contributions to thermal conductivity from different parts of the phonon spectrum. The double-hatched region represents reduced thermal conductivity due to boundary scattering, the single-hatched region that due to alloy scattering in a solid solution.

Sec. 4 · **PREDICTION OF THE LATTICE THERMAL CONDUCTIVITY**

vibrations can be represented by the dimensionless quantity $\alpha_T \gamma T$, and it was suggested that the phonon mean free path should be approximately equal to the lattice constant a divided by this quantity. Thus,

$$\kappa_L = \frac{c_V a v}{3 \alpha_T \gamma T}. \tag{50}$$

The expansion coefficient is given by the Debye equation of state:

$$\alpha_T = \frac{\chi \gamma c_V}{3}, \tag{51}$$

where χ is the compressibility. Also, the speed of sound and the compressibility are related to the Debye temperature through the equation

$$v = (\rho \chi)^{-1/2} = \frac{2 k_B a \theta_D}{h}, \tag{52}$$

where ρ is the density.

When these ideas were first considered, they were applied to simple crystals having small numbers of atoms per unit cell. No distinction was drawn between the acoustic and optical modes and a was set equal to the cube root of the atomic volume. It was then found that

$$\kappa_L = 8 \left(\frac{k_B}{h} \right)^3 \frac{M V^{1/3} \theta_D^3}{\gamma^2 T}. \tag{53}$$

This differs from Eq. (49) only in the value of the numerical constant.

Notice that Eq. (53) and (49) require a knowledge of the Debye temperature and, therefore, might not be as useful as they otherwise would be. Using the Lindemann melting rule to estimate the compressibility, Keyes[26] obtained a formula that does not have this objection. Thus,

$$\chi = \frac{\varepsilon_m V}{R T_m}, \tag{54}$$

where R is the gas constant and T_m is the melting temperature. This rule makes use of the assumption that a solid melts when the amplitude of the lattice vibrations reaches a fraction ε_m of the lattice constant, ε_m being approximately the same for all substances. Then, from Eq. (50) through (52) we find

$$\kappa_L T = B \frac{T_m^{3/2} \rho^{2/3}}{A_m}, \tag{55}$$

where

$$B = \frac{R^{3/2}}{3 \gamma^2 \varepsilon_m^3 N_A^{1/3}}, \tag{56}$$

N_A is Avogadro's number, and A_m is the mean atomic weight.

The advantage of Keyes' Eq. (55) is that it involves the quantities T_m, ρ, and A_m, which are known as soon as a new material is synthesized, and a quantity B that should not vary much from one substance to another in a given system. For example, Figure 5 shows the values of $\kappa_L T$ plotted against $T_m^{3/2} \rho^{2/3} A_m^{-7/6}$ for semiconductors with covalent or partly covalent bonds. As expected from Eq. (55), the plot is approximately linear.

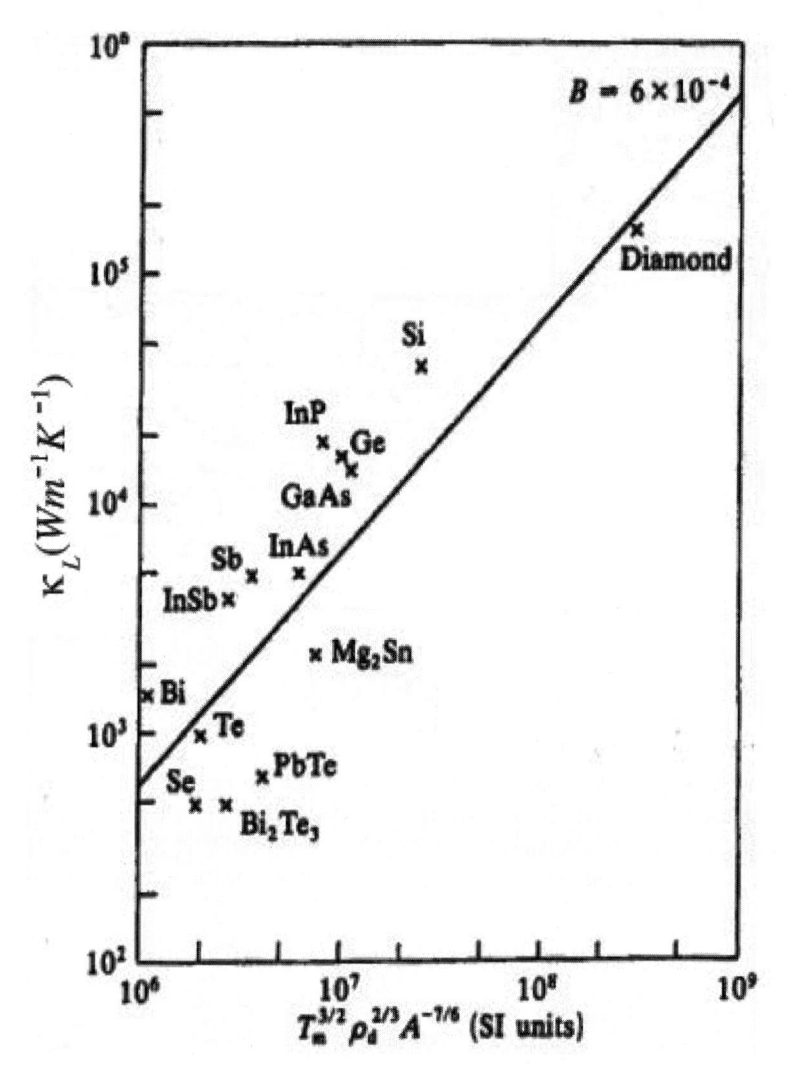

FIGURE 5 The values of $\kappa_L T$ plotted against $T_m^{3/2} \rho^{2/3} A_m^{-7/6}$ for various semiconductors.

Some of the newer semiconductors that have been proposed as possible thermoelectric materials have relatively complex crystal structures with many atoms per unit cell. Such materials will have three acoustic modes and a very large number of optical modes. Moreover, the group velocity of the phonons is what is really important in Eq. (40), and this property varies greatly from mode to mode and from low to high frequency. It is generally likely to be much smaller for optical modes than for acoustic modes, so it is likely that most of the heat will be transported by acoustic phonons. It is, in fact, these phonons that are best approximated by the Debye model. Thus, in applying the Keyes relation to complex crystals, it is probably appropriate to regard a as the cube root of the volume of the unit cell. Also, A_m should be regarded as the molecular weight. However, it is, perhaps, stretching Eq. (55) beyond the realm that Keyes had in mind when we apply it to complex materials. Nevertheless, it is still of some use when we are trying to select semiconductors of low thermal conductivity.

5. REFERENCES

1. G. S. NOLAS, J. SHARP, AND H. J. GOLDSMID *Thermoelectrics: Basic Principles and New Materials Developments* (Springer, Berlin, 2001).
2. A. SOMMERFELD, Z. Phys. **47**, 1, 43 (1928).
3. J. MCDOUGALL AND E. C. STONER Phil. Trans. A **237**, 67 (1938).
4. P. RHODES Proc. R. Soc. A **204**, 396 (1950).
5. A. C. BEER, M. N. CHASE, AND P. F. CHOQUARD Helv. Phys. Acta **28**, 529 (1955).
6. H. FRÖHLICH AND C. KITTEL Physica **20**, 1086 (1954).
7. H. J. GOLDSMID *The Thermal Conductivity of Bismuth Telluride* Proc. Phys. Soc. B **69**, 203–209 (1956).
8. J. W. SHARP, E. H. VOLCKMANN, AND H. J. GOLDSMID *The Thermal Conductivity of Polycrystalline $Bi_{88}Sb_{12}$*, Phys. Stat. Sol. (a) **185**, 257–265 (2001).
9. H. J. GOLDSMID *Heat Conduction in Bismuth Telluride* Proc. Phys. Soc. **72**, 17–26 (1958).
10. YA. KORENBLIT, M. E. KUZNETSOV, V. M. MUZHDABA, AND S. S. SHALYT Soviet Physics:JETP **30**, 1009 (1969).
11. C. UHER AND H. J. GOLDSMID *Separation of the Electronic and Lattice Thermal Conductivities in Bismuth Crystals* Phys. Stat. Sol. (b) **65**, 765–771 (1974).
12. P. DEBYE Ann. Phys. **39**, 789 (1912).
13. R. E. PEIERLS Ann. Phys. **3**, 1055 (1929).
14. A. EUCKEN Ann. Phys. **34**, 185 (1911).
15. J. CALLAWAY *Model for Lattice Thermal Conductivity at Low Temperatures* Phys. Rev. **113**, 1046–1051 (1959).
16. J. E. PARROTT *The High Temperature Thermal Conductivity of Semiconductor Alloys* Proc. Phys. Soc. **81**, 726–735 (1963).
17. T. H. GEBALLE AND G. W. HULL *Isotopic and Other Types of Thermal Resistance in Germanium* Phys. Rev. **110**, 773–775 (1958).
18. G. S. NOLAS, T. J. R. WEAKLEY, J. L. COHN, AND R. SHARMA *Structural Properties and Thermal Conductivity of Crystalline Ge Clathrates* Phys. Rev. B **61**, 3845–3850 (2000).
19. D. G. CAHILL, S. K. WATSON, AND R. O. POHL *Lower Limit to the Thermal Conductivity of Disordered Crystals* Phys. Rev. B **46**, 6131–6139 (1992).
20. G. A. SLACK *The Thermal Conductivity of Nonmetallic Crystals*, in *Solid State Physics*, edited by F. Seitz, D. Turnbull, and H. Ehrenreich (Academic Press, New York, 1979) pp. 1–17.
21. H. B. G. CASIMIR Physica, **5**, 495 (1938).
22. H. J. GOLDSMID AND R. W. PENN *Boundary Scattering of Phonons in Solid Solutions* Phys. Lett. A **27**, 523–524 (1968).
23. H. J. GOLDSMID, H. B. LYON, AND E. H. VOLCKMANN Proc. 14th Int. Conf. on Thermoelectrics (St. Petersburg, 1955) pp. 16–19.
24. G. LEIBFRIED AND E. SCHLÖMANN Nachr. Akad. Wiss. Göttingen, Kl. 2 Math-Phys. **2**, 71 (1954).
25. J. S. DUGDALE AND D. K. C. MACDONALD *Latttice Thermal Conductivity*, Phys. Rev. **98**, 1751–1752 (1955).
26. R. W. KEYES *High Temperature Thermal Conductivity of Insulating Materials: Relationship to the Melting Point* Phys. Rev. **115**, 564-567 (1959).

Chapter 1.5

SEMICONDUCTORS AND THERMOELECTRIC MATERIALS

G. S. Nolas

Department of Physics, University of South Florida, Tampa, FL, USA

J. Yang

Materials and Processes Laboratory, General Motors R&D Center, Warren, MI, USA

H. J. Goldsmid

School of Physics, University of New South Wales, Sydney, NSW, Australia

1. INTRODUCTION

The theory of heat conduction in semiconductors was reviewed in Chapter 1.4 and some of the experimental observations have been used as illustrations. In this chapter the main emphasis is the behavior of materials emerging for practical applications, particularly in the field of thermoelectric energy conversion. The thermoelectric figure of merit Z is defined as $\alpha^2\sigma/\kappa$, where α is the Seebeck coefficient, σ is the electrical conductivity, and κ is the thermal conductivity. Very often, the figure of merit is used in its dimensionless form, ZT. Also, the figure of merit is seen to be the ratio of the so-called power factor, $\alpha^2\sigma$, to the thermal conductivity, κ. The electronic contribution to the thermal conductivity, κ_e, is related to the electrical conductivity, as shown in Chap. 1.4. Thus, a good thermoelectric material combines a high power factor with a small value for the lattice conductivity, κ_L.[1] It is not surprising, therefore, that a great deal of effort has been devoted to the search for materials with a low lattice conductivity. The substances reviewed in this chapter largely reflect the results of this search.

In the next section we briefly summarize the observations made on thermoelectric materials already used in Peltier coolers and electric generators. These materials include alloys of bismuth and antimony, which are useful at low temperatures; alloys based on bismuth telluride, which are the mainstay of the thermoelectrics industry; and the IV–VI compounds and the Si–Ge alloys, which are used in thermoelectric generation.

In subsequent sections we discuss skutterudites, clathrates and half-Heusler compounds that continue to be of interest for thermoelectrics, and review some novel oxides and chalcogenides.

2. ESTABLISHED MATERIALS

2.1. Bismuth Telluride and Its Alloys

Bismuth telluride (Bi_2Te_3), has a layered structure with strongly anisotropic mechanical properties. There are two equivalent a axes and a mutually perpendicular c axis. In the direction of the c axis, the atomic planes follow the sequence Te–Bi–Te–Bi–Te, which is then repeated. The Te–Te layers are held together by weak van der Waals forces, while the remaining atoms are linked by strong ionic–covalent bonds[2]. Single crystals of bismuth telluride can be cleaved readily in the plane of the a axes. Not surprisingly, the transport properties, including the electronic and lattice thermal conductivities, are anisotropic.

The electronic component of the thermal conductivity is anisotropic because the electrical conductivity is different in the a and c directions. The ratio of this quantity in the a direction to that in the c direction is larger for n-type material than for p-type material. It is somewhat larger than the ratio of the lattice conductivities in the two directions, which is reliably reported to be equal to 2.1, although the ready cleavage of bismuth telluride makes measurements of any transport properties in the c direction rather difficult.[3]

In all samples of bismuth telluride, the two components of the thermal conductivity are of comparable magnitude. It is possible to calculate the electronic component from the measured electrical conductivity. However, the calculation is not so simple as it would be for a metal because the samples are usually only partly degenerate, so account must be taken of both the scattering law for the charge carriers and the position of the Fermi level (see Chapter 1.4). This being so, it is probably best to determine the lattice conductivity by extrapolation of the total thermal conductivity for samples of differing carrier concentrations. Such an extrapolation must make use only of extrinsic samples because of the large bipolar contribution for mixed or intrinsic conduction.

Figure 1 shows how the lattice conductivity of bismuth telluride along the cleavage planes varies with temperature.[4–6] At high-temperatures, κ_L varies as $1/T$, but as the temperature falls there is a maximum value of about 60 W/m-K at about 10 K. It has been suggested that the position of the maximum depends on boundary scattering, but it may also be associated with point–defect scattering. Even in chemically pure samples, there are numerous defects due to the inherent nonstoichiometry of bismuth telluride.

Bismuth telluride has been largely replaced as a material for thermoelectric refrigeration by its alloys with isomorphous compounds, bismuth selenide (Bi_2Se_3) and antimony telluride (Sb_2Te_3). The study of these materials followed the sugges-

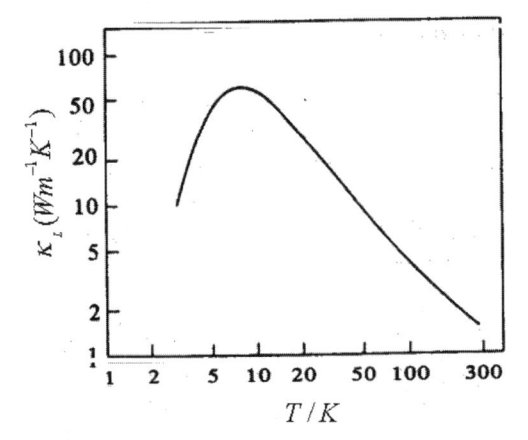

FIGURE 1. Lattice conductivity plotted against temperature for bismuth telluride in the plane of the a axes.[3-5]

tion by Ioffe et al.[7] that the formation of a solid solution could reduce the lattice conductivity without lowering the carrier mobility. It was indeed found that this was the case for p-type material in the bismuth-antimony telluride system. Figure 2 shows plots obtained by 3 sets of workers for the lattice conductivity in the a direction at 300 K against composition.[8-10] The difference between the results is undoubtedly due to the techniques used to determine the electronic thermal conductivity, but in each case there is a clear minimum at a composition close to $Bi_{0.5}Sb_{1.5}Te_3$. Since the power factor for this alloy is virtually the same as it is for bismuth telluride, its advantage for thermoelectric applications is obvious.

An even greater reduction in lattice conductivity occurs when bismuth selenide is added to bismuth telluride and the resultant solid solutions are commonly used as

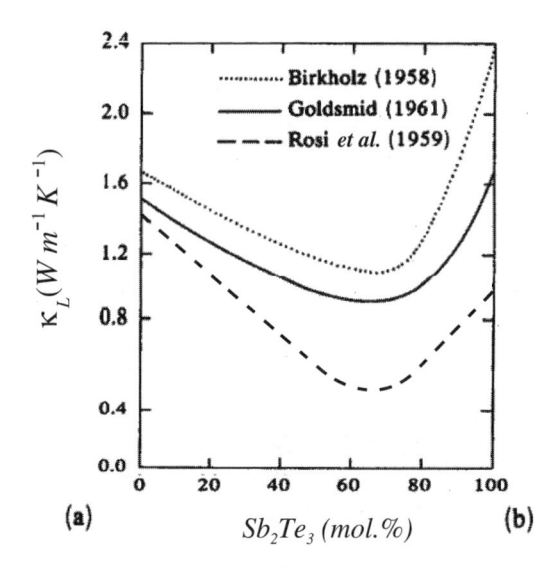

FIGURE 2. Lattice thermal conductivity for solid solutions between Bi_2Te_3 and Sb_2Te_3 at 300 K. The measurements were made along the cleavage planes. (Reprinted from Ref. 1. Copyright 2001, Springer-Verlag.)

n-type thermoelements. However, in this case, there is a reduction in the electron mobility, and for this reason the useful alloys are restricted to the compositions Bi$_2$Te$_{3-x}$Se$_x$ with $x < 0.6$.

2.2. Bismuth and Bismuth–Antimony Alloys

Alloys based on bismuth telluride may be used for thermoelectric refrigeration at low-temperature, but the dimensionless figure of merit falls off quite rapidly as the temperature is reduced. This would be the case even if Z were independent of temperature, but in fact, this quantity itself becomes smaller as T becomes less because of an increase in the lattice conductivity. The lattice conductivity does not vary so rapidly with temperature for the solid solutions as it does for the pure compound, but ZT nevertheless varies more rapidly than $1/T$. There are actually significantly superior *n*-type materials below about 200 K, namely the alloys of bismuth with antimony.

Bismuth itself is a semimetal so that, although it displays a large power factor, the total thermal conductivity is also large and the figure of merit is not particularly high. However, the addition of up to about 15% of antimony causes a band gap to appear.[11] The gap is too small to allow the alloys to be used in thermoelectric applications at ordinary temperatures, but it becomes the best *n*-type material as the temperature is reduced.[12] The lattice conductivity is less for Bi–Sb alloys than it is for pure bismuth, which also helps increase the figure of merit.

Bismuth and Bi–Sb crystals have the same structure as those of bismuth telluride and display a similar anisotropy of the thermal conductivity and other properties. They are also characterized by high values of the electron mobility, especially at low

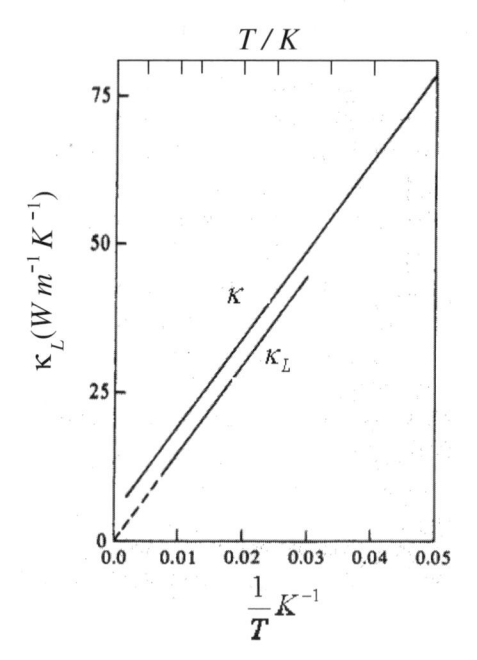

FIGURE 3. Total thermal conductivity and the lattice contribution in the binary direction plotted against the reciprocal of the temperature for pure bismuth. The electronic component has been determined by the procedure outlined in Chap. 1.4.[14] (Reprinted from Ref. 1. Copyright 2001, Springer-Verlag.)

Sec. 2 • ESTABLISHED MATERIALS

temperatures. The application of a transverse magnetic field may be used not only to improve the thermoelectric figure of merit[13] but also in the separation of the lattice and electronic components of the thermal conductivity,[14] as described in Chap. 1.4. Figure 3 shows how the total thermal conductivity and the lattice component vary with temperature for pure bismuth.

Some indication of the way in which the lattice conductivity falls as the concentration of antimony is increased in Bi–Sb alloys is given in Figure 4. Here the variation of the total thermal conductivity in the binary direction is plotted against the strength of the magnetic field in the bisectrix direction for three different alloys.[15] The measurements were made at 80 K, at which temperature the field effects are much stronger than at 300 K. In a field of 1 T there is very little remaining of the electronic contribution to the thermal conductivity. The Nernst–Ettingshausen figure of merit is not particularly large for the given field orientation, so the thermal conductivity does indeed tend toward the lattice component at high magnetic field strengths. Note that the lattice conductivity of pure bismuth at 80 K is about 17 W/m-K and that the addition of no more than 3% antimony is sufficient to reduce κ_L by a factor of more than 3.

2.3. IV–VI Compounds

The bismuth telluride solid solutions become less effective as thermoelectric materials when the temperature is raised much above, say, 400 K. This is mainly because the energy gap is only about 0.3 eV so that there are a significant number of minority carriers present at higher temperatures. The minority carriers give rise to a thermoelectric effect, which opposes that of the majority carriers. Lead telluride (PbTe) and other compounds between Group IV and Group VI elements have somewhat larger energy gaps, and, therefore, the onset of mixed conduction occurs at rather larger temperatures.

In many ways lead telluride is an easier material to work with than bismuth telluride. In particular, it has the cubic sodium chloride structure and, as a result,

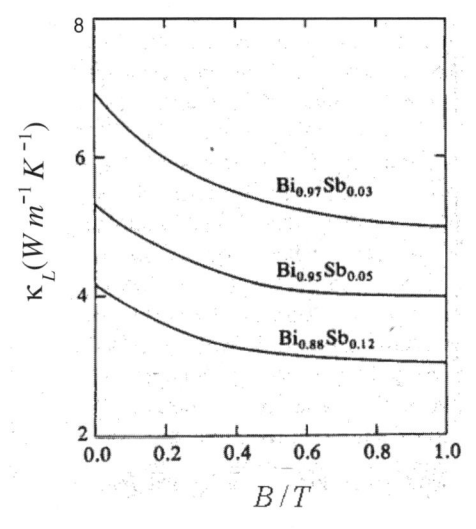

FIGURE 4. Thermal conductivity plotted against magnetic field for certain bismuth–antimony alloys at 80 K.[15] The temperature gradient is in the binary direction, and the magnetic field is in the bisectrix direction. (Reprinted from Ref. 1. Copyright 2001, Springer-Verlag.)

the basic transport properties are the same in all directions. The combination of a rather high mobility with a small effective mass for both electrons and holes[16] gives a power factor of the same order as that for bismuth telluride. However, at room temperature the lattice conductivity is 2.0 W/m-K and the figure of merit is smaller for the lead compound. Above 200°C the larger energy gap is sufficient compensation for the higher lattice conductivity, and lead telluride is the superior thermoelectric material.

In fact, it is preferable to use lead telluride solid solutions rather than the simple compound. Airapetyants et al.[17] observed a reduction of the lattice conductivity on adding tin telluride (SnTe) or lead selenide (PbSe) to lead telluride. $Pb_{1-y}Sn_yTe$ is still regarded as one of the most useful p-type thermoelectric materials up to perhaps 800 K, but quaternary alloys have been used as an n-type material in preference to a solid solution based on lead telluride. The quaternary alloys have been given the acronym TAGS since they contain the elements Te, Ag, Ge, and Sb. They are essentially alloys between $AgSbTe_2$ and GeTe, and for certain ranges of composition they share the sodium chloride structure with PbTe. When the concentration of GeTe falls below about 80%, the alloys take on a rhombohedral structure. For compositions lying close to that of the phase transformation there is considerable lattice strain, and it is believed that this may be the reason for exceptional low values of thermal conductivity in this region.[18] This is illustrated by the plots of lattice conductivity against the reciprocal of the temperature for PbTe–SnTe and so-called TAGS-85 (Fig. 5). The latter material contains 85% GeTe.

2.4. Silicon, Germanium, and Si–Ge Alloys

Silicon is remarkable in that, though it is nonmetallic, it has a thermal conductivity comparable with that of many metals. The need for large and perfect crystals of

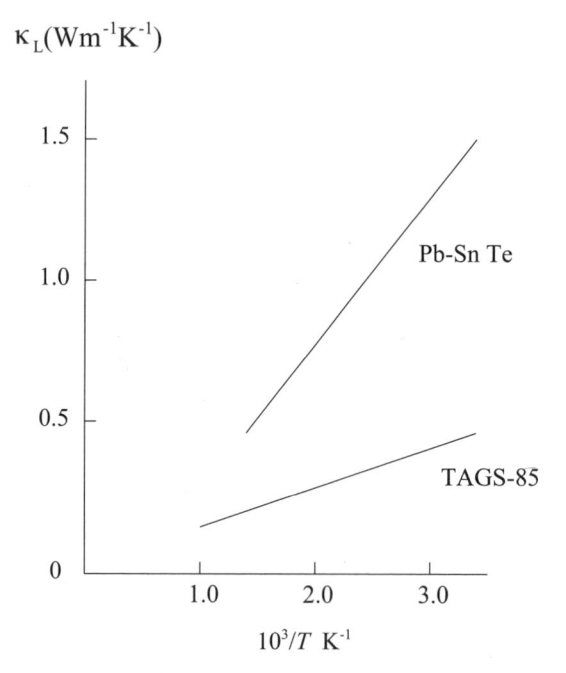

FIGURE 5. Plots of lattice conductivity against the reciprocal of the temperature for PbTe-SnTe and TAGS-85.[18]

high-purity silicon for the electronics industry has made available excellent samples for basic thermal conductivity studies.

Its high thermal conductivity makes silicon of little use as a thermoelectric material. Germanium, also has a rather large thermal conductivity, but silicon–germanium alloys have been used for high-temperature thermoelectric generation. The lattice conductivities of silicon and germanium at 300 K are 113 and 63 W/m-K, respectively, but $Si_{0.7}Ge_{0.3}$ has a value of only about 10 W/m-K.[19]

Silicon, germanium, and Si–Ge alloys all possess the cubic diamond structure, and their thermal and electrical conductivities do not depend on crystal orientation. Thus, in principle, randomly oriented polycrystalline samples should be just as good as single crystals in thermoelectric applications. In fact, there have been claims that fine-grained polycrystals are superior because of a reduction of the thermal conductivity through boundary scattering of the phonons.[20] This exemplifies the principles discussed in Sec. 2.3 of Ch. 1.4. Although the mean free path of the charge carriers is undoubtedly higher than that of the phonons in Si–Ge alloys, the preferential scattering of the low-frequency phonons by the grain boundaries can lead to a greater reduction of the lattice conductivity than of the carrier mobility.[21] Boundary scattering at room temperature has been observed for thin samples of single-crystal silicon, the effect being enhanced by introducing, through neutron irradiation, phononscattering point defects.[22]

One of the most interesting experiments on the thermal conductivity of solids was carried out by Geballe and Hull.[23] They were able to obtain a single crystal of germanium in which the isotope Ge^{74} had been enriched to 96%. They compared the thermal conductivity of this crystal with that of another crystal in which the isotopes had their naturally occurring distribution. As shown in Fig. 6, the thermal conductivity of the isotopically enriched sample was significantly higher at all temperatures and about 3 times as large at 20 K. The results are remarkable in that the two crystals were virtually indistinguishable in most of their physical properties. Nevertheless, the behavior is in substantial agreement with the theory of mass-defect scattering of phonons that was outlined in Sec. 2.2.

3. SKUTTERUDITES

Skutterudite compounds have been extensively studied as potential high-efficiency thermoelectric materials in recent years. The chemical formula for binary skutterudites is MX_3, where the metal atom M can be Co, Ir, or Rh, and the pnicogen atom X can be P, As, or Sb. Binary skutterudite compounds crystallize in a body-centered-cubic structure with space group *Im3*, and the crystal structure contains large voids at the a positions (12-coordinated); each M atom is octahedrally surrounded by X atoms, thus forming a MX_6 octahedron. They are semiconductors with small band gaps (~100 meV), high carrier mobilities, and modest thermopowers. Detailed structural and electronic properties can be found in two reviews.[24,25] $CoSb_3$-based skutterudites have particularly been the focal point of research mainly because of the abundance of the constituent elements. Despite their excellent electronic properties, they possess thermal conductivities (~10 W/m-K at room temperature) that are too high to compete with state-of-the-art thermoelectric materials. Partially filled skutterudites $G_yM_4X_{12}$ are formed by inserting small guest ions into the large 12-coordinated sites of binary skutterudites, where G represents a guest ion and y is its filling fraction. The maximum filling fraction y_{max} varies

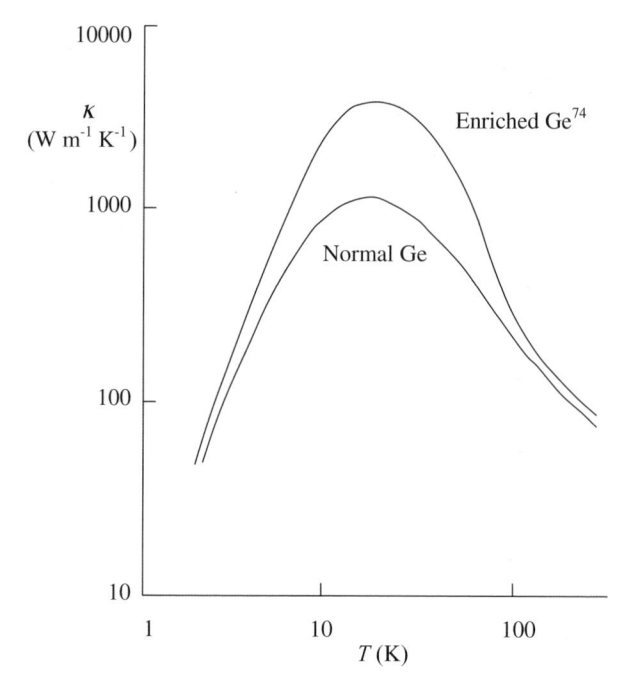

FIGURE 6. Thermal conductivity of normal germanium and enriched Ge[74].[23]

depending on the valence and the size of the guest ion. By replacing M or X with electron-deficient elements, one can attain full filling of the voids. These guest ions often have large thermal parameters. It was suggested that these guest ions or "rattlers" may strongly scatter the heat carrying lattice phonons in the low-frequency region, thereby, very effectively reducing the lattice thermal conductivities of the parent binary skutterudite compounds.[26] These low-frequency phonons are otherwise difficult to be scattered by conventional methods. Low lattice thermal conductivities over a wide temperature range subsequently observed in filled skutterudites led to ZT well above 1.0 between 500 K and 1000 K.[27–33] It is evident that filling the crystal structure voids results in a dramatic decrease of the lattice thermal conductivity of skutterudites; the effect of doping is equally dramatic and beneficial.[28,29,34,35] A very important feature of skutterudites is the large number of different isostructural compositions that can be synthesized. The diversity of potential compositional variants remains one of the key reasons why this material system continues to be investigated by many research groups.

3.1. Binary (Unfilled) Skutterudites

Figure 7 shows lattice thermal conductivity κ_L as a function of temperature for single-crystal $CoSb_3$, polycrystalline $CoSb_3$, polycrystalline $IrSb_3$, and polycrystalline $Ru_{0.5}Pd_{0.5}Sb_3$.[36] Also included are the room temperature data for $PtSb_2$ and $RuSb_2$ and the calculated minimum thermal conductivity for $IrSb_3$. The minimum thermal conductivity κ_{min} is calculated by taking the minimum mean free path of the acoustic phonons as one half of the phonon wavelength. For single-crystal $CoSb_3$[37] at low temperatures, κ_L is about two orders of magnitude higher than that of polycrystalline $CoSb_3$.[38] This is attributed to the differences in the grain size

Sec. 3 · SKUTTERUDITES

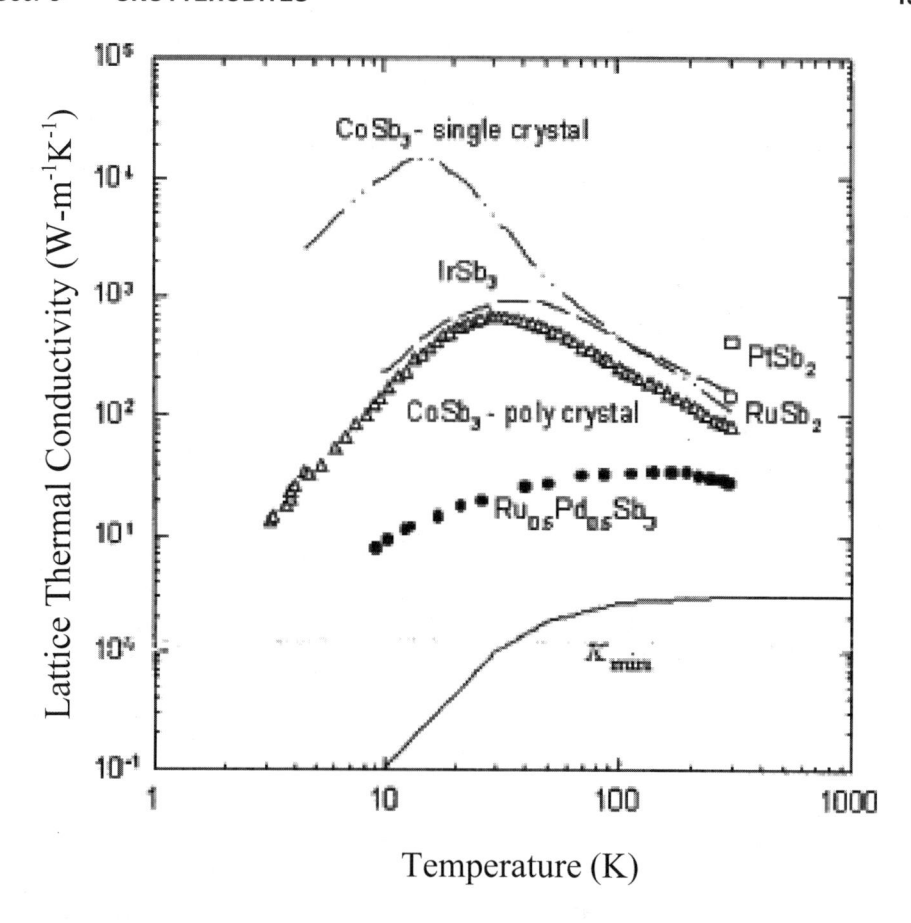

FIGURE 7. Lattice thermal conductivity versus temperature for single-crystal CoSb$_3$, polycrystalline CoSb$_3$, polycrystalline IrSb$_3$, and polycrystalline Ru$_{0.5}$Pd$_{0.5}$Sb$_3$.[36] Also included are the room temperature data for PtSb$_2$ and RuSb$_2$, and the calculated minimum thermal conductivity λ_{min} for IrSb$_3$.[36,38] (Reprinted from Ref. 36. Copyright 1996, American Institute of Physics.)

and the amount of defects between the two. At high temperatures κ_L for single-crystal CoSb$_3$ decreases exponentially with increasing temperature, indicative of the predominant phonon–phonon umklapp scattering and sample quality. At room temperature κ_L's of the single-crystal and polycrystalline CoSb$_3$ approach a common value. The overall κ_L for polycrystalline IrSb$_3$ is somewhat higher than that of polycrystalline CoSb$_3$, despite Ir having a higher atomic mass than Co. The room temperature κ_L of Ru$_{0.5}$Pd$_{0.5}$Sb$_3$ is about a factor of 5 smaller than the binary skutterudite compounds, and the difference is even greater at low temperatures. The room temperature data for PtSb$_2$ and RuSb$_2$ are also plotted for comparison. None of these binary compounds has room temperature κ_L as low as Ru$_{0.5}$Pd$_{0.5}$Sb$_3$. A NEXAFS (near-edge extended absorption fine structure) study and a simple electron count reveal that Ru is in a mixed valence state, and a correct composition can be written as $\square_v \text{Ru}_x^{2+} \text{Ru}_y^{4+} \text{Pd}_z^{2+} \text{Sb}_3$, where $\square$ represents a vacancy on the metal site.[36] A mass fluctuation with strain field correction estimation for κ_L is not sufficient to reproduce the data at room temperature. The low κ_L is ascribed to additional phonon scattering by a rapid transfer of electrons between the d shells of Ru^{2+} and Ru^{4+}.

3.2. Effect of Doping on the Co Site

The conductivity κ_L can also be significantly suppressed upon doping transition metal elements on the Co site.[38–40] Figure 8 shows the temperature dependence of κ_L between 2 K and 300 K for $Co_{1-x}Fe_xSb_3$. Iron doping dramatically decreases the overall κ_L. At the κ_L peak a reduction of about an order of magnitude is observed from $CoSb_3$ to $Co_{0.9}Fe_{0.1}Sb_3$. The additional strain (6%) and mass (5%) fluctuations introduced by alloying Fe on the Co site simply could not account for the observed κ_L reduction. The experimental data are modeled by the Debye approximation, including phonon-boundary, phonon–point-defect, and phonon–phonon umklapp scatterings.[38] In Fig. 8 the solid and dashed lines represent theoretical fits for $CoSb_3$ and $Co_{0.9}Fe_{0.1}Sb_3$, respectively. The agreement between calculations and experimental data is very good. It is found that phonon–point-defect scattering prefactor increases by a factor of about 25 from $CoSb_3$ to $Co_{0.9}Fe_{0.1}Sb_3$, indicating a rapid increase of point-defect concentration upon Fe alloying on the Co site. Based on the thermal conductivity and study of other low-temperature transport properties, the increased Fe doping is believed to significantly increase the amount of vacancies with severed atomic bonds on the Co site which leads to strong scattering of the lattice phonons. This is reasonable in light of the eventual instability of the skutterudite structure at higher Fe concentration and the lack of existence of a $FeSb_3$ phase. Recently reported electron tunneling experiments on $Co_{1-x}Fe_xSb_3$ suggest that the observed strong zero-bias conductance anomaly arises from a structural disorder produced by vacancies on the Co sites, in agreement with analysis.[41]

Figure 9 displays the temperature dependence of κ_L from 2 K to 300 K for $Co_{1-x}Ni_xSb_3$ samples, and it is clear that Ni doping strongly suppresses κ_L. For very small Ni concentrations ($x \leq 0.003$), κ_L decreases rapidly with increasing x, and this effect is especially manifested at low-temperature ($T < 100$ K). As the Ni

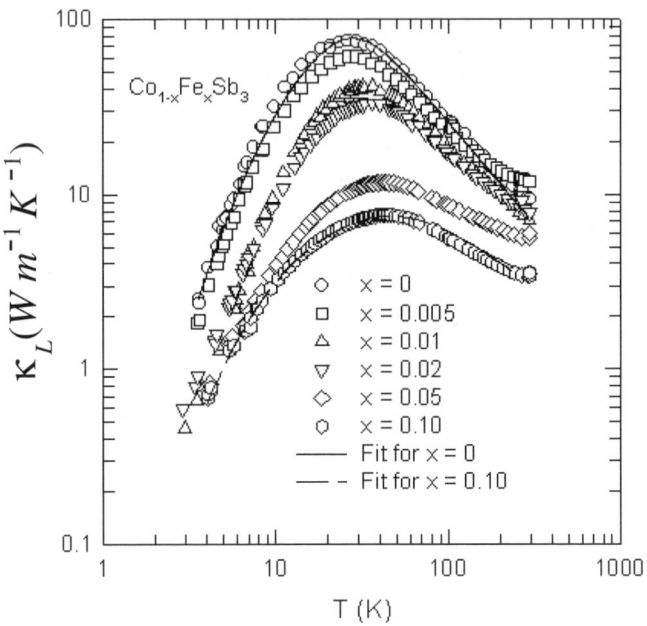

FIGURE 8. Lattice thermal conductivity of $Co_{1-x}Fe_xSb_3$ versus temperature. The solid and dashed lines represent calculations based on the Debye approximation.[39] (Reprinted from Ref. 39. Copyright 2001, American Physical Society.)

Sec. 3 · SKUTTERUDITES

concentration increases, the suppression of κ_L seems to saturate. The lines are drawn to illustrate the T^a dependence of κ_L at low-temperature ($T < 30$ K), where $a = 2.08$, 1.22, and 1 for $x = 0$, $x = 0.001$, and $x \geq 0.003$, respectively. The observation that $\kappa_L \propto T^1$ for $T < 30$ K for all $x \geq 0.003$ samples is a strong indication of the presence of electron–phonon interaction.[38] From analysis of electrical transport and magnetic data, it is found that electrical conduction at low-temperature (less than 30 K) is dominated by hopping of electrons among the impurity states.[42] The electron mean free path is estimated to be about 10^{-9} m, which is much shorter than the estimated phonon wavelength $\sim 10^{-8}$ m for $T < 30$ K.[38] The electron–phonon interaction is included in the theoretical fits (Debye approximation) for the data in addition to the phonon-boundary, phonon–point-defect, and phonon–phonon umklapp scatterings. The theoretical model fits the experimental data very well from 2 K to 300 K.[38] It is concluded that the reduction in κ_L (especially at low temperature) is the result of strong electron–phonon interaction between very-heavy-impurity electrons and lattice phonons.

Doping on the Co site of $CoSb_3$ with transition-metal elements significantly suppresses heat conduction in these binary skutterudites. This partially accounts for the high values of the figure of merit obtained in both *p*-type and *n*-type filled skutterudites.[28,42]

3.3. Filled Skutterudites

One of the most important findings of the recent skutterudite studies is the effect of guest ion filling of the large interstitial voids on κ_L. The guest ions are enclosed in an irregular dodecahedral cage of X atoms. Based on the large x-ray thermal parameters of these ions, Slack suggested that, if smaller than the voids, the guest

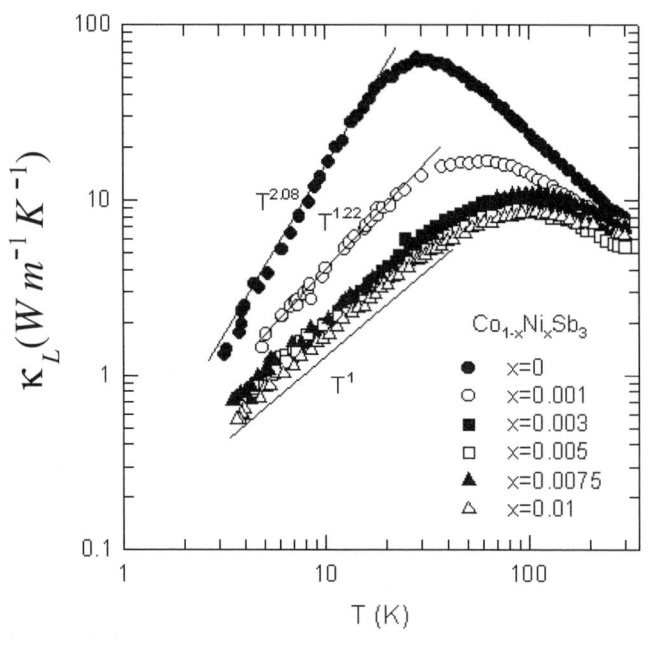

FIGURE 9. Temperature dependence of the lattice thermal conductivity of $Co_{1-x}Ni_xSb_3$ between 2 K and 300 K. The lines indicate $\kappa_L \propto T^{2.08}$, $T^{1.22}$, and T^1 between 2 K and 30 K for $x = 0$, 0.001, and 0.003, respectively.[38] (Reprinted from Ref. 38. Copyright 2002, American Physical Society.)

ions may rattle and therefore interact with lattice phonons, resulting in substantial phonon scattering.[26] The most direct evidence of rattling comes from the atomic displacement parameters (ADPs, a.k.a. thermal parameters) obtained from either x-ray or neutron scattering of single crystals. The isotropic ADP, U_{iso}, represents the mean square displacement average over all directions.[43] Figure 10 shows the temperature dependence of U_{iso} between 10 K and 300 K for RFe_4Sb_{12} (R = Ce, La, and Yb). The ADP values for Fe and Sb are typical for these elements in compounds with similar coordination numbers, whereas those for Ce, La, and Yb are anomalously large, particularly at high-temperatures. These results suggest that the guest ions are loosely bonded in the skutterudite structure and rattle about their equilibrium positions. It is also interesting to notice that U_{iso} increases as the ion size decreases from Ce to Yb.

The demonstration of thermal conductivity reduction over a wide temperature range due to fillers (or rattlers) was first realized in Ce-filled skutterudites.[27] Subsequently, similar effects were observed in skutterudites filled with La, Tl, Yb, Ba, and Eu.[29-33] Figure 11 shows the temperature dependence of κ_L for unfilled $IrSb_3$, La, Sm, and Nd-filled $IrSb_3$.[44] The calculated κ_{min} for $IrSb_3$ is also plotted. Upon rare-earth filling, κ_L is reduced by more than an order of magnitude over most of the temperature range. The temperature dependence of κ_L is significantly altered by rare-earth filling as well. The dielectric peak of κ_L that is evident for $IrSb_3$ completely disappears for the filled skutterudite samples. The Sm and Nd ions are smaller than La ions and therefore are freer to rattle inside the voids. They are believed to interact with lower-frequency phonons as compared to La ions, leading to a larger κ_L reduction. Furthermore, the low-lying $4f$ electronic energy levels of Sm and Nd ions also produce additional phonon scattering. Since the ground state of Nd ions splits into more levels of smaller energy separation than that of Sm ions,

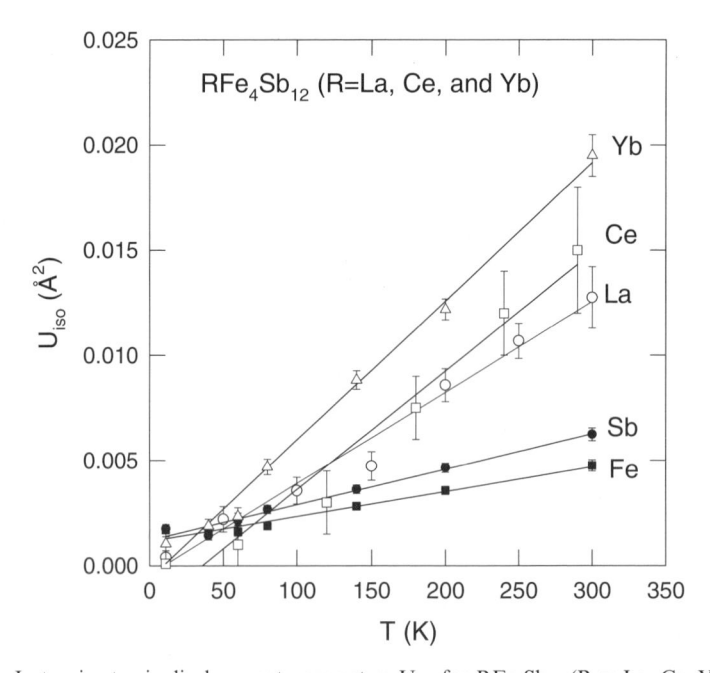

FIGURE 10. Isotropic atomic displacement parameters U_{iso} for RFe_4Sb_{12} (R = La, Ce, Yb) alloys.

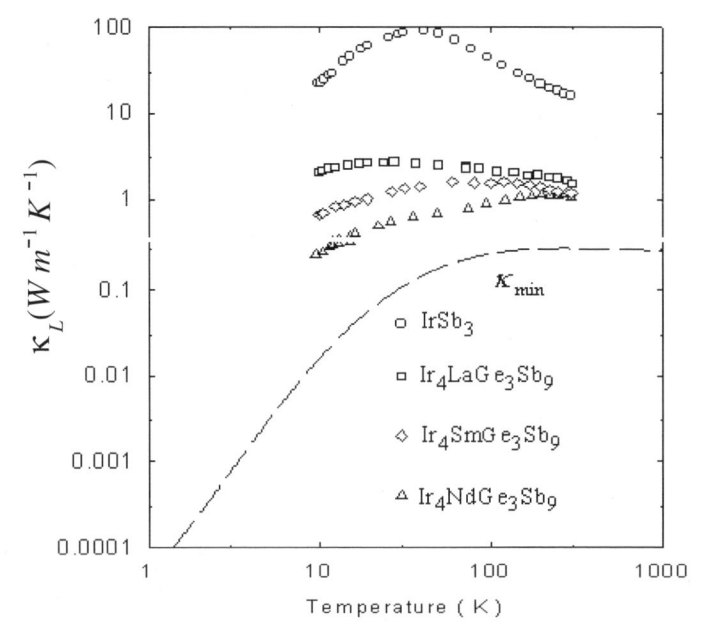

FIGURE 11. Lattice thermal conductivity versus temperature for the La-, Sm-, and Nd-filled skutterudite samples as well as the unfilled IrSb$_3$ sample. The calculated values of κ_{min} for IrSb$_3$ are also included.[44] (Reprinted from Ref. 44. Copyright 1996, American Institute of Physics.)

Nd ions will scatter a larger spectrum of phonons than will Sm ions. This effect should be manifested at low temperatures because these phonons have long wavelengths. As shown in Fig. 11, at room temperature, κ_L of Nd-filled and Sm-filled samples are about the same. At 10 K, κ_L of the former is only about half that of the latter.

The filling fraction y of the filled skutterudites can normally be increased by replacing electron-deficient elements on the M or X sites. In the case of Ce-filled skutterudites, Fe alloying on the Co site not only alters the optimal Ce filling fraction, the location of the Fermi level, carrier type, and carrier concentration,[45] but also strongly affects the nature of the phonon scattering of the optimally Ce-filled skutterudites.[46] In a series of optimally Ce-filled skutterudite samples, Meisner et al.[46] found that κ_L first decreases with increasing Ce filling fraction, reaches a minimum, and then increases for higher Ce filling fractions. This anomalous effect is justified by considering these Ce-filled skutterudites as solid solutions of CeFe$_4$Sb$_{12}$ (fully filled) and □ Co$_4$Sb$_{12}$ (unfilled), where □ denotes a vacancy. The predominant phonon–point–defect scattering does not arise between Co and Fe, but between Ce and □. Figure 12 plots the room temperature thermal resistivity of (CeFe$_4$Sb$_{12}$)$_\alpha$(□ Co$_4$Sb$_{12}$)$_{1-\alpha}$ as a function of α. The solid line represents a theoretical calculation for the thermal resistivity of these solid solutions. The dashed line includes additional thermal resistivity arising from other phonon scatterings. This solid solution model explains the variation of κ_L very well without even a single adjustable parameter.[46]

The interaction between the lattice phonons and the rattlers can be modeled by a phonon resonance scattering term in the Debye approximation. Figure 13 shows the experimental κ_L data with theoretical fits (solid lines) for CoSb$_3$, Yb$_{0.19}$Co$_4$Sb$_{12}$, and Yb$_{0.5}$Co$_4$Sb$_{11.5}$Sn$_{0.5}$.[47] The theoretical fit for CoSb$_3$ is calculated with the De-

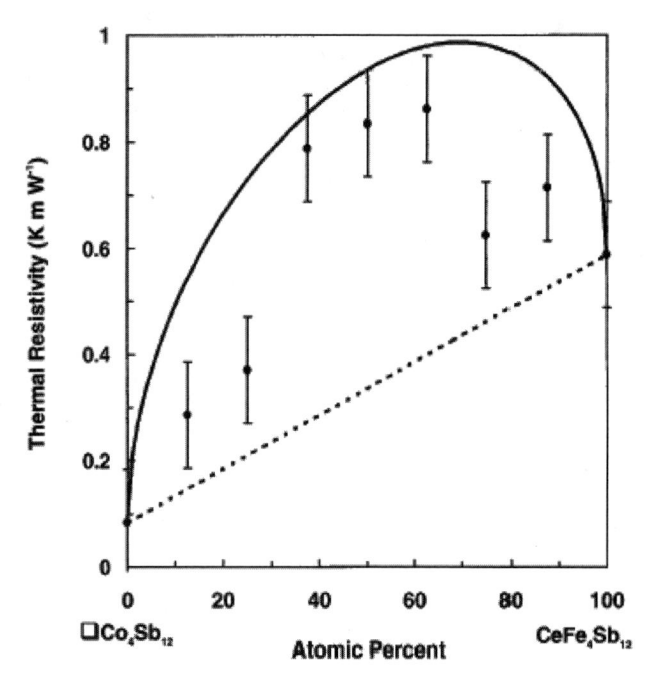

FIGURE 12. Variation of the thermal resistivity of $(CeFe_4Sb_{12})_\alpha(\square\ Co_4Sb_{12})_{1-\alpha}$ solid solutions as a function of α at room temperature. Dashed line represents variation from the rule of mixtures. Solid line includes calculated additional thermal resistivity due to the formation of solid solutions.[46] (Re-printed from Ref. 46. Copyright 1998, American Physical Society.)

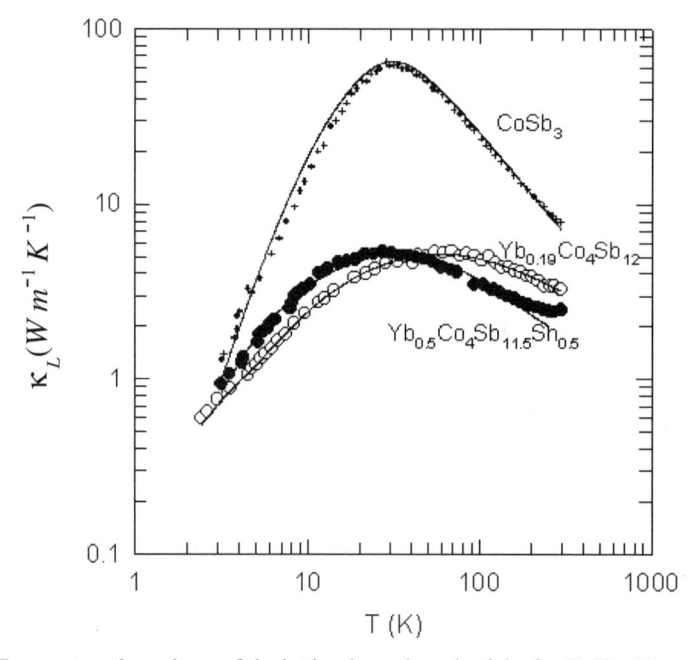

FIGURE 13. Temperature dependence of the lattice thermal conductivity for $CoSb_3$, $Yb_{0.19}Co_4Sb_{12}$, and $Yb_{0.5}Co_4Sb_{11.5}Sn_{0.5}$. The symbols are the experimental data, and the solid lines represent a calculation based on the Debye approximation.[47]

bye approximation and assuming only the boundary, point-defect and umklapp scatterings. An additional phonon–resonant scattering term is included in the model for the calculations of the Yb-filled samples. The solid lines fit the experimental data very well for all three samples over the entire two orders of magnitude temperature span. Yang et al.[47] find that the predominant phonon–point-defect scattering in these samples is on the Yb sites between Yb and □, and the phonon-resonant scattering increases linearly with increasing Yb filling fraction in agreement with theory.

4. CLATHRATES

Compounds with the clathrate crystal structure display an exceedingly rich number of physical properties, including semiconducting behavior, superconductivity, and thermal properties reminiscent of amorphous materials. All of these properties are a direct result of the nature of the structure and bonding of these materials.[48]

Clathrate compounds form in a variety of different structure types. They are basically Si, Ge, and Sn network structures or three-dimensional arrays of tetrahedrally bonded atoms built around various guests. The majority of work thus far on the transport properties of clathrates has been on two structure types that are isotypic with the clathrate hydrate crystal structures of type I and type II. The type I clathrate structure can be represented by the general formula $X_2Y_6E_{46}$, where X and Y are alkali-metal, alkaline earth, or rare-earth metal "guest" atoms that are encapsulated in two different polyhedra and E represents the Group IV elements Si, Ge, or Sn (although Zn, Cd, Al, Ga, In, As, Sb, or Bi can be substituted for these elements to some degree). Similarly, the type II structure can be represented by the general formula $X_8Y_{16}E_{136}$.[48] As shown in Fig. 14, these structures can be thought of as being constructed from two different face-sharing polyhedra; two pentagonal dodecahedra (E_{20}) and 6 tetrakaidecahedra (E_{24}) per unit cell in the case of the type I structure, and 16 dodecahedra (E_{20}) and 8 hexakaidecahedra (E_{28}) for type II compounds.

The guest–host interaction is one of the most conspicuous aspects of these compounds and directly determines the variety of interesting and unique properties

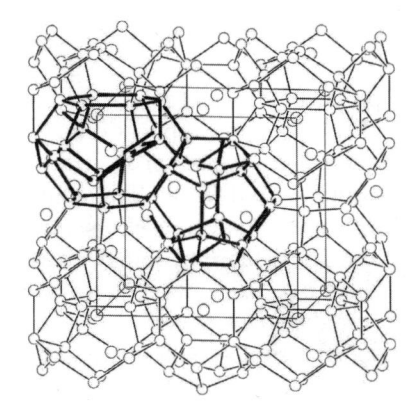

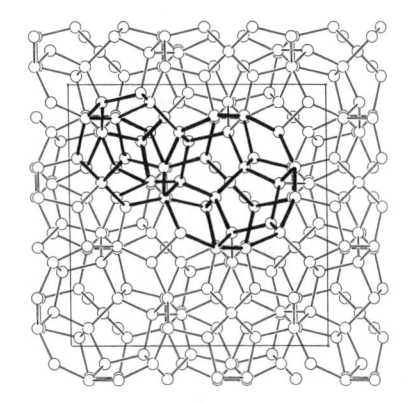

FIGURE 14. Type I (left) and type II (right) clathrate crystal structures. Only group 14 elements are shown. Outlined are the two different polyhedra that form the unit cell. (Reprinted from Ref. 1. Copyright 2001, Springer-Verlag.)

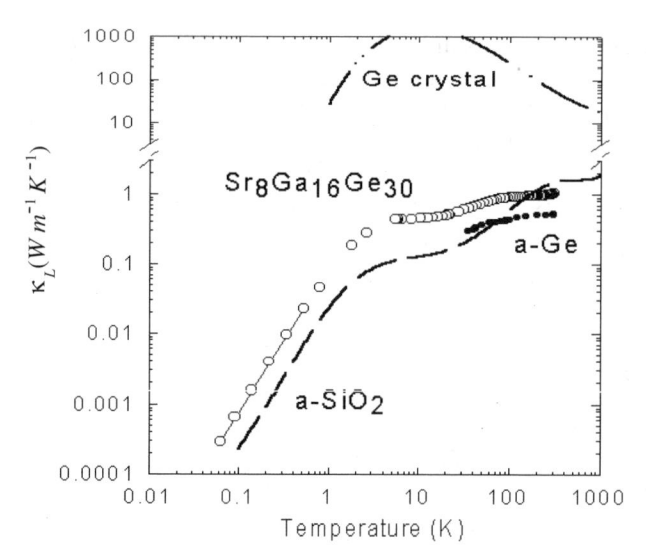

FIGURE 15. Lattice thermal conductivity versus temperature for polycrystalline $Sr_8Ga_{16}Ge_{30}$, single-crystal Ge, amorphous SiO_2, and amorphous Ge. The straight-line fit to the $Sr_8Ga_{16}Ge_{30}$ data below 1 K produces a T^2 temperature dependence characteristic of glasses.

these materials possess. One of the more interesting of these properties is a very distinct thermal conductivity. For example, the thermal conductivity of semiconductor $Sr_8Ga_{16}Ge_{30}$ shows a similar magnitude and temperature dependence to that of amorphous materials.[49–51] In fact at room temperature the thermal conductivity of these compounds is lower than that of vitreous silica and very close to that of amorphous germanium (Fig. 15). The low-temperature (< 1 K) data indicate a T^2 temperature dependence, as shown by the straight-line fit to the data in Fig. 15. Higher-temperature data show a minimum, or dip, indicative of resonance scattering. This resonance "dip" is more pronounced in single-crystal specimens, since scattering from grain boundaries is not a factor in these specimens.[49–51] It is clear from these results that in the $Sr_8Ga_{16}Ge_{30}$ compound the traditional alloy phonon scattering, which predominantly scatters the highest-frequency phonons, has been replaced by one or more much lower frequency scattering mechanisms. The scattering of the low-frequency acoustic phonons by the "rattle" modes of the encapsulated Sr atoms results in low thermal conductivity. We note that this glasslike thermal conductivity is found in semiconducting compounds that can be doped, to some extent, in order to vary their electronic properties.[49] This result indicates why these materials are of specific interest for thermoelectric applications. The transport properties are closely related to the structural properties of these interesting materials, as discussed later. Figure 16 shows the lattice thermal conductivity of several type I clathrates. Semiquantitative fits to these data as well as recent resonant ultrasound spectroscopic studies[52] indicate the existence of low-frequency vibrational modes of the "guest" atoms inside their oversized polyhedra. Low-frequency optic "rattle" modes were theoretically predicted to be well within the framework acoustic phonon band.[53] A Raman spectroscopic analysis of several type I clathrates experimentally verified these results.[54] The correlation between the thermal, ultrasound, and optical properties with the structural properties, as well as recent theoretical analysis,[55] is strong evidence that the resonance scattering

Sec. 4 · CLATHRATES

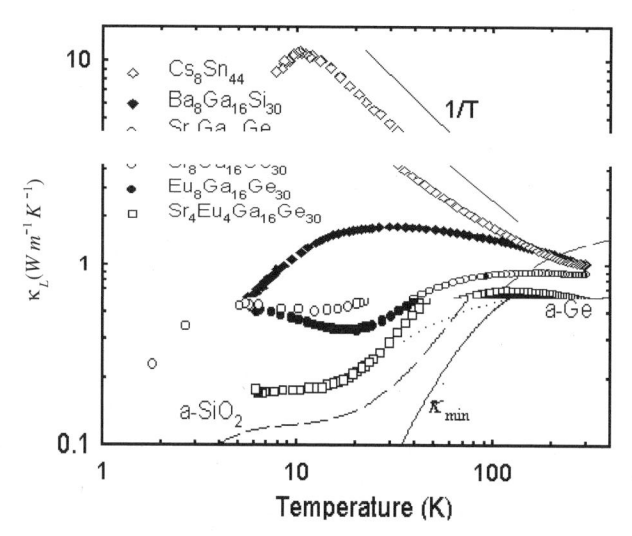

FIGURE 16. Lattice thermal conductivity of five representative polycrystalline type I clathrates. The dashed and dotted curves are for amorphous SiO$_2$ and amorphous Ge, respectively, and the solid curve is the theoretical minimum thermal conductivity calculated for diamond-Ge. (Reprinted from Ref. 50. Copyright 1999, American Physical Society.)

of acoustic phonons with the low-frequency optical "rattle" modes of the "guest" atoms is a strong phonon scattering process in clathrates.

Room temperature structural refinements from single-crystal and powder neutron scattering and x-ray diffraction reveal large atomic displacement parameters (ADPs) for atoms inside the tetrakaidecahedra as compared to the framework atoms of the type I clathrates. This is an indication of localized disorder within these polyhedra beyond typical thermal vibration. This is illustrated in Fig. 17a, where the ADPs of the Sr(2) (i.e., Sr inside the tetrakaidecahedra) exhibit an anisotropic ADP that is almost an order of magnitude larger than those of the other constituents in Sr$_8$Ga$_{16}$Ge$_{30}$.[51,56] Large and anisotropic ADPs are typically observed for relatively small ions in the tetrakaidecahedra in this crystal structure. The enormous ADP for the Sr(2) site implies the possibility of a substantial dynamic or "rattling" motion. It is also possible to correlate these ADPs with a static disorder in addition to the dynamic disorder. The electrostatic potential within the polyhedra is not everywhere the same, and different points may be energetically preferred. The ADP data can then also be described by splitting the Sr(2) site into four equivalent positions, as shown in Fig. 17b. This would suggest that these Sr atoms may tunnel between the different energetically preferred positions. The possibility of a "freeze-out" of the "rattle" modes of Sr in Sr$_8$Ga$_{16}$Ge$_{30}$ was initially postulated by Cohn et al.[50] from low-temperature thermal conductivity data (see Fig. 15). The "split-site" model was very eloquently revealed by Chakoumakos et al.[56], who employed temperature-dependent neutron diffraction data on single-crystal and powder Sr$_8$Ga$_{16}$Ge$_{30}$, where the split-site (static + dynamic) model and the single-site (dynamic disorder) model are clearly distinguished. These results clearly illustrate how the specific framework and thermal parameters associated with the atoms in these compounds are an important aspect of this structure and have an effect on the transport properties.[48] It also shows the unique properties thus far revealed in type I clathrates. Moreover, many different compositions have been synthesized in investigating the thermal properties of these interesting materials.

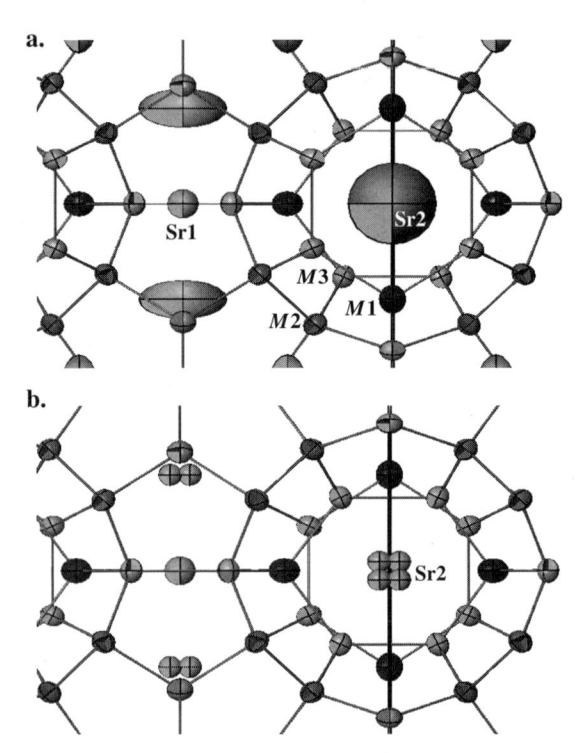

FIGURE 17. Crystal structure projection on a (100) plane of $Sr_8Ga_{16}Ge_{30}$ illustrating the large anisotropic atomic displacement parameters for the single-site model (a) and the split-site model (b) where the combined static and dynamic disorder of the Sr(2) crystallographic site with isotropic atomic displacement parameters is indicated. (Reprinted from Ref. 56. Copyright 1999, Elsevier Science.)

Fig. 17 illustrates this by showing the thermal conductivity of several type I clathrates.

Unlike type I clathrate compounds, type II clathrates can be synthesized with no atoms inside the polyhedra that form the framework of the crystal structure. Gryko et al.[57] have synthesized Si_{136} and demonstrated it is a clathrate form of elemental silicon with a 2-eV band gap and semiconducting properties. This experimental result verified the theoretical prediction made by Adams et al.[58] The thermal conductivity of Si_{136} shows a very low thermal conductivity, as shown in Fig. 18.[59] The room temperature thermal conductivity is comparable to that of amorphous silica and 30 times lower from that of diamond structure Si. The temperature dependence, however, is quite different from that of a-SiO_2. Indeed the temperature dependence of Si_{136} at low temperatures does not follow the glasslike behavior characteristic of a-SiO_2. This again confirms that the low value of κ_L is an intrinsic property of Si_{136}. The solid line in Fig. 18 illustrates the trend toward a T^3 dependence at the lower temperatures, where scattering from grain boundaries will dominate. As seen in the figure the lowest-temperature κ_L data begin to follow this temperature dependence. Again, this is different from $Sr_8Ga_{16}Ge_{30}$, for example, where a T^2 dependence is observed at low temperatures due to the localized disorder of Sr inside the (Ga,Ge) polyhedra.

The low thermal conductivity suggests that Si_{136} is another low-thermal-conductivity semiconductor. It may be that in this framework optic modes play an important role in the thermal conduction of this compound and, indeed, type II

Sec. 5 • HALF-HEUSLER COMPOUNDS

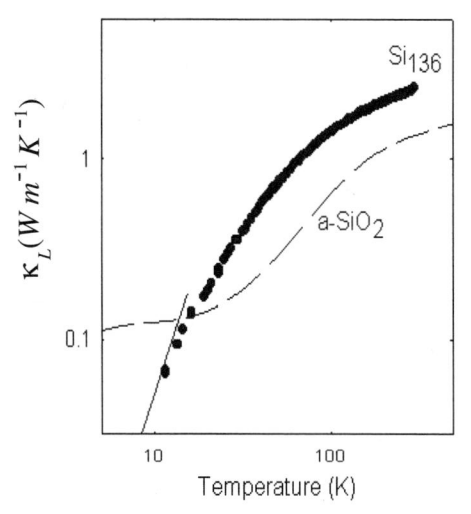

FIGURE 18A. Lattice thermal conductivity as a function of temperature for polycrystalline Si$_{136}$ and a-SiO$_2$. The solid line indicates a temperature dependence of T^3. (Reprinted from Ref. 59. Copyright 2003, American Institute of Physics.)

clathrates in general. The data imply a strong effect on the thermal transport from the large number of atoms per unit cell in Si$_{136}$ upon damping of the acoustic phonons by zone-edge singularities. The polyhedra are "empty"; therefore, localized disorder is not the reason for the low thermal conductivity. The thermal conductivity for a "guest-free" clathrate (hypothetical type I Ge$_{46}$) was recently calculated by using a Tersoff potential to mimic interatomic interactions.[55] A 10-fold decrease in lattice thermal conductivity compared with the diamond-structure semiconductor was noted due to scattering of heat carrying acoustic phonons by zone-boundary modes "folded back" due to the increase in unit cell size. The result indicates that low-thermal-conductivity values are achieved in clathrate compounds without "rattling" guest atoms, due to the intrinsic vibrational properties of the framework and the enlarged unit cell.

5. HALF-HEUSLER COMPOUNDS

Half-Heusler compounds possess many interesting transport and magnetic properties. These compounds can be semiconductors, semimetals, normal metals, weak ferromagnets, antiferromagnets, or strong half-metallic ferromagnets. In particular, MNiSn (M = Ti, Zr, or Hf) half-Heusler compounds show promising thermoelectric characteristics. The MNiSn compounds are a subset of a much larger family of compounds possessing the so-called MgAgAs-type structure. This structure is cubic and consists of three interpenetrating fcc sublattices with an element of each type on each sublattice displaced by one fourth of the distance along the body diagonal.[60] The fourth site along the body diagonal is not occupied and thus is an ordered array of vacancies. The MgAgAs-type structure is closely related to the MnCuAl$_2$-type structure in which this vacant position is occupied. The fully occupied structures, which in the MNiSn family are of the form MNi$_2$Sn, are metals known as Heusler compounds. Following this nomenclature, the MNiSn compounds are frequently termed half-Heusler compounds. The difference between the Heusler and the half-Heusler crystal structures, however, is not only in the filling of a vacant site, but also involves a change of space group and a reordering of the atoms. The MNiSn compounds have the space group $F\bar{4}3m$ with four crystallographically inequivalent fourfold sites, M, Ni, Sn, and * (the vacancy). In MNiSn, one published structure[60] has the sequence M, Sn, Ni, and * at positions (0,0,0), (1/4,1/4,1/4), (1/2,1/2,1/2), and (3/4,3/4,3/4), respectively. An alternative structure[61] has the M and Sn positions reversed. Indeed, there may be substantial site-exchange disorder between M and Sn, particularly when M is Hf or Zr.

While the full Heusler alloys are metals, many half-Heusler compounds are semiconductors with band gaps of about 0.21–0.24 eV.[62,63] The large Seebeck coefficient, due to the heavy conduction band, and modestly large electrical resistivity make MNiSn compounds ideal candidates for optimization.[64–67] Alloying Zr and Hf, Ti, and Zr on the M sites, we find that Pd on the Ni site leads to an overall reduction of the total thermal conductivity κ.[63–65, 68–70] Doping 1 at.% Sb on the Sn site results in a lowering, by several orders of magnitude, of the resistivity without significantly reducing the Seebeck coefficient, leading to peak power factors in excess of 35 μW/cm-K^2 between 675 K and 875 K for a number of doped half-Heusler compounds.[63,64,68,70,71] The highest ZT value reported was ~0.7 around 800 K for the $Zr_{0.5}Hf_{0.5}Ni_{0.8}Pd_{0.2}Sn_{0.99}Sb_{0.01}$ compound.[70] Doped half-Heusler compounds are prospective materials for high-temperature thermoelectric power generation. Thermal conductivities for these materials have to be further reduced (without lowering power factors) to make it a practical reality.

5.1. Effect of Annealing

Since intermixing between M and Sn sublattices depends on the heat treatment, a series of measurements monitoring transport properties as a function of annealing conditions were made. Especially, the behavior of κ is extremely sensitive to the structural disorder and is thus a very good indicator of the evolving perfection of the crystal lattice.[65] Figure 19 shows the temperature dependence of κ on ZrNiSn (panel a), HfNiSn (panel b), and $Zr_{0.5}Hf_{0.5}NiSn$ (panel c) subjected to up to 1-week annealing periods. The transport properties remain stable beyond 1-week annealing. The electronic thermal conductivity component is estimated to be less than 1% of the total; therefore, the data shown are almost entirely the lattice part. All samples show dramatic changes in κ with prolonged annealing. For ZrNiSn and HfNiSn, annealing enhances κ. This is especially manifested at low temperatures. After a week of annealing, the peak in κ (T) increases by about a factor of 3, suggesting an improvement in the structural quality of the annealed materials. Even though HfNiSn has higher thermal conductivity values at all temperatures and in all corresponding annealing stages than does ZrNiSn, the percentage increase in κ is about the same for both samples. Annealing, however, promotes an opposite trend in κ for $Zr_{0.5}Hf_{0.5}NiSn$: κ decreases upon annealing. This suggests that the as-cast sample does not have the M site completely randomized, and it takes heat treatment to ensure that the Zr and Hf atoms are completely mixed.

5.2. Isoelectronic Alloying on the M and Ni Sites

Figure 19 also shows that the overall κ is much suppressed over the entire temperature range by isoelectronically alloying Zr and Hf on the M site. This is due to the phonon–point-defect scattering between Zr (atomic mass M_{Zr} = 91) and Hf

Sec. 5 · HALF-HEUSLER COMPOUNDS

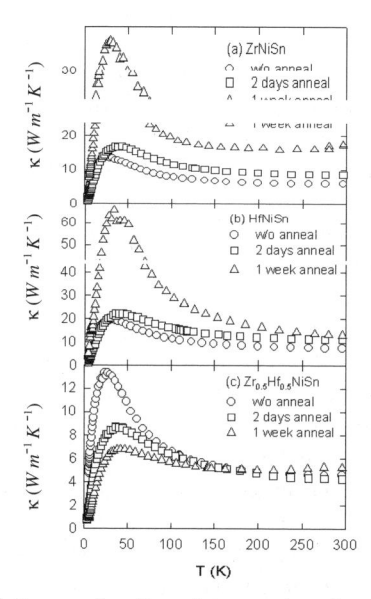

FIGURE 19. Thermal conductivity as a function of temperature for different annealing periods. (a) ZrNiSn; (b) HfNiSn; (c) $Zr_{0.5}Hf_{0.5}NiSn$.[65] (Reprinted from Ref. 65. Copyright 1999, American Physical Society.)

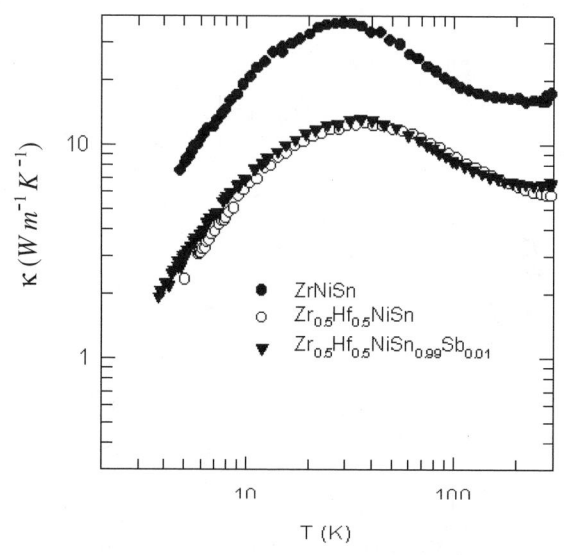

FIGURE 20. Thermal conductivity as a function of temperature for one-week-annealed ZrNiSn, $Zr_{0.5}Hf_{0.5}NiSn$, and $Zr_{0.5}Hf_{0.5}NiSn_{0.99}Sb_{0.01}$.[65] (Reprinted from Ref. 65. Copyright 1999, American Physical Society.)

(atomic mass M_{Hf} = 179). This effect is better illustrated in Fig. 20, where the temperature dependence of κ for one-week-annealed ZrNiSn, $Zr_{0.5}Hf_{0.5}NiSn$, and $Zr_{0.5}Hf_{0.5}NiSn_{0.99}Sb_{0.01}$ is plotted between 2 K and 300 K.[65] At room temperature, κ decreases from 17.2 W/m-K for ZrNiSn to 5.3 W/m-K for $Zr_{0.5}Hf_{0.5}NiSn$, more than a factor of 3. There is also a slight thermal conductivity increase near room temperature by doping 1 at.% of Sb on the Sn site of $Zr_{0.5}Hf_{0.5}NiSn$. This is attributed to the increased electronic thermal conductivity as a consequence of an increased electron concentration upon Sb-doping.[65]

The thermal conductivity of ZrNiSn-based half-Heusler compounds is further reduced by simultaneously alloying Hf on the Zr site and Pd on the Ni site. A recent high-temperature thermoelectric property study of these materials shows that upon alloying 20 at.% of Pd on the Ni site, there is a reduction of κ by about a factor of 1.5 over the entire temperature range.[70] This is the result of the phonon–point-defect scattering between Ni (atomic mass M_{Ni} = 58.7) and Pd (atomic mass M_{Pd} = 106.4). Further reduction of κ is accomplished by alloying more (50 at.%) Pd on the Ni site, as illustrated in Fig. 21. At room temperature, κ = 4.5 and 3.1 W/m-K for $Zr_{0.5}Hf_{0.5}Ni_{0.8}Pd_{0.2}Sn_{0.99}Sb_{0.01}$ and $Zr_{0.5}Hf_{0.5}Ni_{0.5}Pd_{0.5}Sn_{0.99}Sb_{0.01}$, respectively. Increasing Pd alloying on the Ni site, however, decreases the overall power factor of these compounds.[70] The highest ZT values are observed for n-type $Zr_{0.5}Hf_{0.5}Ni_{0.8}Pd_{0.2}Sn_{0.99}Sb_{0.01}$, not necessarily the compound with the lowest λ.

5.3. Effect of Grain Size Reduction

At 800 K the highest ZT value achieved by half-Heusler compounds is 0.7. This is at least as good as the conventional high-temperature thermoelectric materials at

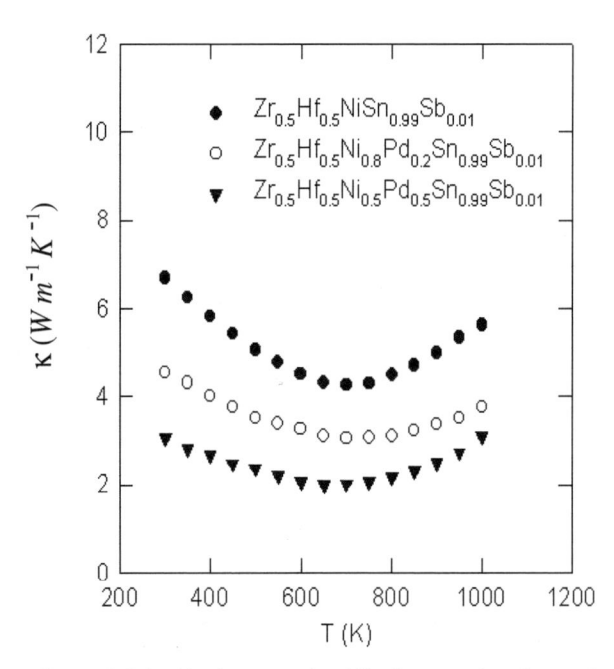

FIGURE 21. The effect of Pd alloying on the Ni site on the thermal conductivity of $Zr_{0.5}Hf_{0.5}NiSn_{0.99}Sb_{0.01}$.[70] (Reprinted from Ref. 70. Copyright 2001, American Institute of Physics.)

Sec. 6 · NOVEL CHALCOGENIDES AND OXIDES

the same temperature; i.e., $ZT \approx 0.7$ and 0.55 for PbTe and SiGe alloys, respectively. Future improvement in ZT will have to come from further reduction of κ. It was predicted that some improvement of ZT may be realized by reducing the grain sizes of the polycrystalline half-Heusler compounds.[72] This is based on theoretical results suggesting that the lattice thermal conductivity decreases much faster than the carrier mobility with decreasing grain size for these materials. A recent experimental study shows a linear correlation between the room temperature lattice thermal conductivity and the average diameter of grains for Ti-based half-Heusler compounds,[73] including ball-milled and shock-compacted samples with < 1 μm grain sizes. The results are plotted in Fig. 22. The lattice thermal conductivity of these materials decreases with decreasing grain size. At room temperature, the ball-milled and shock-compacted sample exhibits a total thermal conductivity of about 5.5 W/m-K with a lattice component of about 3.7 W/m-K. Even though a subsequent study reports a disappointing power factor decrease upon decreasing the grain size of a $TiNiSn_{0.95}Sb_{0.05}$ sample,[74] a systematic investigation on the effect of the grain size dependence of the power factor is much needed.

6. NOVEL CHALCOGENIDES AND OXIDES

The thermal properties of complex chalcogenide and oxide compounds have recently been investigated in an effort to expand the understanding of thermal transport in these materials to develop new materials for thermoelectric applications. In the last few years new compounds with interesting phonon scattering mechanisms that result in a low thermal conductivity have been investigated. In this section we review several of these compounds in terms of their thermal conduction.

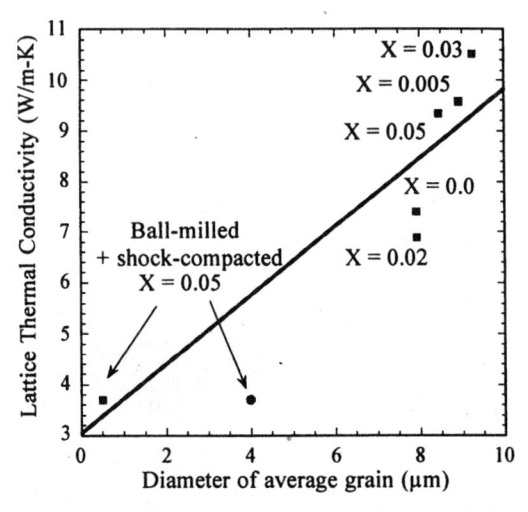

FIGURE 22. Lattice thermal conductivity versus average grain diameter of $TiNiSn_{1-x}Sb_x$ including ball-milled and shock-compacted $TiNiSn_{0.95}Sb_{0.05}$. The line is a linear fit to the data.[73] (Reprinted from Ref. 73. Copyright 1999, American Institute of Physics.)

6.1. Tl$_9$GeTe$_6$

Tl$_9$GeTe$_6$ belongs to a large group of ternary compounds that may be thought of as derived from the isostructural compound Tl$_5$Te$_3$. There are no Te–Te bonds in the structure, and the Te nearest neighbors are Tl atoms. A simple crystal chemistry analysis indicates that tellurium is in the Te^{2-} valence state, thallium is in the Tl$^+$ state, and bismuth is in the Bi^{3+} valence state. The 4c crystallographic site is equally occupied with Tl$^+$ and Bi^{3+} and octahedrally coordinated with Te atoms.[75] This mixed valence site makes for a strong phonon scattering center, as shown in Fig. 23. This type of phonon scattering has been documented in Ru$_{0.5}$Pd$_{0.5}$Sb$_3$[36,76] and Fe$_3$O$_4$;[77] however, in this case the valence disorder is from two different atoms occupying the same crystallographic site, Tl$^+$ and Bi^{3+}, and results in a very low thermal conductivity with κ_L = 0.39 W/m-K at room temperature.[75] These results strongly indicate how mixed valence results in a very low thermal conductivity and may be a useful phonon scattering mechanism in the search for new thermoelectric materials. The thermoelectric figure of merit of Tl$_9$GeTe$_6$ was estimated to exceed unity above 450 K. The very low thermal conductivity for this compound is the key for obtaining this figure of merit.[75]

6.2. Tl$_2$GeTe$_5$ and Tl$_2$SnTe$_5$

The Tl–Sn–Te and Tl–Ge–Te systems both contain several ternary compounds, including a 2-1-5 composition in both systems.[78] Both compounds are tetragonal and contain columns of Tl ions along the crystallographic c axis. Transverse to the c axis are alternate columns with chains of composition (Sn/Ge)Te$_5$. These compounds can be thought of as polytypes of one another, with different stacking sequences of (Ge/Sn)Te$_4$ tetrahedra and TeTe$_4$ square planar units linked into chains. Large ADP values confirm that a portion of the Tl atoms is loosely bound in these structures. Again, large ADP values are associated with low values of κ_L, as previously discussed. The thermal conductivity is very low for both Tl$_2$SnTe$_5$ and Tl$_2$GeTe$_5$. Apparently κ_L in these polycrystalline samples is approximately 5 W/m-K at room temperature. This is less than one third of the value for pure Bi$_2$Te$_3$. At

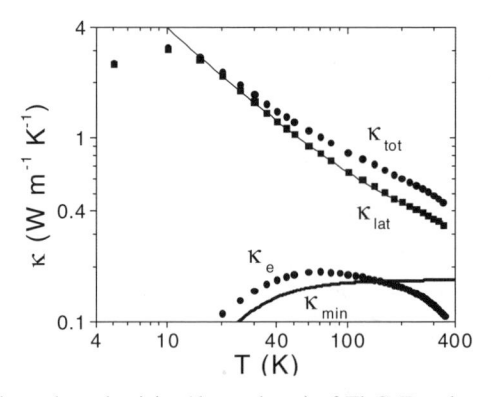

FIGURE 23. The total thermal conductivity (denoted κ_{tot}) of Tl$_9$GeTe$_6$ along with the calculated electrical, κ_e, and phonon, κ_{lat}, contributions. The fit to the data employs the expression $AT^{-0.93} + \kappa_{min}$, where A is a constant, T is the absolute temperature, and κ_{min} is the minimum thermal conductivity, also plotted in the figure.[75] (Reprinted from Ref. 75. Copyright 2001, American Physical Society.)

room temperature the electronic contribution to κ was estimated to be about 20% for Tl_2SnTe_5 but less than 10% for Tl_2GeTe_5, due to the higher resistivity.[78]

6.3. $CsBi_4Te_6$

One of the more interesting new bulk materials that have recently been investigated for thermoelectric refrigeration is $CsBi_4Te_6$.[79] This compound has a layered anisotropic structure composed of Bi_4Te_6 layers alternating with layers of Cs^+ ions. The Cs^+ ions lie between the Bi_4Te_6 layers and their atomic displacement parameters (ADPs) are almost twice as large as those of the Bi and Te atoms, indicating that Cs^+ may display dynamic disorder. This disorder, together with the complexity of the crystal structure, is presumably the cause of the relatively low thermal conductivity in this material. The thermal conductivity of dense polycrystalline pellets was measured to be 1.5 W/m-K at room temperature.

6.4. $NaCo_2O_4$

Recently, dimensionless figures of merit of ~1.0 at 800°K and ~0.8 at 800°C were reported for single-crystal and polycrystalline $NaCo_2O_4$, respectively,[80,81] making this a promising high-temperature thermoelectric material, particularly since this is an oxide made up of light atoms with large electronegativity differences, contradicting the general guidelines for good thermoelectric materials.[82] $NaCo_2O_4$ possesses a layered hexagonal structure that consists of two-dimensional triangular lattices of Co. The cobalt atoms are octahedrally surrounded by O atoms above and below each Co sheet. The resulting CoO_2 layers are stacked along the hexagonal c-axis. Na atoms are in planes sandwiched by adjacent CoO_2-layers and randomly occupy 50% of the regular lattice sites in those planes. The material shows metallic-like transport properties as measured in single-crystal samples, with highly anisotropic behavior.[80] Because of the low electrical resistivity and large thermopower, a large power factor of 50 $\mu W\ cm^{-2}\ K^{-2}$ at room temperature was reported in the in-plane direction.[83] This power factor value exceeds that of Bi_2Te_3 despite the much lower mobility estimated for $NaCo_2O_4$. Terasaki et al. suggested that strong electron–electron correlation in this layered oxide plays an important role in the unusually large thermopower observed.[83]

Another important characteristic of $NaCo_2O_4$ is its low thermal conductivity. At room temperature, a total thermal conductivity of 2 W/m-K with lattice component of 1.8 W/m-K is reported for polycrystalline samples. Figure 24 shows the temperature dependence of κ_L for polycrystalline $(NaCa)Co_2O_4$ between 10 K and 300 K.[84] For both samples in Fig. 24, κ_L increases monotonically with increasing temperature. The solid curve A in the figure is a Callaway model fit to the data, assuming only boundary, point-defect, and phonon–phonon umklapp scatterings are present, whereas the dashed curve B represent the calculated minimum thermal conductivity. Based on the fit, Takahata et al. conclude that these amorphous-like κ_L are predominantly due to strong point-defect phonon scattering on the Na site between the Na atoms and the vacancies,[84] even though the value of L (a fitting parameter which represents an inelastic scattering length) determined from the fit does not seem to have a clear physical origin.

Recently, Rivadulla et al.[85] reported a very different temperature dependence and room temperature value of κ_L for polycrystalline $NaCo_2O_4$, compared to those in Fig. 24. The data are plotted in Fig. 25 and show typical temperature dependence

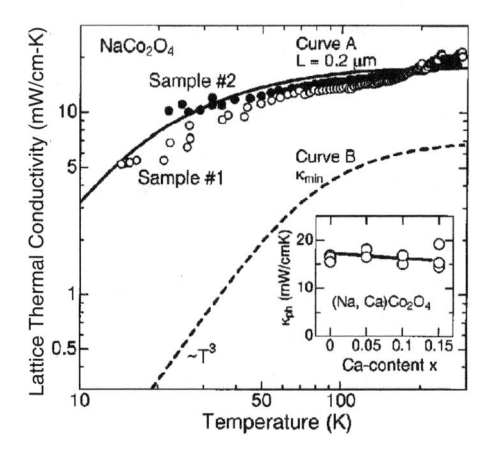

FIGURE 24. Lattice thermal conductivity of $NaCo_2O_4$. Open and closed circles represent data for samples no. 1 and no. 2, where the electron thermal conductivity was estimated from the Wiedemann–Franz law. Curve A is the calculation based on the Debye approximation, Curve B is the calculated minimum thermal conductivity, as discussed in Ref. 84. The inset shows data at 200 K.

for crystalline materials. A κ_L peak is observed at ~40 K, and at room temperature $\kappa_L \approx 4.0$ W/m-K, twice as large as that reported by Takahata et al.[84] The origin of phonon scattering mechanisms that govern the thermal conductivity for this type of compound is an open question, the answer to which awaits further investigation. It is a prerequisite for assessing the thermoelectric properties of these materials.

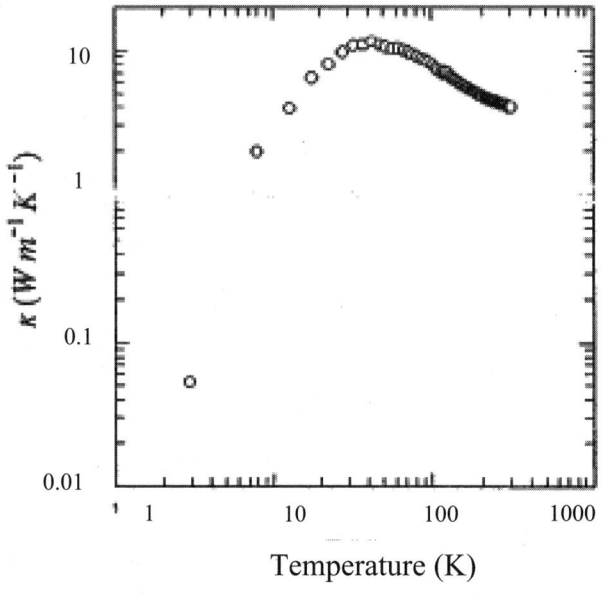

FIGURE 25. Lattice thermal conductivity of polycrystalline $NaCo_2O_4$.[85]

7. SUMMARY

We have reviewed thermal conductivities of conventional and newly developed thermoelectric materials. For most conventional materials and half-Heusler alloys, low lattice thermal conductivities are achieved via alloying between isomorphous compounds without deteriorating the electrical properties. In the cases of filled skutterudites, clathrates, and some of the novel chalcogenides, guest ions can be inserted into the large interstitial voids of the crystal structures. These guest ions rattle inside the voids and therefore interact resonantly with low-frequency acoustic phonons, leading to significant thermal conductivity reduction. Phonon scattering by valence disorder is responsible for the low thermal conductivity observed in Tl_9GeTe_6. In-depth investigation on the $NaCo_2O_4$-based oxides, potential high-temperature thermoelectric materials, is much needed to clarify the exact phonon scattering mechanisms in these compounds.

8. REFERENCES

1. G. S. NOLAS, J. SHARP, AND H. J. GOLDSMID *Thermoelectrics: Basic Principles and New Materials Developments* (Springer, Berlin, 2001).
2. J. R. DRABBLE AND C. H. GOODMAN *Chemical Bonding in Bismuth Telluride* J. Phys. Chem. Solids **5**, 142–144 (1958).
3. H. J. GOLDSMID *The Thermal Conductivity of Bismuth Telluride* Proc. Phys. Soc. London, Sect. B **69**, 203–209 (1956).
4. D. K. C. MACDONALD, E. MOOSER, W. B. PEARSON, I. M. TEMPLETON, AND S. B. WOODS *On the Possibility of Thermoelectric Refrigeration at Very Low Temperatures* Phil. Mag. **4**, 433–446 (1959).
5. P. A. WALKER *The Thermal Conductivity and Thermoelectric Power for Bismuth Telluride at Low Temperatures* Proc. Phys. Soc. London **76**, 113–126 (1960).
6. H. J. GOLDSMID *Heat Conduction in Bismuth Telluride* Proc. Phys. Soc. London **72**, 17–26 (1958).
7. A. F. IOFFE, S. V. AIREPETYANTS, A. V. IOFFE, N. V. KOLOMOETS, AND L. S. STIL'BANS, Dokl. Akad. Nauk. SSSR **106**, 981 (1956).
8. U. BIRKHOLZ *Untersuchung Der Intermetallischen Verbindung Bi_2Te_3 Sowie Der Festen Losungen $Bi_{2-X}Sb_XTe_3$ Und $Bi_2Te_{3-X}Se_X$ Hinsichtlich Ihrer Eignung Als Material Fur Halbleiter-Thermoelemente* Z. Naturforsch A **13**, 780–792 (1958).
9. H. J. GOLDSMID *Recent Studies of Bismuth Telluride and Its Alloys* J. Appl. Phys. **32**, 2198 Suppl. S (1961).
10. F. D. ROSI, B. ABELES, and R. V. JENSEN *Materials for Thermoelectric Refrigeration* J. Phys. Chem. Solids **10**, 191–200 (1959).
11. A. L. JAIN *Temperature Dependence of the Electrical Properties of Bismuth-Antimony Alloys* Phys. Rev. **114**, 1518–1528 (1959).
12. G. E. SMITH AND R. WOLFE *Thermoelectric Properties of Bismuth-Antimony Alloys* J. Appl. Phys. **33**, 841 (1962).
13. R. WOLFE AND G. E. SMITH *Effects of a Magnetic Field on the Thermoelectric Properties of a Bismuth-Antimony Alloy* Appl. Phys. Lett. **1**, 5–7 (1962).
14. C. UHER AND H. J. GOLDSMID *Separation of the Electronic and Lattice Thermal Conductivities in Bismuth Crystals* Phys. Stat. Solidi B **65**, 765–772 (1974).
15. K. F. CUFF, R. B. HORST, J. L. WEAVER, S. R. HAWKINS, C. F. KOOI, AND G. W. ENSLOW *The Thermomagnetic Figure of Merit and Ettingshausen Cooling in Bi-Sb Alloys* Appl. Phys. Lett. **2**, 145–146 (1963).
16. R. S. ALLGAIER AND W. W. SCANLON *Mobility of Electrons and Holes in PbS, PbSe, and PbTe Between Room Temperature and 4.2-Degrees-K* Phys. Rev. **111**, 1029–1037 (1958).
17. S. V. AIRAPETYANTS, B. A. EFIMOVA, T. S. STAVITSKAYA, L. S. STIL'BANS, AND L. M. SYSOEVA, Zh. Tekh. Fiz. **27**, 2167 (1957).
18. E. A. SKRABEK AND D. S. TRIMMER *Properties of the General TAGS System*, in *CRC Handbook of Thermoelectrics*, edited by D. M. Rowe (CRC Press, Boca Raton, FL, 1995), pp. 267–285.

19. M. C. Steele and F. D. Rosi *Thermal Conductivity and Thermoelectric Power of Germanium-Silicon Alloys* J. Appl. Phys. **29**, 1517–1520 (1958).
20. D. M. Rowe and C. M. Bhandari *Effect of Grain-Size on the Thermoelectric Conversion Efficiency of Semiconductor Alloys at High-Temperature* Appl. Energy **6**, 347–351 (1980).
21. H. J. Goldsmid and R. W. Penn *Boundary Scattering of Phonons in Solid Solutions* Phys. Lett. A **27**, 523 (1968).
22. N. Savvides and H. J. Goldsmid *Boundary Scattering of Phonons in Silicon Crystals at Room-Temperature* Phys. Lett. A **41**, 193 (1972).
23. T. H. Geballe and G. W. Hull *Isotopic and Other Types of Thermal Resistance in Germanium* Phys. Rev. **110**, 773–775 (1958).
24. C. Uher *Skutterudites: Prospective Novel Thermoelectrics*, in *Semiconductors and Semimetals*, Vol. **69**, edited by T. M. Tritt (Academic Press, San Diego, CA, 2000), pp. 139–253.
25. G. S. Nolas, D. T. Morelli, and T. M. Tritt *Skutterudites: A Phonon-Glass-Electron-Crystal Approach to Advanced Thermoelectric Energy Conversion Applications* Ann. Rev. Mater. Sci. **29**, 89–116 (1999).
26. G. A. Slack and V. G. Tsoukala *Some Properties of Semiconducting IrSb$_3$* J. Appl. Phys **76**, 1665-1671 (1994).
27. D. T. Morelli and G. P. Meisner *Low Temperature Properties of the Filled Skutterudite CeFe$_4$Sb$_{12}$* J. Appl. Phys. **77**, 3777–3781 (1995).
28. J. -P. Fleurial, A. Borshchevsky, T. Caillat, D. T. Morelli, and G. P. Meisner *High Figure in Ce-filled Skutterudites*, in *Proc. 15th Intl. Conf. on Thermoelectrics* (IEEE, Piscataway, NJ, 1996), pp. 91–95.
29. B. C. Sales, D. Mandrus, and R. K. Williams *Filled Skutterudite Antimonides: A New Class of Thermoelectric Materials* Science **272**, 1325–1328 (1996).
30. B. C. Sales, B. C. Chakoumakos, and D. Mandrus *Thermoelectric Properties of Thallium-Filled Skutterudites* Phys. Rev. B **61**, 2475–2481 (2000).
31. G. S. Nolas, M. Kaeser, R. T. Littleton IV, and T. M. Tritt *High Figure of Merit in Partially Filled Ytterbium Skutterudite Materials* Appl. Phys. Lett. **77**, 1855–1857 (2000).
32. L. D. Chen, T. Kawahara, X. F. Tang, T. Hirai, J. S. Dyck, W. Chen, and C. Uher *Anomalous Barium Filling Fraction and n-Type Thermoelectric Performance of Ba$_y$Co$_4$Sb$_{12}$* J. Appl. Phys. **90**, 1864–1868 (2001).
33. G. A. Lamberton, Jr., S. Bhattacharya, R. T. Littleton IV, M. A. Kaeser, R. H. Tedstrom, T. M. Tritt, J. Yang, and G. S. Nolas *High Figure of Merit in Eu-Filled CoSb$_3$-Based Skutterudites* Appl. Phys. Lett. **80**, 598–600 (2002).
34. H. Anno and K. Matsubara *A New Approach to Research for Advanced Thermoelectric Materials: CoSb$_3$ and Yb-Filled Skutterudites* Recent Res. Dev. Appl. Phys. **3**, 47–61 (2000).
35. J. S. Dyck, W. Chen, C. Uher, L. Chen, X. Tang, and T. Hirai *Thermoelectric Properties of the n-Type Filled Skutterudites Ba$_{0.3}$Co$_4$Sb$_{12}$ Doped with Ni* J. Appl. Phys. **91**, 3698–3705 (2002).
36. G. S. Nolas, V. G. Harris, T. M. Tritt, and G. A. Slack *Low Temperature Transport Properties of the Mixed-Valence Semiconductor Ru$_{0.5}$Pd$_{0.5}$Sb$_3$* J. Appl. Phys. **80**, 6304–6308 (1996).
37. D. T. Morelli, T. Caillat, J. -P. Fleurial, A. Borshchevsky, J. Vandersnade, B. Chen, and C. Uher *Low-Temperature Transport Properties of p-Type CoSb$_3$* Phys. Rev. B **51**, 9622–9628 (1995).
38. J. Yang, D. T. Morelli, G. P. Meisner, W. Chen, J. S. Dyck, and C. Uher *Influence of Electron-Phonon Interaction on the Lattice Thermal Conductivity of Co$_{1-x}$Ni$_x$Sb$_3$* Phys. Rev. B **65**, 094115 (2002).
39. J. Yang, G. P. Meisner, D. T. Morelli, and C. Uher *Iron Valence in Skutterudites: Transport and Magnetic Properties of Co$_{1-x}$Fe$_x$Sb$_3$* Phys. Rev. B **63**, 014410 (2001).
40. H. Anno, K. Matsubara, Y. Notohara, T. Sakakibara, and H. Tashiro *Effects of Doping on the Transport Properties of CoSb$_3$* J. Appl. Phys. **86**, 3780–3786 (1999).
41. J. Nagao, M. Ferhat, H. Anno, K. Matsubara, E. Hatta, and K. Mukasa *Electron Tunneling Experiments on Skutterudite Co$_{1-x}$Fe$_x$Sb$_3$ Semiconductors* Appl. Phys. Lett. **76**, 3436–3438 (2000).
42. J. S. Dyck, W. Chen, J. Yang, G. P. Meisner, and C. Uher *Effect of Ni on the Transport and Magnetic Properties of Co$_{1-x}$Ni$_x$Sb$_3$* Phys. Rev. B **65**, 115204 (2002).
43. B. C. Sales, D. Mandrus, and B. C. Chakoumakos *Use of Atomic Displacement Parameters in Thermoelectric Materials Research*, in *Semiconductors and Semimetals* Vol. **70**, edited by T. M. Tritt (Academic Press, San Diego, CA, 2001), pp. 1–36.
44. G. S. Nolas, G. A. Slack, D. T. Morelli, T. M. Tritt, and A. C. Ehrlich *The Effect of Rare-Earth Filling on the Lattice Thermal Conductivity of Skutterudites* J. Appl. Phys. **79**, 4002–4008 (1996).

45. D. T. Morelli, G. P. Meisner, B. Chen, S. Hu, and C. Uher *Cerium Filling and Doping of Cobalt Triantimonide* Phys. Rev. B **56**, 7376–7383 (1997).
46. G. P. Meisner, D. T. Morelli, S. Hu, J. Yang, and C. Uher *Structure and Lattice Thermal Conductivity of Fractionally Filled Skutterudites: Solid Solutions of Fully Filled and Unfilled End Members* Phys. Rev. Lett. **80**, 3551–3554 (1998).
47. J. Yang, D. T. Morelli, G. P. Meisner, W. Chen, J. S. Dyck, and C. Uher *Effect of Sn Substituting Sb on the Low Temperature Transport Properties of Ytterbium-Filled Skutterudites*, Phys. Rev. B **67**, 165207 (2003).
48. For two recent reviews see Chapter 6 of G. S. Nolas, J. Sharp, and H. J. Goldsmid, *Thermoelectrics: Basics Principles and New Materials Developments* (Springer-Verlag, New York, 2001), and G. S. Nolas, G. A. Slack, and S. B. Schujman *Semiconductor Clathrates* in *Semiconductors and Semimetals*, Vol. **69**, edited by T. M. Tritt (Academic Press, San Diego, 2000), pp. 255–300, and references therein.
49. G. S. Nolas, J. L. Cohn, G. A. Slack, and S. B. Schujman *Semiconducting Ge-Clathrates: Promising Candidates for Thermoelectric Applications* Appl. Phys. Lett. **73**, 178–180 (1998).
50. J. L. Cohn, G. S. Nolas, V. Fessatidis, T. H. Metcalf, and G. A. Slack *Glass-like Heat Conduction in High-Mobility Crystalline Semiconductors* Phys. Rev. Lett. **82**, 779–782 (1999).
51. G. S. Nolas, T. J. R. Weakley, J. L. Cohn, and R. Sharma *Structural Properties and Thermal Conductivity of Crystalline Ge-Clathrates* Phys. Rev. B **61**, 3845–3850 (2000).
52. V. Keppens, B. C. Sales, D. Mandrus, B. C. Chakoumakos, and C. Laermans *When Does a Crystal Conduct Heat Like a Glass?* Phil. Mag. Lett. **80**, 807–812 (2000).
53. J. Dong and O. F. Sankey *Theoretical Study of Two Expanded Phases of Crystalline Germanium: Clathrate-I and clathrate-II* J. Phys. Condensed Matter **11**, 6129–6145 (1999).
54. G. S. Nolas and C. A. Kendziora *Raman Scattering Study of Ge and Sn Compounds with Type-I Clathrate Hydrate Crystal Structure* Phys. Rev. B **62**, 7157–7161 (2000).
55. J. Dong, O. F. Sankey, and C. W. Myles *Theoretical Study of the Lattice Thermal Conductivity in Ge Framework Semiconductors* Phys. Rev. Lett. **86**, 2361–2364 (2001).
56. B. C. Chakoumakos, B. C. Sales, D. G. Mandrus, and G. S. Nolas *Structural Disorder and Thermal Conductivity of the Semiconducting Clathrate $Sr_8Ga_{16}Ge_{30}$* J. Alloys Comp. **296**, 80–86 (1999).
57. J. Gryko, P. F. McMillan, R. F. Marzke, G. K. Ramachandran, D. Patton, S. K. Deb, and O. F. Sankey *Low-Density Framework Form of Crystalline Silicon with a Wide Optical Band Gap* Phys. Rev. B **62**, R7707–7710 (2000)
58. G. B. Adams, M. O'Keeffe, A. A. Demkov, O. F. Sankey, and Y.-M. Huang *Wide-Band-Gap Si in Open Fourfold-Coordinated Clathrate Structures* Phys. Rev. B **49**, 8048–8053 (1994).
59. G. S. Nolas, M. Beekman J. Gryko, G. Lamberton, T. M. Tritt, and P. F. McMillan *Thermal Conductivity of the Elemental Crystalline Silicon Clathrate Si_{136}* Appl. Phys. Lett. **82**, 901–903 (2003).
60. W. Jeitschko *Transition Metal Stannides with MgAgAs and $MnCu_2Al$-Type Structure* Metall. Trans. **1**, 3159, 1970.
61. R. V. Skolozdra, Y. V. Stadnyk, and E. E. Starodynova *The Crystal Structure and Magnetic Properties of Me'Me"Sn Compounds (Me' = Ti, Zr, Hf, Nb; Me" = Co, Ni)* Ukrain. Phys. J. **31**, 1258–1261, 986.
62. F. G. Aliev, N. B. Brandt, V. V. Moshchalkov, V. V. Kozyrkov, R. V. Skolozdra, and A. I. Belogorokhov *Gap at the Fermi Level in the Intermetallic Vacancy System RNiSn (R = Ti, Zr, Hf)* Z. Phys. B **75**, 167–171 (1989).
63. B. A. Cook, G. P. Meisner, J. Yang, and C. Uher *High Temperature Thermoelectric Properties of MNiSn (M = Zr, Hf)* Proc. 18th Intl. Conf. on Thermoelectrics (IEEE, Piscataway, NJ, 1999), pp. 64–67.
64. C. Uher, J. Yang, and G. P. Meisner *Thermoelectric Properties of Bi-Doped Half-Heusler Alloys* Proc. 18th Intl. Conf. on Thermoelectrics (IEEE, Piscataway, NJ, 1999), pp. 56–59.
65. C. Uher, J. Yang, S. Hu, D. T. Morelli, and G. P. Meisner *Transport Properties of Pure and Doped MNiSn (M = Zr, Hf)* Phys. Rev. B **59**, 8615–8621 (1999).
66. H. Hohl, A. P. Ramirez, C. Goldmann, G. Ernst, B. Wolfing, and E. Bucher *Efficient Dopants for ZrNiSn-Based Thermoelectric Materials* J. Phys.: Condensed Matter **11**, 1697–1709 (1999).
67. B. A. Cook and J. L. Harringa *Electrical Properties of Some (1,1,1) Intermetallic Compounds* J. Mat. Sci. **34**, 323–327 (1999).
68. S. Bhattacharya, V. Ponnambalam, A. L. Pope, Y. Xia, S. J. Poon, R. T. Littleton IV, and T. M. Tritt *Effect of Substitutional Doping on the Thermal Conductivity of Ti-based Half-Heusler Compounds* Mat. Res. Soc. Symp. **626**, Z5.2.1–z5.2.6 (2000).

69. V. M. Browning, S. J. Poon, T. M. Tritt, A. L. Pope, S. Bhattacharya, P. Volkov, J. G. Song, V. Ponnambalam, and A. C. Ehrlich *Thermoelectric Properties of the Half-Heusler Compound (Zr,Hf)(Ni,Pd)Sn* Mat. Res. Soc. Symp. **545**, 403–412 (1998).
70. Q. Shen, L. Chen, T. Goto, T. Hirai, J. Yang, G. P. Meisner, and C. Uher *Effects of Partial Substitution of Ni by Pd on the Thermoelectric Properties of ZrNiSn-Based Half-Heusler Compounds* Appl. Phys. Lett. **79**, 4165–4167 (2001).
71. S. J. Poon *Electronic and Thermoelectric Properties of Half-Heusler Alloys*, in *Semiconductors and Semimetals*, Vol. **70**, edited by T. M. Tritt (Academic Press, San Diego, CA, 2001), pp. 37–75.
72. J. W. Sharp, S. J. Poon, and H. J. Goldsmid *Boundary Scattering and the Thermoelectric Figure of Merit* Phys. Stat. Sol. (a) **187**, 507–516 (2001).
73. T. M. Tritt, S. Bhattacharya, Y. Xia, V. Ponnambalam, S. J. Poon, and N. Thadhani *Effects of Various Grain Structure and Sizes on the Thermal Conductivity of Ti-based Half-Heusler Alloys* Proc. 20th Intl. Conf. on Thermoelectrics (IEEE, Piscataway, NJ, 2001), pp. 7–12.
74. S. Bhattacharya, Y. Xia, V. Ponnambalam, S. J. Poon, N. Thadhani, and T. M. Tritt *Reductions in the Lattice Thermal Conductivity of Ball-Milled and Shock Compacted $TiNiSn_{1-x}Sb_x$ Half-Heusler Alloys* Mat. Res. Soc. Symp. **691**, G7.1.1–G7.1.6 (1998).
75. B. Wolfing, C. Kloc, J. Teubner, and E. Boucher *High Performance Thermoelectric Tl_9GeTe_6 with an Extremely Low Thermal Conductivity* Phys. Rev. Lett. **86**, 4350–4353 (2001).
76. T. Caillat, J. Kulleck, A. Borshchevsky, and J. P. Fleurial *Preparation and Thermoelectric Properties of Skutterudite-Related Phase $Ru_{0.5}Pd_{0.5}Sb_3$* J. Appl. Phys. **79**, 8419–8426 (1996).
77. G. A. Slack *Thermal Conductivity of MgO, Al_2O_3, $MgAl_2O_4$ and Fe_3O_4 Crystals from 3 to 300 K* Phys. Rev. **126**, 427–441 (1962).
78. J. W. Sharp, B. C. Sales, D. G. Mandrus, and B. C. Chakoumakos *Thermoelectric Properties of Tl_2GeTe_5 and Tl_2SnTe_5* Appl. Phys. Lett. **74**, 3794–3797 (1999).
79. D. Y. Chung, T. Hogan, P. Brazis, M. Rocci-Lane, C. Kannewurf, M. Bastea, C. Uher, and M. G. Kanatzidis *$CsBi_4Te_6$: A High-Performance Thermoelectric Material for Low-Temperature Applications* Science **287**, 1024–1027 (2000).
80. K. Fujita, T. Mochida, and K. Nakamura *High-Temperature Thermoelectric Properties of $NaCo_2O_{4-y}$ Single Crystals* Jpn. J. Appl. Phys., Part 1 **40**, 4644 (2001).
81. M. Ohtaki, Y. Nojiri, and E. Maeda *Improved Thermoelectric Performance of Sintered $NaCo_2O_4$ with Enhanced 2-Dimensional Microstructure* Proc. 19th Intl. Conf. on Thermoelectrics (Babrow, Wales, 2000), pp. 190–195.
82. G.A. Slack *New Materials and Performance Limits for Thermoelectric Cooling*, in *CRC Handbook of Thermoelectrics*, edited by D. M. Rowe (CRC Press, Boca Raton, FL, 1995), pp. 407–440.
83. I. Terasaki, Y. Sasago, and K. Uchinokura *Large Thermoelectric Power in $NaCo_2O_4$ Single Crystals* Phys. Rev. B **56**, R12685–R12687 (1997).
84. K. Takahata, Y. Iguchi, D. Tanaka, T. Itoh, and I. Terasaki *Low Thermal Conductivity of the Layered Oxides $(NaCa)Co_2O_4$: Another Example of a Phonon Glass and an Electron Crystal* Phys. Rev. B **61**, 12551–12555 (2000).
85. F. Rivadulla and J. B. Goodenough, unpublished.

Chapter 1.6

THERMAL CONDUCTIVITY OF SUPERLATTICES

G. D. Mahan

Department of Physics and Materials Research Laboratory
Pennsylvania State University, University Park, PA, USA

1. INTRODUCTION

The word *superlattice* describes a solid composed of alternate materials. Usually the materials are different semiconductors and are grown by molecular-beam epitaxy (MBE) or by chemical vapor deposition (CVD). Usually each material is a single-crystalline semiconductor, and usually the superlattice is composed of alternate layers of two different materials. Usually the system is periodic, so material *A* has *n* layers, material *B* has *m* layers, and this pattern repeats. It is an advantage to have the lattice constants for each layer be the same within a fraction of a percent. Otherwise severe strains result near the interfaces. A typical superlattice has alternate layers of GaAs and AlAs and is represented as GaAs/AlAs. AlAs has a larger energy gap than GaAs, so conduction electrons are confined to GaAs. These superlattices have been widely studied for their electrical properties. The investigation of their thermal properties is just beginning.[1-8] Recently a number of superlattices have been grown to investigate their properties as thermoelectric devices.[9-12] In some cases they have achieved a high figure of merit.[11] In thermoelectric applications it is desirable to have the thermal conductivity be as small as possible. For electronic applications one wants the thermal conductivity to be large in order to remove the Joule heat.

We have repeatedly used the word "usually." Superlattices do not have to be composed of only two materials, and the alternate layers do not need to be a periodic system. Indeed, for reducing the thermal conductivity along the growth axis ("cross-plane" direction), it might be useful to intentionally introduce some disorder by making the layer thickness have a degree of randomness. Such randomness would induce Anderson localization of the phonons.[13] Also, the superlattices do not have to be composed of semiconductors. Some measurements have been done on metal–metal and metal–semiconductor superlattices.

A superlattice is anisotropic, with a different thermal conductivity along the layers and in the cross-plane direction. It is useful to divide the topic into these two categories. The experiments and the theories for these two directions are very different. The experimental methods have been reviewed.[14–15]

2. PARALLEL TO LAYERS

The initial measurements and theories were for transport parallel to the layers. The simple viewpoint is that heat transport should be quite efficient parallel to the layers. If the interfaces are atomically smooth, then phonons would specularly reflect from these interfaces. Each layer becomes a phonon waveguide that efficiently channels heat along each layer. Recent experiments on superlattices with atomically smooth surfaces confirms this expectation. The thermal conductivity parallel to the layers is similar to the bulk value.

The initial measurements of thermal conductivity parallel to the layers found values smaller than bulk values by a factor of 4. Yao[2] found in GaAs/AlAs superlattices that the thermal conductivity was similar to that of a random alloy of the two materials. This was explained in Ref. 16 as due to the interfaces boundaries not being perfectly planar. Atomic scale defects at the interface are efficient scattering centers for the phonons.

3. PERPENDICULAR TO LAYERS

The thermal conductivity of a superlattice along the growth axis is a complicated topic. The experiments are hard because the samples are quite small. Different groups, using different techniques, do not get the same results on the same sample. Generally, it is found that the thermal conductivity in the cross-plane direction is quite small. It is usually smaller than the thermal conductivity of the bulk materials in the layers, and it is smaller than a random alloy of the same materials.

The theory has similar problems. Several theoretical issues are unresolved. The earliest theories found that the thermal conductivity of the superlattice was similar to the bulk material, and the interfaces had little effect. This conclusion is now known to be wrong, and the earliest theories are now ignored.[17] Before describing the different theoretical approaches, we discuss some major issues.

3.1. Thermal Boundary Resistance

Thermal boundary resistance is also called Kapitza resistance. Kapitza found that when heat flowed between copper and superfluid helium, a temperature step developed at the interface that was proportional to the heat flow:

$$\sigma_B \Delta T = J_Q. \qquad (1)$$

Heat flow J_Q has units of W/m^2, so the boundary conductance σ_B has units of W/m^2-K. The boundary resistance $R_B = 1/\sigma_B$ has units of m^{2}K/W. The same phenomena is now known to exist at every interface, even at a twin boundary of the same material. The boundary resistance has now been measured at a variety of interfaces.[18,19] For twin boundaries it is quite low.[20] It can be much larger for an interface between two different materials.

Bartkowiak and Mahan noted that there are a whole range of interface transport parameters.[21,22] Start with the equation for transport of electricity (J) and heat (J_Q) in a bulk thermoelectric,

$$J = -\sigma\left(\frac{dV}{dx} + S\frac{dT}{dx}\right), \tag{2}$$

$$J_Q = -\sigma T S \frac{dV}{dx} - K'\frac{dT}{dx}, \tag{3}$$

$$J_Q = JST - K\frac{dT}{dx}, \tag{4}$$

$$K = K' - \sigma T S^2, \tag{5}$$

where K' is the thermal conductivity at zero electric field, which is different than the thermal conductivity K at zero current. By analogy, similar equations for the boundary impedances are

$$J = -\sigma_B(\Delta V + S_B \Delta T), \tag{6}$$

$$J_Q = -\sigma_B T S_B \Delta V - K'_B \Delta T, \tag{7}$$

$$J_Q = J S_B T - K_B \Delta T, \tag{8}$$

$$K_B = K'_B - \sigma_B T S_B^2. \tag{9}$$

Here we have introduced the boundary impedances: S_B (V/K) and K_B (W/m^2-K). The flow of current has impedances if there is a temperature difference ΔT across the boundary, or if there is a potential difference ΔV, or both. The heat current behaves similarly. An Onsager relation proves that the same Seebeck coefficient S_B enters both terms.

These equations are not the most general result. Stoner and Maris[18] measured the Kapitza resistance between diamond and other crystals. Between diamond and lead the heat flow was much higher than could be explained by a purely phonon conduction. Huberman and Overhauser[23] suggested that phonons in diamond carried heat to the interface, and electrons in lead carried it away. A theory by Sergeev[24] assumed that there was an abrupt change in temperature at the boundary between diamond and Pb. His theory agrees with experiment. If the superlattice carries heat by electrons and phonons, then the interface is a place of conversion between these two kinds of heat flow. That would be an additional term in the preceding equations.

The existence of a temperature step at a boundary is controversial. Many think that temperature differences cannot be localized that rapidly. A temperature step was found in the first molecular dynamic simulation of heat flow through a twin boundary in silicon,[20] as shown in Fig.1. The theoretical system was a slab that was periodic in the xy plane and about 200 atoms thick in the z-direction. A low-index

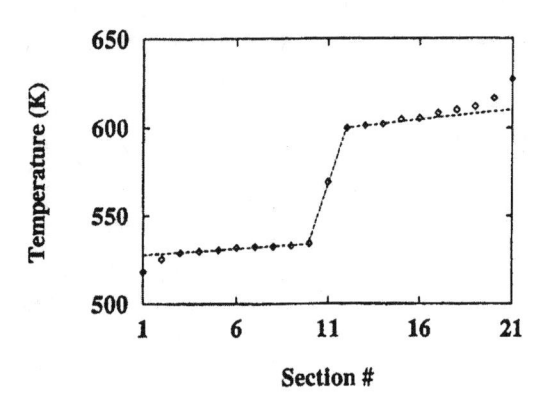

FIGURE 1 Molecular dynamics simulation of heat flow through a twin boundary in silicon. Vertical axis is temperature and horizontal axis is position in a slab. The temperature step is at the twin boundary. [From Maiti, Mahan and Pantelides, Sol. State Comm. **102**, 517 (1997).]

twin boundary is in the middle of the slab. Heat was inserted on one side of the slab and removed from the other. This was the first calculation of Kapitza resistance that used purely numerical methods. The molecular dynamics simulation was entirely classical, so a classical definition of temperature was used. The simulation calculated the position and velocity of each atom through millions of time steps. It is quite easy to define a classical temperature for each site:

$$k_B T(\vec{R}_i) = \frac{m}{3} \langle v_i^2 \rangle. \tag{10}$$

The bracket $\langle \cdots \rangle$ denotes time averaging. There was a temperature step at the twin boundary. Since large heat flows were used in the simulation, the temperature step was as large as 50°C. The calculation was done for several twin boundaries in silicon. A consistent value was found of $\sigma_B = 0.7$ GW-m^2/K. Recent calculations by Schelling et al.[25] show similar results.

Heat flow in the cross-plane direction travels through a periodic array of interfaces. The thermal boundary resistance at each interface makes a large contribution to the thermal resistance of the device. This effect is severe for superlattices with short periods. One theoretical problem is that the boundary resistance is unknown for most pairs of materials.

3.2. Multilayer Interference

Most students learn about a multilayer interferometer in optics.[26] A stack of glass plates that alternate between two different materials, each with a different refractive index, will filter out most of the light. It is an interference effect. The same phenomena of interference is found in superlattices. A typical superlattice also has alternate layers of two different materials. In this case the phonons in each layer have a diffferent sound velocity. That achieves the same kind of multilayer interference. However, the filtering is effective even for phonons of shorter wavelength.

This feature was first discovered by Narayanamurti et al.[1] They measured the transmission of high frequency phonons through a superlattice and showed that only phonons of selected frequencies could pass through. Phonons of a few frequencies carry very little heat, so the interference significantly reduces the thermal

Sec. 3 • PERPENDICULAR TO LAYERS

conductivity. Narayanamurti et al.[1] showed the interference only for phonons going normal to the layers. There is a greater filtering effect for phonons going at oblique angles.[27] It is important to evaluate all of the phonons in doing the calculation of heat transport in the superlattice.

This interference is the primary reason that superlattices have a low value of thermal conductivity. The calculation of the thermal conductivity, including interference, was first done by Hyldgaard and Mahan[27] for a Si/Ge superlattice, by Maris's group,[28] by Simkin and Mahan[29] and by Bies et al.[30] for a variety of superlattices. They found that the thermal conductivity in the cross-plane direction is typically a factor of 10 smaller than it is in either of the layer materials. This wave interference is not included in calculations using the Boltzmann equation, which is why earlier calculations failed to notice the effect. One actually has to calculate all of the phonon modes of the superlattice. There are quite a few modes when the layers are thick. Simkin and Mahan noted that the phonon mean free path λ_q was an important length. When $q\lambda_q \gg 1$, interference is important. When $q\lambda_q < 1$, interference subsides, and Kapitza resistance at the interfaces was the primary phenomena.

3.3. What Is Temperature?

The usual definition of temperature is related to the average energy of a system of particles. This definition is for a system in equilibrium and works even for nanoscale systems. However, our interest is in the transport of heat through nanoscale systems. Can temperature, which is an equilibrium concept, still be invoked in a nonequilibrium process such as heat flow? The answer is affirmative for macroscopic systems. They are so large that one can define a local temperature in each region in space. This local temperature will vary from region to region. Then one finds, for example, that the heat current is proportional to $-\vec{\nabla}T$. The question what is temperature? is really about the size of the regions over which a local temperature can be defined. In many semiconductor superlattices, the layer thickness is 2–5 nm. Are these regions large enough to define a temperature?

There have been several molecular dynamic simulations of heat flow through a twin boundary in silicon.[20,25,31] These simulations raise new problems which needed to be addressed. The most important problem is to define the temperature scale at different planes in the slab. The MD calculates the position $\vec{R}_i(t_n)$ and velocity $\mathbf{v}_i(t_n)$ at each time step in the simulation. It is relatively simple to store the velocities, and compute an averate kinetic energy over N steps in time by using Eq.(10). The time averaging must be done over very long times.

Is the classical formula an adequate temperature scale? Another possible definition of temperature is provided by quantum mechanics. The collective excitations of the atomic motions are phonons. The phonons have a frequency $\omega_\lambda(\vec{q})$ that depends upon the wave vector $\vec{q}$ and polarization λ of the phonon. In quantum mechanics the average kinetic energy of the phonons is

$$\left\langle \frac{1}{2}mv_i^2 \right\rangle = \frac{1}{4N} \sum_{\lambda,\vec{q}} \hbar\omega_\lambda(\vec{q}) \xi^{(i)2} \left[\frac{2}{e^{\hbar\omega_\lambda(q)/k_B T_i} - 1} + 1 \right], \qquad (11)$$

where N is the number of unit cells, which is also the number of $\vec{q}$ points. The polarization vectors are $\xi^{(i)2}$. They are normalized so that $\sum_\lambda \xi^{(i)2} = 3$. In the limit of high temperature, where $k_B T \gg \hbar\omega_\lambda(\vec{q})$, the Bose–Einstein occupation factors are approximated as

$$\frac{2}{e^{\hbar\omega_\lambda(q)/k_B T_i} - 1} + 1 \approx 2\frac{k_B T_i}{\hbar\omega_\lambda(\vec{q})}. \tag{12}$$

In this high-temperature limit the quantum relation (Eq. 11) becomes identical to the classical one. (Eq. 10) The problem with many simulations is that they are not in the high-temperature limit. Silicon has optical phonons of very high energy (62 meV $\approx$ 750 K). For any temperature between 300 and 1000 K, the two definitions of temperature [Eqs. (10)–(11)] gave very different values for T_i.

What size of region is required to define a temperature? The classical definition is entirely local, and one can define a temperature for each atom or plane of atoms. For the quantum definition the length scale is defined by the mean free path $\ell_{\lambda\vec{p}}$ of the phonon. If two regions of space have a different temperature, then they have a different distribution of phonons. The phonons can change their distribution by scattering. The most important scattering is the anharmonic process, where one phonon divides into two, or two combine to one. This process occurs on the length scale of the mean free path. A local region with a designated temperature must be larger than the phonon scattering distance. However, phonons of different frequency have very different values of $\ell_{\lambda\vec{p}}$. Low-frequency phonons have a long $\ell_{\lambda\vec{p}}$, and high-frequency phonons have a short $\ell_{\lambda\vec{p}}$. For phonons which carry most of the heat, one can plausibly define an aveage mean free path. It is typically larger than the numerical slab used in the MC calculation. This case is called the Casimir limit, where the phonon mpf is larger than the size of the system.

The phonon definition of temperature requires that temperature not be defined for a particular atom or a plane of atoms. There should be not abrupt variation in temperature between a plane of atoms. Although this definition seems quite reasonable, it makes the numerical results in Fig. 1 quite puzzling. The numerical simulation by different groups do show an abrupt change in the kinetic energy of a plane of atoms at the twin boundary. Regardless of which temperature scale is adopted[10], or[11], a graph of temperature versus distance will show an abrupt change. A possible resolution for this puzzle is that a grain or twin boundary may form a natural boundary for a region of temperature. The statement that temperature cannot be defined within a scale of distance given by $\ell_{\lambda\vec{p}}$ may not apply across grain boundaries. Even if one adopts this hypothesis, it still means that temperature cannot vary within a grain, or within a superlattice layer, on a scale smaller than $\ell_{\lambda\vec{p}}$. If the layer thickness of the superlattice is less than $\ell_{\lambda\vec{p}}$, then the whole layer is probably at the same temperature. This point is emphasized, since all theories of heat transport in superlattices have assumed that one could define a local temperature $T(z)$ within each layer. This theoretical treatment is wrong for superlattices with short periods.

Although the discussion of temperature has been cast in the framework of MD simulations, the issues are more general. One of the major issues of thermal transport in nanoscale systems is whether temperature can be defined locally. If it cannot, then how is transport calculated?

Another approach is to adopt a Landauer formalism, assume there are two thermal reservoirs at known temperatures, and consider the ballistic flow of heat between them. This does not work for nanoscale devices since the values of $\ell_{\lambda\vec{p}}$ are much smaller than the distance between the reservoirs. Hence, heat flow is diffusive

Sec. 3 • PERPENDICULAR TO LAYERS

rather than ballistic. How do you calculate diffusive heat flow without a local temperature?

The theory of thermal conductivity is simple if the system satisfies two conditions: (1) the layers are thick enough that one can ignore interference, and (2) the layers are thin enough that each layer is at the same temperature. In that case assume the system has n interfaces. Consider the heat flow at any interface between two materials 1 and 2. The heat flow from 1 to 2 is $\sigma_{12}T_1$. The heat flow from 2 to 1 is $\sigma_{21}T_2$. The principle of detailed balance requires that $\sigma_{12} = \sigma_{21}$. If two layers were at the same temperature and $\sigma_{12} \neq \sigma_{21}$, then the layers would not exchange the same amount of heat. One layer would heat up, the other would cool down, which violates the second law of thermodynamics. So $\sigma_{12} = \sigma_{21} \equiv \sigma_B$. Then the net heat flow in that one interface is $\sigma_B(T_1 - T_2)$. One can extend this argument to n-interfaces and find that the net heat flow is

$$J_Q = \frac{\sigma_B}{n}(T_H - T_C). \tag{13}$$

The same reasoning can be applied at any temperature. At low temperatures the heat flow at one interface goes as $f(T_1) - f(T_2) = \sigma'(T_1^4 - T_2^4)$. The T^4 law for phonons is the same as for photons. In that case the same reasoning gives

$$J_Q = \frac{\sigma'}{n}(T_H^4 - T_C^4). \tag{14}$$

The general result, valid at any temperature in a superlatttice of alternate materials, is

$$J_Q = \frac{1}{n}[f(T_H) - f(T_C)], \tag{15}$$

where $f(T)$ is the general function that goes as T^4 at low temperatures and as T at high temperatures. For thicker superlattices one is tempted to modify this formula with the thermal resistance of the bulk of the layers, resulting in

$$J_Q = \frac{T_H - T_C}{n/\sigma_B + L_1/K_1 + L_2/K_2}, \tag{16}$$

where K_i and L_i are the thermal conductivity and total thickness of the two materials.

3.4. Superlattices with Thick Layers

Superlattices with thick layers are an important subtopic. In these cases one can ignore wave interference and treat the boundary effects, using Boltzmann equations for electrons and phonons. Kapitza resistance is the primary phenomenon. Nevertheless, the topic has interesting features. In thermoelectric devices it is important to have a flow of electric current and heat. Both electrons and phonons go through the interface, and both have boundary resistances.[22] They differ. In this case the foregoing treatment for the phonon part of the thermal conductivity of a superlattice is wrong, since the electrons and phonons invariably exchange energy. This exchange is not due to phonon drag, but is an interface phenomenon.

Most of the calculations have been done with a formulism pioneered in Russia.[32–34] They assume each layer has different temperature for electrons $T_e(z)$ and phonons $T_p(z)$. They become different since electrons and phonons have dif-

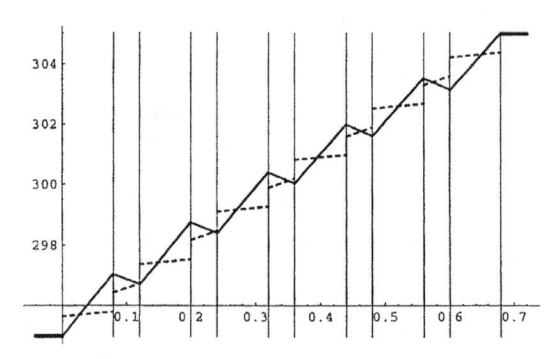

FIGURE 2 Temperatures of electrons (solid curve) and phonons (dashed curve) for heat flow in the cross-plane direction in a superlattice composed of Bi_2Te_3/Sb_2Te_3. Vertical scale is temperature, horizontal axis is position in microns. Calculations by Bartkowiak and Mahan.[21,22]

ferent boundary resistances and different values of ΔT at each interface. There is a healing length over which the two interacting systems (electrons and phonons) return to local equilibrium. The healing lengths are many nanometers. In a superlattice with layer thicknesses of a few nanometers, the electron and phonon temperatures are totally different in each layer. These calculations predict some wierd behavior. A typical result is shown in Fig. 2 from Bartkowiak and Mahan.[21,22]

Actual results depend on the boundary resistances, for which we used theoretical estimates. Whether these predicted effects are real depends on whether our estimates are correct and whether the temperature can be defined on a local length scale.

The coupled equations for the heat exchange between electrons and phonons[22,33] are

$$-K_e \frac{d^2 T_e}{dx^2} = \rho J^2 - P\delta T, \tag{17}$$

$$-K_p \frac{d^2 T_p}{dx^2} = P\delta T, \tag{18}$$

$$\delta T = T_e - T_p, \tag{19}$$

where $K_{e,p}$ are the bulk thermal conductivity of electrons and phonons. Joule heating (ρJ^2) generates electronic heat in the layer. The terms $P\delta T$ exchange heat between the electron and phonon systems. The constant P is well known for metals,[35] and we have calculated it for semiconductors. The length scale (λ) over which the electron and phonon systems come to the same temperature is given by

$$\lambda^2 = \frac{K_e K_p}{P(K_e + K_p)}. \tag{20}$$

Since P depends on the density of conduction electrons, each material will have a different length. The scale of λ is many nanometers. Short-period superlattices may never have the electron and phonon temperatures in agreement at any point.

4. "NON-KAPITZIC" HEAT FLOW

A measurement of electrical resistance always uses four probes on the sample. Soldering a wire onto the sample causes "non-Ohmic" contacts, so the voltage change must be measured with separate wire leads. Until recently, similar measurements on thermal conductance used only two contacts. The interface did not cause nonlinear behavior analogous to non-Ohmic electrical contacts. Recently there has been an appreciation of the Kapitza resistance at the interface. However, it is included by assuming a linear behavior. The Kaptiza resistance is another thermal resistance in series with that of the sample.

Here we wish to discuss the possibility that there may be nonlinear behavior in the heat flow at interfaces, which cannot be described simply by a boundary resistance. This nonlinear heat flow is analogous to the non-Ohmic electrical contacts. We have been searching for a name for this process. I propose we name surface resistance after Kapitza (Kz=m^2-K/W) and call this behavior "non-Kapitzic."

It is likely that non-Kapitzic heat flow has been present in all measurements of heat flow. However, it has been on a length scale too small to be observed. Nonlinear effects are on the nanoscale. They should play a role in the measurement and theory of heat transport in nanoscale devices.

There have been numerous MD calculations of heat transport in one-dimensional lattices.[36–43] A chain of atoms connected only by harmonic springs will have heat conduction only by ballistic pulses. Some mechanism of scattering the phonons is required for heat diffusion. Usually that is included by a nonlinearity in the springs or by the addition of impurities or isotope scattering. The usual procedure has been to introduce some anharmonic component into the spring constant, such as a bit of quartic potential energy.

Nearly all of these calculations fail to find ordinary diffusion of heat. The heat flow fails to have the right scaling with the number of atoms in the chain ($J_Q \propto 1/N$),[29,40–42] and the temperature along the chain is not linear. An example from my own group,[29] is shown in Fig. 3.

This phenomenon is explained in Ref.(43) as due to non-Kapitzic effects. The results are all based on how one puts in the heat at one end of the chain and takes it out on the other side. The usual method, due to Hoover,[44] used by us[20] and

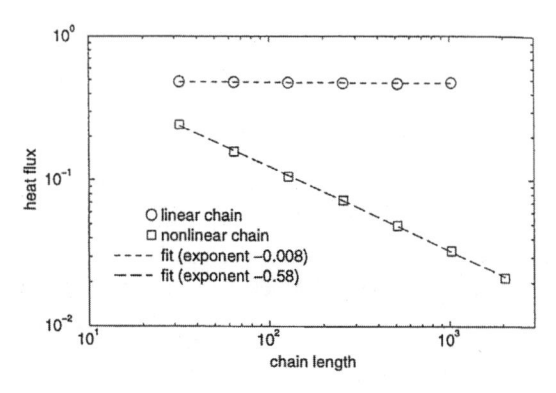

FIGURE 3 One-dimensional heat flow as a function of chain length N for harmonic and anharmonic lattices. The same temperature is at the ends. The nonlinear chain does not scale as $1/N$. Calculations by Simkin and Mahan (unpublished).

nearly everyone, causes nonequilibrium effects to persist far into the interior of the chain. The earlier studies did not have chains long enough to get the interior of the chain away from the surface effects.

The message from these very different numerical studies is that surface effects persist far into the chain. Even very anharmonic lattices take a long distance to relax into the diffusive regime, which has a temperature which varies linearly with distance. In the surface region the distribution of phonons cannot be described by a single parameter such as temperature. Since many simulations, including ours, show similar behavior, it seems to be a universal phenomenon. Do experimental systems show the same non-Kapitzic behavior? No one knows. Answering this question is a future task for experimentalists.

4.1. Analytic Theory

Most theories of transport in solids employ the Boltzmann equation (BE). For both electron and phonon transport, the form of the equation, and the form of the various scattering mechanisms are very well known. This theory can explain, in bulk homogeneous materials, the dependence of the electrical conductivity, the thermal conductivity, and the Seebeck coefficient on temperature, impurity content, isotope scattering, and quantum confinement.

There have been several calculations of thermal conductivity of the superlattice which have solved the Boltzmann equation. That of Ren and Dow[16] did not predict a large reduction of thermal conductivity and disagreed with experiments. Several calculations from Chen's group[45,46] agree much better with experiment.

However, the Boltzmann equation treats electrons and phonons as classical particles. One is solving, say for phonons, for the density $f(\mathbf{r}, \omega_\lambda(\mathbf{q}), t)$ of excitations with polarization λ, wave vector $\mathbf{q}$, and frequency $\omega_\lambda(\vec{q})$ at point $\mathbf{r}$ at time t. The wave nature of the excitation is neglected, as is any interference phenomena caused by the wave nature of the phonons. Furthermore, the scattering rates in the BE are calculated under the assumption that the system is only slightly perturbed from equilibrium. The solutions to the BE assume the existence of a local temperature. Neither of these two assumptions may be valid in nanoscale devices.

Wave interference becomes important in nanoscale devices. The wavelengths of the phonons are similar to the length scale of the microstructure. At room temperature, in most solids, all of the phonon states in the Brillouin zone are involved in the transport. Their wavelengths span the range from atomic dimensions to the size of the sample.

One method is to calculate the phonon states for the actual microstructure, including interference, and use those states in the BE. For transport in a superlattice, this approach puts in some of the wave phenomena. The phonon states depend significantly on the interference due to scattering from the multiple-layer boundaries of the superlattice. Solving this problem by using the BE is also complicated by phonon band folding. There are many superlattice bands, phonons scatter between them, and the BE becomes a matrix equation of large dimension. This calculation is quite ambitious and has never been done. Even doing it does not include all wave interference phenomena.

Another possibility is to use the quantum Boltzmann equation[47] to solve for the distribution $f(\vec{R}, t; \omega, \mathbf{q})$. Compared to the BE there is one more vector variable in the argument that makes the solutions more complex. This more fundamental equation does include wave information. However, it is difficult to solve and is seldom used.

There have been many calculations of thermal conductivity in the cross-plane direction. Most are based on the standard formula for thermal conductivity

$$K = -\hbar \sum_\lambda \int \frac{d^3q}{(2\pi)^3} v_{z\lambda}(\mathbf{q})^2 \omega_\lambda(\vec{q}) \tau_\lambda(\vec{q}) \left(\frac{dn_B(\omega_\lambda(\vec{q}))}{dT} \right), \qquad (21)$$

where $v_{z\lambda}(\mathbf{q})$ is the velocity of the phonon, $\tau_\lambda(\mathbf{q})$ is the lifetime, and $n_B(\omega_\lambda(\vec{q}))$ is the Boson occupation function. This formula is derived from the BE. Calculations by Kato et al.,[48] Hyldgaard and Mahan,[27] Tamura et al.,[28] Simkin and Mahan,[29] Bies et al.,[30] and Kiselev et al.[49] just include Eq.(21) the actual phonon modes of the superlattice and evaluate. Usually, the lifetime $\tau_\lambda(\mathbf{q})$ or the mean free path $\ell_{\lambda\vec{p}}$ is selected to be the same as in the homogeneous material. These calculations show that thermal conductivity in the superlattice is reduced significantly, sometimes by a factor of 10, compared to that of the constituents of the superlattice. This large reduction agrees with experimental findings. Simkin and Mahan[29] used a complex wave vector to compute the superlattice modes. The imaginary part of the wave vector was a phenomenological way of including the mean free path in the calculation of the phonon modes. This calculation showed a minimum in the thermal conductivity as a function of superlattice period.

Most alternate theories are purely numerical. Molecular dynamics (MD) is far more suitable for a phonon system than for an electron system. The phonon modes can be calculated quite accurately with classical methods. Quantum effects are important at low temperatures but relatively unimportant in most solids for phonon effects at room temperature or above. Wave effects are included automatically. Using an anharmonic potential between neighboring atoms includes effects such as one phonon dividing in two, or two phonons combining into one. One limitation on MD is that it is limited to insulators. In order to model an actual nanoscale device, thousands of atoms have to be included in the simulation. This calculation is beginning to be practical with modern parallel computers.

5. SUMMARY

The calculation of the thermal conductivity of a superlattice is rather easy in some cases. If the layers are thick and all of the heat is carried by phonons, then one can use Kapitza resistances at interfaces and Boltzmann equations for the interior of the layers. These calculations are rather easy and predict that the main influence of the superlattice is the thermal boundary resistance at the interfaces.

There is much interest in superlattices where the layers have a thickness of a few nanometers. In this case calculating thermal conductivity runs into difficulties. One has to include (i) wave interference and (ii) nonlocal temperature scales. Further complications arise if a significant amount of heat is carried by electrons. The electrons and phonons exchange heat at the interface and in the interior of the layers. They have different boundary resistances. The electrons and phonon systems are not in equilibrium with each other throughout the superlattice. MD cannot accurately describe the transport by electrons, nor can it describe the exchange of energy between the phonon and electron systems.

If one throws out the concept of a local temperature, so the BE cannot be used, then there is no known way to solve for the heat transport and thermal conductivity. Many semiconductor devices depend on the electric currents provided by the electrons. The heat currents have components from both electrons and phonons. Presently there is no accurate way to model the heat flows in these nanoscale systems while including the exchange of heat between electrons and phonons. A major theoretical challenge is to invent a new method of solving this and related transport problems.

6. REFERENCES

1. V. NARAYANAMURTI, H. L. STÖRMER, M. A. CHIN, A. C. GOSSARD, AND W. WIEGMANN *Selective Transmission of High-Frequency Phonons by a Superlattice* Phys. Rev. Lett. **43**, 2012–2016 (1979).
2. T. YAO *Thermal Properties of AlAs/GaAs Superlattices* Appl. Phys. Lett. **51**, 1798 (1987).
3. X. Y. YU, G. CHEN, A. VERMA, AND J. S. SMITH *Temperature Dependence of Thermophysical Properties of GaAs/AlAs Periodic Structure* Appl. Phys. Lett. **67**, 3554–6 (1995).
4. W. S. CAPINSKI AND H. J. MARIS *Thermal Conductivity of GaAs/AlAs Superlattices* Physic B **219 220**, 699–701 (1996).
5. S. M. LEE, D. G. CAHILL, AND R. VENKATASUBRAMANIAN *Thermal Conductivity of Si-Ge Superlattices* Appl. Phys. Lett. **70**, 2957–2959 (1997).
6. W. S. CAPINSHKI, H. J. MARIS, T. RUF, M. CARDONA, K. PLOOG, AND D. S. KATZER *Thermal-Conductivity Measurements of GaAs/AlAs Superlattices Using Picosecond Optical Pump-and-Probe Technique* Phys. Rev. B **59**, 8105–8113 (1999).
7. S. T. HUXABLE, A. R. ABRAMSON, C. L. TIEN, A. MAJUMDAR, C. LABOUNTY, X. FAN, G. H. ZENG, J. E. BOWERS, A. SHAKOURI, AND E. T. CROKE *Thermal Conductivity of Si/SiGe and SiGe/SiGe Superlattices* Appl. Phys. Lett. **80**, 1737–1739 (2002).
8. B. YANG, J. L. LIU, K. L. WANG AND G. CHEN *Simultaneous Measurements of Seebeck Coefficient and Thermal Conductivity across Superlattices* Appl. Phys. Lett. **80**, 1758–1760 (2002).
9. G. D. MAHAN AND H. B. LYON *Thermoelectric Devices Using Semiconductor Quantum Wells* J. Appl. Phys. **76**, 1899–1901 (1994).
10. J. O. SOFO AND G. D. MAHAN *Thermoelectric Figure of Merit of Superlattices* Appl. Phys. Lett. **63**, 2690 (1994).
11. R. VENKATASUBRAMANIAN, E. SIVOLA, T. COLPITTS, AND B. O'QUINN *Thin-Film Thermoelectric Devices with High Room-Temperature Figures of Merit* Nature **413**, 597 (2001).
12. G. CHEN AND A. SHAKOURI *Heat Transfer in Nanostructures for Solid-State Energy Conversion* J. Heat Transfer **124** (2), 242–252 (2002).
13. M. J. MCKENNA, R. L. STANLEY, AND J. D. MAYNARD *Effects of Nonlinearity on Anderson Localization* Phys. Rev. Lett. **69**, 1807–1810 (1992).
14. D. G. CAHILL, K. GOODSON, AND A. MAJUMDAR *Thermometry and Thermal Transport in Micro/Nanoscale Solid State Devices and Structures* J. Heat Transfer **124** (2), 223–241 (2002).
15. D. G. CAHILL, W. K. FORD, K. E. GOODSON, G. D. MAHAN, A. MAJUMDAR, H. J. MARIS, R. MERLIN, AND S. R. PHILLPOT *Nanoscale Thermal Transport* J. Appl. Phys. **93**, 793–818 (2003).
16. P. HYLDGAARD AND G. D. MAHAN *Phonon Knudsen Flow in GaAs/AlAs Superlattices*, Thermal Conductivity **23**, 171–182 (1996).
17. S. Y. REN AND J. D. DOW *Thermal Conductivity of Superlattices* Phys. Rev. B **25**, 3750–3755 (1982).
18. R. J. STONER AND H. J. MARIS *Kapitza Conductance and Heat Flow between Solids at Temperatures from 50 to 300 K* Phys. Rev. B **48**, 16373–16387 (1993).
19. D. G. CAHILL, A. BULLEN AND S. M. LEE *Interface Thermal Conductance and the Thermal Conductivity of Multilayer Thin Films* High Temp.–High Pressure **32**, 135–142 (2000).
20. A. MAITI, G. D. MAHAN, AND S. T. PANTELIDES *Dynamical Simulations of Nonequilibrium Processes–Heat Flow and the Kapitza Resistance Across Grain Boundaries* Sol. State. Comm. **102**, 517–521 (1997).
21. G. D. MAHAN AND M. BARTKOWIAK *Wiedemann–Franz Law at Boundaries* Appl. Phys. Lett., **74**, 953–954 (1999).
22. M. BARTKOWIAK AND G. D. MAHAN *Heat and Electricity Transport Through Interfaces*, Semiconductors and Semimetals Vol. **71** (Academic Press, 2001), pp. 245–273.

Sec. 6 · REFERENCES

23. M. L. HUBERMAN AND A. W. OVERHAUSER *Electronic Kapitza Conductance at Diamond-Pb Interface* Phys. Rev. B **50**, 2865 (1994).
24. A. V. SERGEEV *Electronic Kapitza Conductance due to Inelastic Electron-Boundary Scattering* Phys. Rev. B **58**, R10199 (1998).
25. P. K. SCHELLING, S. R. PHILLPOT, AND P. KEBLINSKI *Comparison of Atomic-level Simulation Methods for Computing Thermal Conductivity* Phys. Rev. B **65**(14), 4306 (2002).
26. P. HARIHARAN *Optical Interferometry* (Academic, Sydney, 1985).
27. P. HYLDGAARD AND G. D. MAHAN *Phonon Superlattice Transport* Phys. Rev. B **56**, 10754–10757 (1997).
28. S. I. TAMURA, Y. TANAKA, AND H. J. MARIS *Phonon Group Velocity and Thermal Conductivity in Superlattices* Phys. Rev. B **60**, 2627–2630 (1999).
29. M. V. SIMKIN AND G. D. MAHAN *Minimum Thermal Conductivity of Superlattices* Phys. Rev. Lett. **84**, 927–930 (2000).
30. W. E. BIES, H. EHRENREICH, AND E. RUNGE, *Thermal Conductivity in HgTe/CdTe Superlattices* J. Appl. Phys. **91**, 2033–2036 (2002).
31. A. R. ABRAMSON, C. L. TIEN, AND A. MAJUMDAR *Interface and Strain Effects on the Thermal Conductivity of Heterostructures* J. Heat Transfer **124** (5), 963–970 (2002).
32. L. P. BULAT AND V. G. YATSYUK *Theory of Heat Conduction in Solids* Sov. Phys. Solid State **24**, 1994–1995 (1982).
33. L. P. BULAT AND V. G. YATSYUK *Thermal Effects at Boundaries of Solids* Sov. Phys. Semicond. **18**, 383–384 (1984).
34. L. I. ANATYCHUK, L. P. BULAT, D. D. NIKIRSA, AND V. G. YATSYUK *Influence of Size Effects on the Properties of Cooling Thermoelements* Sov. Phys. Semicond. **21**, 206–207 (1986).
35. P. B. ALLEN *Theory of Thermal Relaxation of Electrons in Metals* Phys. Rev. Lett. **59**, 1460–1463 (1987).
36. R. BOURBONNAIS AND R. MAYNARD *Energy Transport in One- and Two-Dimensional Anharmonic Lattices with Isotropic Disorder* Phys. Rev. Lett. **64**, 1397–1400 (1990).
37. S. WANG *Localized Vibrational Modes in an Anharmonic Chain* Phys. Lett. A **182**, 105–108 (1993).
38. S. WANG *Influence of Cubic Anharmonicity on High Frequency Modes* Phys. Lett. A **200**, 103–108 (1995).
39. S. SEN, R. S. SINKOVITS, AND S. CHAKRAVARTI *Algebraic Relaxation Laws for Classical Particles in 1D Anharmonic Potentials* Phys. Rev. Lett. **17**, 4855–4859 (1996).
40. S. LEPRI, R. LIVI, AND A. POLITI *Heat Conduction in Chains of Nonlinear Ascillators* Phys. Rev. Lett. **78**, 1896–1899 (1997).
41. S. LEPRI, R. LIVI, AND A. POLITI *On the Anomalous Thermal Conductivity of One-Dimensional Lattices* Europhy. Lett. **43**, 271–276 (1998).
42. C. GIARDINA, R. LIVI, A. POLITI, AND M. VASSALLI *Finite Thermal Conductivity in 1D Lattices* Phys. Rev. Lett. **84**, 2144–2147 (2000).
43. O. GENDELMAN AND A.V. SAVIN *Normal Heat Conductivity of the One-Dimensional Lattice with Periodic Potential of Nearest Neighbor Interaction* Phys. Rev. Lett. **84**, 2381 (2000).
44. W. G. HOOVER *Canonical Dynamics: Equilibrium Phase-Space Distributions* Phys. Rev. A **31**, 1693–1697 (1985).
45. G. CHEN *Size and Interface Effects on Thermal Conductivity of Superlattices and Periodic Thin Film Structures* J. Heat Transfer, **119**, 220–229 (1997).
46. G. CHEN AND M. NEAGU *Thermal Conductivity and Heat Transfer in Superlattices* Appl. Phys. Lett. **71**, 2761–2763 (1997).
47. G. D. MAHAN *Many-Particle Physics* 3rd ed. (Kluwer-Plenum, 2000).
48. H. KATO, S. TAMURA, AND H. J. MARIS *Resonant-Mode Conversion and Transmission of Phonons in Superlattices* Phys. Rev. B **53**, 7884–7889 (1996).
49. A. A. KISELEV, K. W. KIM, AND M. A. STROSCIO *Thermal Conductivity of Si/Ge Superlattices: A Realistic Model with a Diatomic Unit Cell* Phys. Rev. B **62**, 6896–6899 (2000).

Chapter 1.7

EXPERIMENTAL STUDIES ON THERMAL CONDUCTIVITY OF THIN FILMS AND SUPERLATTICES

Bao Yang

Department of Mechanical Engineering
University of Maryland
College Park, MD, USA

Gang Chen

Department of Mechanical Engineering
Massachusetts Institute of Technology
Cambridge, MA, USA

1. INTRODUCTION

The thermal conductivity of thin films is a very important parameter for a wide range of applications, such as microelectronic devices, photonic devices, thermoelectric devices, and optical and thermal barrier coatings.[1-4] Experimental studies on the thermal conductivity of thin films can be traced back to the mid-1960s when size dependence of the thermal conductivity in metal films was first reported.[5,6] Since the 1980s, the thermal properties of thin films have drawn increasing attention, due to the demands of thermal management in the rapidly growing microelectronics and optoelectronics industries[1-4,7-9] and to the resurgence of thermoelectrics in the early 1990s.[4,10-12] In microelectronics, integrated circuits (ICs) employ various insulating, semiconducting, and metallic thin films. Because the power density and speed of ICs keep increasing, thermal management becomes more challenging, and the thermal conductivity of the constituent thin films becomes more important for the device design. Semiconductor lasers, often made of heterostructures to control electron and photon transport, encounter more severe thermal management problems because they are more sensitive to temperature. The maximum optical output power of these semiconductor lasers depends significantly on the operating temperature, and the lasing wavelength shifts with temperature fluctuation. The thin films used in microelectronic and photonic devices need to have high thermal conductivity in order to dissipate heat as efficiently as possible. On the other hand, thermoelectric devices call for materials or structures with low thermal conductivity because the performance of thermoelectric devices is determined by the figure of merit $Z=S^2\sigma/k$, where k is thermal conductivity, σ is electrical conductivity, and S is the Seebeck coefficient.[13] Nanostructured materials, such as superlattices, can have drastically reduced thermal conductivity in comparison to the corresponding bulk values, and thus have become promising candidates in the search for high-efficiency thermoelectric materials.[10-12,14] Other applications calling for thin films with low thermal conductivity are high-temperature coatings for engines.

The increasing interest in thermal conductivity of thin films and superlattices is coincident with advances in microfabrication technology and measurement techniques. Measurements of the thermal conductivity of thin films and superlattices have proven very challenging, and the conventional methods for bulk materials may not apply to these thin films. New measurement techniques for thin-film thermophysical properties have been reported. Among these methods, the 3ω method,[15,16] developed by Cahill *et al.* in late 1980s, may be the most commonly used technique for measuring thermal conductivity in the direction perpendicular to thin films. This method was later extended to measure the thermal conductivity in the perpendicular and parallel directions simultaneously.[17] Other microfabrication-based methods have also been developed. The optical pump–probe method[18] and optical calorimetry method[19] are also widely used. Chapter 2.2 reviews experimental techniques for the measurement of thermal conductivity in thin films.

The size effects on thermal conductivity become extremely important when the film thickness shrinks to be comparable to the mean free path or wavelength of the heat carriers (i.e., phonons in semiconducting and dielectric materials and electrons in metals).[20-25] Scattering at boundaries and interfaces imposes additional resistance to thermal transport and reduces the thermal conductivity of thin films. When the film thickness is comparable to the wavelength of heat carriers, the quantum size effects step in. As a consequence, the fundamental properties, such as velocity and density of states of heat carriers, will be modified, which may contribute to the reduction in thermal conductivity in thin films. Furthermore, microstructure and stoichiometry in thin films and superlattices strongly depend on the film growth process, and any variations in structure and stoichiometry may significantly influence the thermal conductivity.[1,2] The complexity of structures, stoichiometry, and scattering on interfaces, boundaries, and imperfection in thin films yield many degrees of freedom that challenge the understanding and modeling of thermophysical properties of thin films. The existing models for thermal transport in thin films and superlattices are still not totally satisfactory. Chapter 1.5 reviews the modeling of thermal transport in thin films and superlattices.

In this review we will summarize some experimental results on the thermal conductivity of thin films and superlattices. We start with a brief review of thermal conductivity in metallic thin films. The thermal conductivity of dielectric, semiconducting, and semimetal thin films will be discussed next. Emphasis will be on those

2. THERMAL CONDUCTIVITY OF METALLIC THIN FILMS

Metallic thin films are widely used in microelectronics and micromachined sensors and actuators.[26–30] Thermal transport in metal thin films has been studied since the 1960s because the kinetics of growth and electromigration in metal thin films is related to their thermal properties. In the 1960s the thickness-dependent thermal conductivity, along the film-plane direction (the in-plane direction), was observed at low temperatures in Al foils by Amundsen and Olsen,[5] and in Ag films by Abrosimov et al.[6] In the 1970s, Chopra and co-workers did a series of experiments on the temperature dependence and film thickness dependence of the in-plane thermal conductivity in Cu thin films.[26,27,31] In their work a steady-state method was employed for the measurements at high temperatures, whereas a transient method was used for low-temperature measurements. Later experiments by Wachter and Volklein considered radiation loss in thin metal films, so thermal conductivity and emissivity could be determined.[32] In this section Cu thin films are used as an example for thermal transport in metallic films. Copper thin films are important in modern CMOS technology, replacing Al thin films to reduce the RC delay in interconnect networks.

In metals the heat conduction is dominated by electrons, but the contribution from phonons is very small. Electron scattering at boundaries of thin films may impose additional resistance on the electron transport, and thus size effects on thermal conductivity can be observed. Figure 1 shows the dependence of the in-plane thermal conductivity on film thickness for two different temperatures, 100 K and 325 K, along with the thermal conductivity of bulk Cu.[26,27,31] The thermal conductivity of thin films is reduced in comparison to their bulk values, and ther-

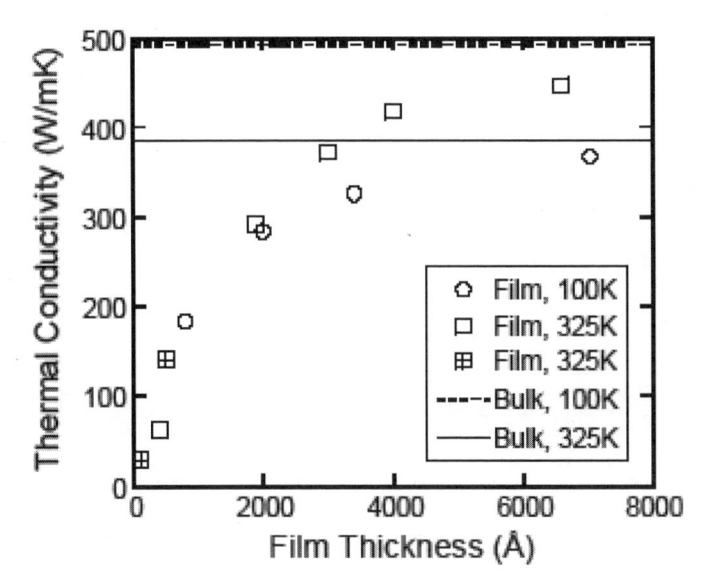

FIGURE 1 Dependence of the in-plane thermal conductivity on film thickness for Cu thin films deposited on a mica slice [data from refs 26, 27, and 31].

mal conductivity is a strong function of film thickness below a certain thickness value, above which it almost remains constant. The thickness-dependent region is smaller for 325 K than that for 100 K, and enhanced size effects are observed at lower temperatures as expected. Fuchs theory,[33] which considers the surface scattering of electrons by solving the Boltzmann transport equation, is often employed to explain the size dependence of thermal conductivity in metal thin films. Fuchs theory can reasonably explain the data for thicker films, but the discrepancy between the prediction and the data for thinner films is relatively large, because Fuchs theory considers only surface scattering and neglects grain boundary scattering. Very thin films typically have smaller grains and thus more grain boundary scattering. Models that consider grain boundary scattering lead to much better agreement with experimental results.[34]

The thermal conductivity of metal thin films also shows a different temperature dependence than the bulk form. As seen in Fig. 2, the trend for thicker Cu films is similar to bulk Cu; i.e., the thermal conductivity decreases with increasing temperature because, at high temperatures, electron-phonon scattering becomes stronger. As film thickness decreases, however, the temperature dependence of thermal conductivity also decreases, and ultimately a reversal of the temperature dependence emerges for films thinner than 400 Å. This reversal can be attributed to boundary or grain boundary scattering, which is more important as the film thickness goes down and ultimately outplays the electron-phonon scattering in very thin films. Because boundary scattering and Fermi velocity are relatively independent of temperature, thermal conductivity roughly follows the temperature dependence of the electron specific heat.

Interestingly in the data by Chopra and Nath,[26] the electrical resistivity, contrary to thermal conductivity, always increases with increasing temperature, no matter how thin the film is. The authors suggested that the Lorenz number could not remain constant anymore and should be a function of temperature and film thickness, because the electron scattering at the boundaries has different effects on the energy transport for thermal conductivity and momentum transport for electrical

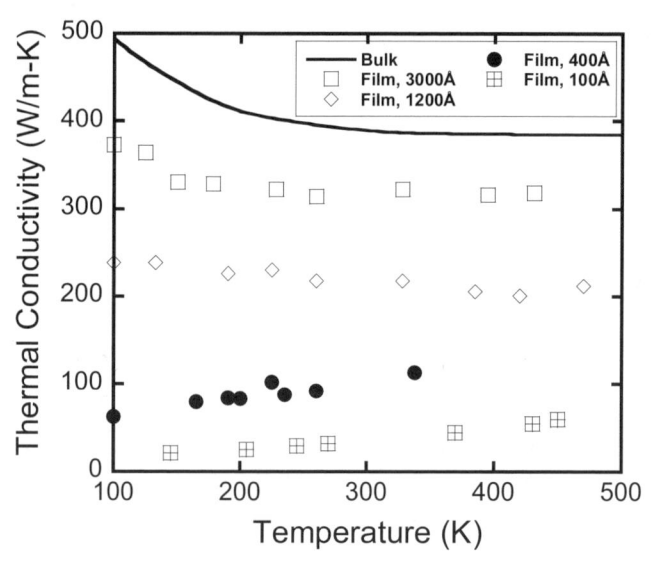

FIGURE 2 Temperature dependence of the in-plane thermal conductivity of Cu thin films with different film thickness [data from Refs. 26, 27, and 31].

conductivity. However, some other experiments did not report significant change in the Lorentz number when the thickness of metal films varied.[35] Hence, it remains inconclusive as to whether the validity of the Lorentz number indeed fails in thin metallic films.

3. THERMAL CONDUCTIVITY OF DIELECTRIC FILMS

Dielectric thin films have wide applications in microelectronics, semiconductor lasers, and optical devices, and may serve as electrical insulators, optical coatings, and thermal barrier coatings. Knowledge of the thermal conductivity of these thin films is essential for the performance and reliability of these structures or devices. These thin films are typically deposited under highly nonequilibrium conditions, so their thermal properties may be very different from those in bulk form, due to the different structure, stoichiometry, and boundary scattering. This section discusses experimental data for the most common dielectric material, amorphous silicon dioxide (a-SiO$_2$), as well as diamond films, thermal barriers, and optical coatings. Other reviews should also be consulted for wider coverage.[1,2,4,20]

3.1. Amorphous SiO$_2$ Thin Films

The thermal conductivity of a-SiO$_2$ thin films has been investigated by many researchers through various experimental methods.[36–41] Figure 3 summarizes some reports on the thermal conductivity of a-SiO$_2$ thin films perpendicular to the film plane (the cross-plane direction). As seen in this figure, the thermal conductivity of a-SiO$_2$ thin films is reduced in comparison to bulk in all cases. The large variation in thermal conductivity is most likely due to the strong dependence of microstructure on the processes used in preparing the samples. For example, the thermal

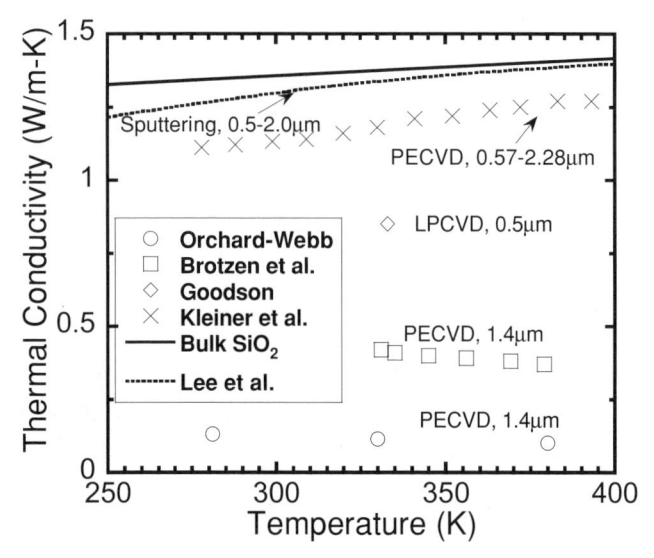

FIGURE 3 Experimental results of the cross-plane thermal conductivity as a function of temperature in amorphous SiO$_2$ thin films (data from Refs. 36-41). The experimental data are labeled by the preparation methods and film thickness in μm.

conductivity of low-pressure chemical vapor deposition (LPCVD) SiO$_2$ films by Goodson et al. is about four times larger than thermal conductivity of plasma-enhanced chemical vapor deposition (PECVD) SiO$_2$ films by Brotzen et al. for films of comparable thickness.[40,41] For a-SiO$_2$ films, the structure variation is often reflected by the mass density variations. Figure 4 lists the representative data on SiO$_2$ films at room temperature for which mass density data are available. In general, higher mass density leads to higher thermal conductivity for a-SiO$_2$ films. For example, SiO$_2$ films grown through thermal oxidation have higher thermal conductivity than PECVD SiO$_2$ films. Another cause for reduced thermal conductivity in a-SiO$_2$ is the interfacial layer, where structural imperfections, such as growth defects, microvoids, lattice strain, and even surface contamination, tend to concentrate.[37–41] Most of the experimental data showing that the thermal conductivity decreases with decreasing film thickness in a-SiO$_2$ films can be explained by the existence of a thermal boundary resistance between the film and the substrate.[37–41]

One more thing to be pointed out is that the trends of thermal conductivity with temperature in Fig. 3 are different. Some data show that the thermal conductivity of a-SiO$_2$ thin films increases with increasing temperature, similar to the behavior of bulk a-SiO$_2$. Yet there are also data suggesting that thermal conductivity of SiO$_2$ decreases with increasing temperature, similar to the behavior of crystalline SiO$_2$ (quartz). There are no satisfactory explanations for this contradiction.

Recently, a new form of SiO$_2$, carbon–doped silicon dioxide (CDO), has attracted growing attention because its low permittivity may increase CMOS performance limited by the RC delay.[42] The permittivity in CDO can be reduced from 4 to 2, and thus CDO is forecast to partially replace SiO$_2$ as the gate material in microelectronics. In CDO the introduction of carbon can reduce mass density and permittivity as well as thermal conductivity. The thermal conductivity of CDO is shown in Fig. 4. This reduced thermal conductivity can impact the thermal management of ICs.

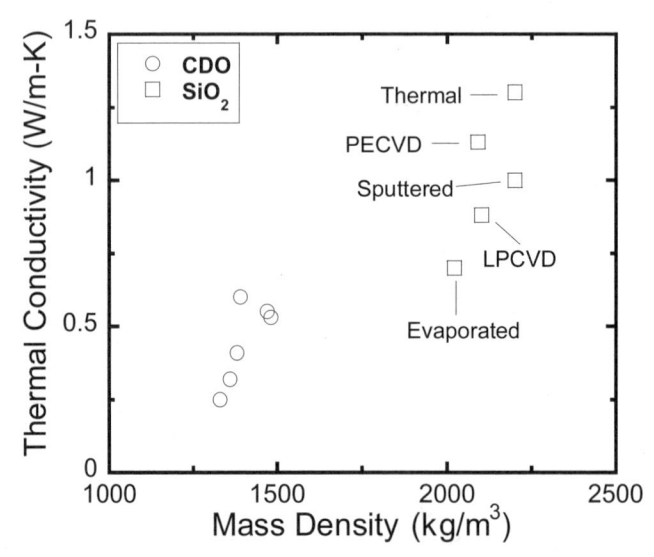

FIGURE 4 Mass density dependence of the cross-plane thermal conductivity of silicon dioxide and carbon-doped silicon dioxide (CDO) films at room temperature.[1,42] The thickness of thermal, PECVD, sputtered, LPCVD and evaporated SiO$_2$ films is 0.99 μm, 0.11 μm, 1.5 μm, 0.5 μm, and 2.18 μm, respectively. The thickness of CDO films is 0.25 μm to 2.0 μm.

Sec. 3 • THERMAL CONDUCTIVITY OF DIELECTRIC FILMS

TABLE I Thermal Conductivity (W/m-K) of Optical Coatings

Film/Substrate	k_{Film}	k_{Bulk}	Direction of the Property	Film Thickness (μm)	Ref.
TiO$_2$/silicon	0.59	7.4-10.4	Cross-plane	0.060-1.246	43
Al$_2$O$_3$/silicon	0.72	20-46	Cross-plane	0.173-0.462	43
MgF$_2$	0.58	14.6-30	Cross-plane	0.209-0.583	43
AlF$_3$/sapphire	0.31		Cross-plane	0.194-0.544	43
ZrO$_2$/sapphire	0.04	1.1-2.2	Cross-plane	0.151-0.465	43
ThO$_2$/sapphire	0.67	16-26	Cross-plane	0.174-0.396	43
CeF$_3$/sapphire	0.08		Cross-plane	0.128-0.357	43
ThF$_4$/sapphire	0.10		Cross-plane	0.162-0.506	43
Si$_3$N$_4$/silicon	0.15	10	Cross-plane	0.5-2.0	46
Si$_{0.7}$Al$_{0.3}$N/silicon	0.82		Cross-plane	0.5-2.0	46
Si$_{0.6}$Al$_{0.4}$NO/silicon	0.83		Cross-plane	0.5-2.0	46
BN/silicon	0.32	62(a axis)	Cross-plane	0.5-2.0	46
SiC/silicon	0.12	25	Cross-plane	0.5-2.0	46
Ta$_2$O$_5$/silicon	0.12		Cross-plane	0.5-2.0	46
TiO$_2$/silica	0.018	7.4-10.4	*[a]	**[b]	44
Ta$_2$O$_3$/silica	0.026		*	**	44
Al$_2$O$_3$/silica	33	20-46	*	**	44
HfO$_2$/silica	0.00077		*	**	44
Al$_2$O$_3$/silicon	1.1-1.7	20-46	Cross-plane	0.5-2.0	36
TiO$_2$/silicon	2.1-6.1	7.4-10.4	Cross-plane	0.5-2.0	36

[a] Assume the thermal conductivity in the cross-plane same as the in-plane.
[b] 4-12 quaterwaves of optical thickness at 1.064μm.

3.2. Thin Film Coatings

A wide variety of dielectric thin films has attracted much attention for applications as optical coatings in various optical components, in which high optical power density may be encountered. The thermal conductivity of optical coatings is an important parameter in optical element design and damage estimation. Table 1 summarizes the thermal conductivity of commonly used optical coatings. Lambropoulos et al. measured oxide and fluoride films and observed a large difference in thermal conductivity between thin films and bulk materials.[43] Ristau and Ebert measured electron–beam-deposited films of Al$_2$O$_3$, TiO$_2$, HfO$_2$, and Ta$_2$O$_5$ on fused-silica substrates and found that only Al$_2$O$_3$ had a value close to the bulk. The other films had from one to several orders of magnitude lower thermal conductivity than the bulk values.[44] Ogden et al. observed that thick (as thick as 85 μm) films of Al$_2$O$_3$ had an average thermal conductivity of 0.73 W/m-K compared with 30 W/m-K for bulk polycrystalline Al$_2$O$_3$.[45] Henager and Pawlewicz measured sputtered oxide and nitride films and showed that thin-film thermal conductivity is typically 10 to 100 times lower than the bulk values.[46] The reduction in thermal conductivity of these films is normally attributed to the varied structure and boundary scattering.

Thermal barrier coatings (TBCs) offer the potential to significantly increase the performance and lifetime of heat engines.[47–50] In contrast to optical coatings, TBCs should have low thermal conductivity, because a high-temperature drop across the

TBC is desired. Today, the standard TBC is partially stabilized zirconia (PSZ) because of its low thermal conductivity and superior bonding to the alloy or superalloy substrate.[49,50] The thermal conductivity of PSZ is 0.4–1.2 W/m-K, depending on the doping oxides, such as Y_2O_3, CeO_2, and MgO_2.[47,48] The PSZ usually exhibits a weak temperature dependence of thermal conductivity, especially at high temperature (>1000°C). One way to reduce the thermal conductivity of PSZ is to replace some zirconium ions with heavier dopants, such as hafnium. It is reported that the thermal conductivity of Hf-doped yttria-stablized zirconia can be about 85% of its yttria-stablized zirconia counterpart.[47]

3.3. Diamond Films

Passive CVD diamond layers have the potential to improve thermal management in optoelectronics and electronic microstructures because of their high thermal conductivity.[1,51,52] The thermal conductivity of polycrystalline diamond films strongly depends on grain size, grain orientation, lattice imperfection, impurities, etc., which are governed by the details of the deposition process. The columnar-grained structure favors heat conduction normal to the diamond films. Verhoeven et al.[52] observed a large degree of anisotropy in the thermal conductivity of polycrystalline diamond films, with the cross-plane thermal conductivity about one order of magnitude higher than its in-plane counterpart. Figure 5c compares the thermal conductivity of a highly-oriented diamond film to that of a random grain film, showing that the former has a much higher thermal conductivity than the latter.[53] Heat conduction in polycrystalline diamond films can be modeled through the introduction of a mean free path caused by grain boundary scattering.[1]

4. THERMAL CONDUCTIVITY OF SEMICONDUCTOR AND SEMIMETAL THIN FILMS

In semiconductors the electron contribution to heat conduction is usually very small, depending on the doping concentration. In semimetals, however, both electrons and phonons may contribute to heat conduction. The most important representative for semiconductor materials is silicon. This section will first discuss the

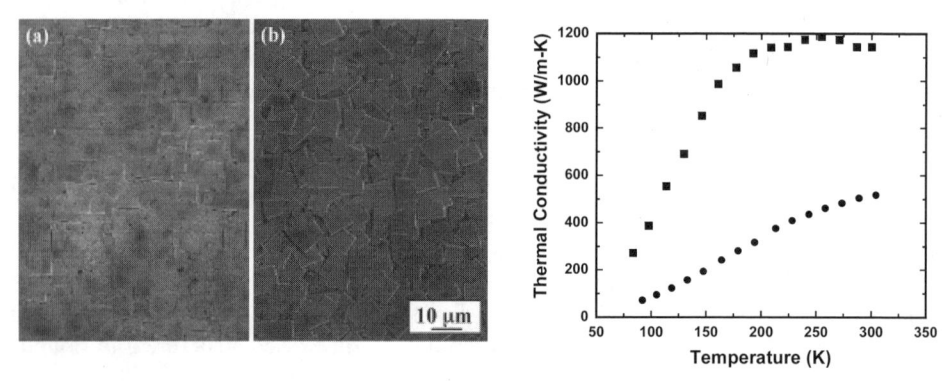

FIGURE 5 SEM pictures of diamond films (a) with highly oriented structure and (b) randomly oriented structure. (c) Thermal conductivity along the diamond film with highly oriented structure (square) and randomly oriented structure (circle). The thickness of both samples is around 70 μm. Data from Ref. 53 and the unpublished paper by D. Borca-Tasciuce et al.

thermal conductivity of silicon thin films in the forms of single-crystalline, polycrystalline, and amorphous states. The second part of this section will be dedicated to several semimetal thin films with potential application as thermoelectric materials.

4.1. Silicon Thin Films

The thermal conductivity of single-crystalline silicon films is an important design parameter for silicon-on-insulator (SOI) ICs and single-crystalline silicon-membrane-based sensors and actuators. Goodson and co-workers have published a series of papers on the in-plane thermal conductivity of single-crystalline silicon films.[54,55] Figure 6 summarizes the in-plane thermal conductivity of silicon thin films in the single-crystal, polycrystalline and amorphous forms. As seen in this figure, the reduction in thermal conductivity of single-crystalline silicon thin films compared to bulk is very large at low temperatures but only moderate at high temperatures. The reduction in thermal conductivity is mainly attributed to interface or diffuse surface scattering.[54–56] Since the interface or surface scattering is not sensitive to temperature, whereas Umklapp scattering grows stronger with temperature, the relative impact of surface scattering is much larger at low temperatures than that at high temperatures. The effects of boundary scattering on thermal conductivity of single-crystalline silicon can be traced to the study by Savvides and Goldsmid,[57] who observed size effects in films of about of 100 μm after subjecting these films to proton irradiation. The onset of size effects in such thick samples was attributed to the fact that short-wavelength phonons are strongly scattered by the implanted ions, while long-wavelength phonons, having a long mean free path, are subject to boundary scattering. At low temperatures impurity scattering, such as ion scattering, will play an important role in thermal transport in thin films. Figure 6 shows that the thermal conductivity of single-crystalline silicon thin film doped to 10^{19} cm^{-3} is reduced by a factor of ~2 at 20 K compared to the undoped sample.

Polysilicon films can be deposited at high quality and are common in MEMS and microelectronics. Muller and co-authors measured heavily doped LPCVD polysilicon films, using microfabricated bridges.[58] They found that the in-plane thermal conductivity ranged from 29 to 34 W/m-K. Volklein and Batles measured the lattice and electronic components of the in-plane thermal conductivity in polysilicon films heavily doped with phosphorus (~5×10^{20} cm^{-3}), using a differential method.[59] They observed that the total thermal conductivity is around 29 W/m-K at temperatures above 200 K and the electronic component is less than 3%. More recently, Goodson's group studied the in-plane thermal conductivity of polysilicon thin films with and without doping, shown in Fig. 6.[60,61] As seen in this figure, the thermal conductivity of polysilicon thin films is strongly reduced at all temperatures, compared to single-crystalline silicon. For a pure polysilicon film 200 nm thick, the thermal conductivity is about two orders of magnitude smaller than that of bulk silicon at temperatures below 100 K, and about 10 times lower than those of undoped single-crystalline silicon thin films. This big reduction in thermal conductivity is due mainly to phonon scattering at the grain boundaries, where the imperfection defects populate, and at interfaces and surfaces.[60,61] Unlike bulk silicon and single-crystalline silicon films, the thermal conductivity of polysilicon films increases with increasing temperature in the plotted temperature range, similar to the trend of specific heat. The undoped polysilicon thin film 200 nm thick has lower

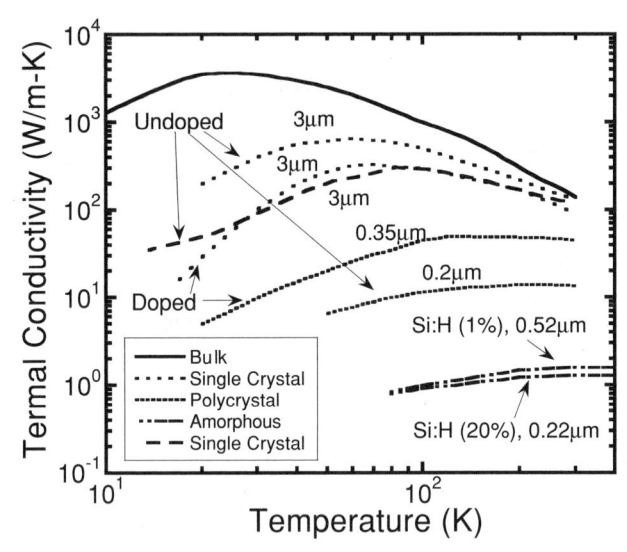

FIGURE 6 Temperature dependence of thermal conductivity of single-crystal bulk silicon, single-crystal silicon thin films,[55,56] polycrystalline silicon thin films,[60,61] and amorphous silicon thin films.[16] These curves are label by film thickness. The thermal conductivity is measured in the in-plane direction for single-crystal and polycrystalline silicon films, and in the cross-plane direction for amorphous silicon thin films. The hydrogen content in the amorphous silicon films is 1% and 20%, respectively.

thermal conductivity than the doped one 350 nm thick, as seen in this figure. This indicates that the impurities, such as ions, are not the primary mechanism for phonon scattering in polysilicon films. However, it does not necessarily mean that boundary scattering is dominant because the microstructure also strongly depends on film thickness.

Amorphous silicon has thermal properties quite different from those of single-crystal or polycrystalline silicon due to its different structure. The introduction of hydrogen to amorphous silicon can greatly modify its electrical properties, and thus amorphous hydrogenated silicon has been widely used in solar cells and thin-film transistors even though it has poor thermal conductivity.[62] Goldsmid et al. measured a 1.15-μm-thick amorphous silicon at room temperature and obtained a rather large thermal conductivity, 2.9 W/m-K.[63] Pompe and Hegenbarth studied the temperature dependence of the thermal conductivity in a 26-μm-thick sputtered amorphous silicon film in the temperature range of 2–50 K and found that the thermal conductivity plateau for the amorphous silicon film occurs at a higher temperature (30 K) than those of other amorphous dielectric films (10–15 K), such as Ge and As.[64] Attaf et al. studied the effects of hydrogen content on the thermal conductivity of amorphous silicon films and found a systematic trend of decreasing thermal conductivity for increasing hydrogen content. Besides the disordered lattice, the mass difference between Si and H seems to further localize the vibration wave.[65] However, Cahill et al. reported a much weaker dependence of the thermal conductivity on the hydrogen content, and part of their data is shown in Fig. 6.[16] In disordered materials, such as amorphous silicon, heat transport by lattice vibration can be separated into two regimes. The lattice vibrations with low energy are wave-like, and thus phonons exist with well-defined wave vector and wave velocity. On the other hand, Orbach[66] proposed that a dominant fraction of high-energy lattice vibrations is localized and unable to contribute to heat transport unless anharmonic forces are present. More recent studies suggested that although high-energy lattice vibrations do not have well-defined wave vector and wave velocity, the dominant mechanism for heat conduction in amorphous silicon thin films should be the coupling between nearly degenerated, extended, but non-propagating vibrations.[16;67]

4.2. Semimetal Thin Films

Semimetal thin films, such as Bi- and Sb-based thin films, have potential application as thermoelectrics, which requires low thermal conductivity. Abrosimov et al.[6,68] and Volklein and Kesseler[69] measured the in-plane thermal conductivity of polycrystalline Bi thin films deposited through thermal evaporation, as well as other thermoelectric properties. Although bulk single-crystal Bi has a fairly small thermal conductivity, $\sim$5–10 W/m-K depending on the crystallographic direction,[70] the measured thermal conductivity shows size dependence for films as thick as $\sim$1000 nm. Abrosimov et al.[68] observed a peak in the thermoelectric figure of merit for Bi films with a thickness $\sim$100 nm and a systematic shift of the peak toward large thickness as temperature decreases. Such a peak behavior, however, was not seen in Volklein and Kesseler's data.[69] Volklein and his co-workers also studied the in-plane thermal conductivity and other thermoelectric thin films, including Sb thin films, Bi_xSb_{1-x} thin films,[71] and $(Bi_{1-x}Sb_x)_2Te_3$ thin films.[72] For both Bi and Sb thin films, after subtracting the electronic contribution to the thermal conductivity according to the Wiedemann–Franz law, the phonon contribution to thermal conductivity increases with increasing temperature, similar to the behavior of many polycrystalline silicon thin films and diamond thin films. This implies that phonon–phonon scattering is overplayed by the temperature-independent boundary scattering in the Bi and Sb thin films, even in the high-temperature range.

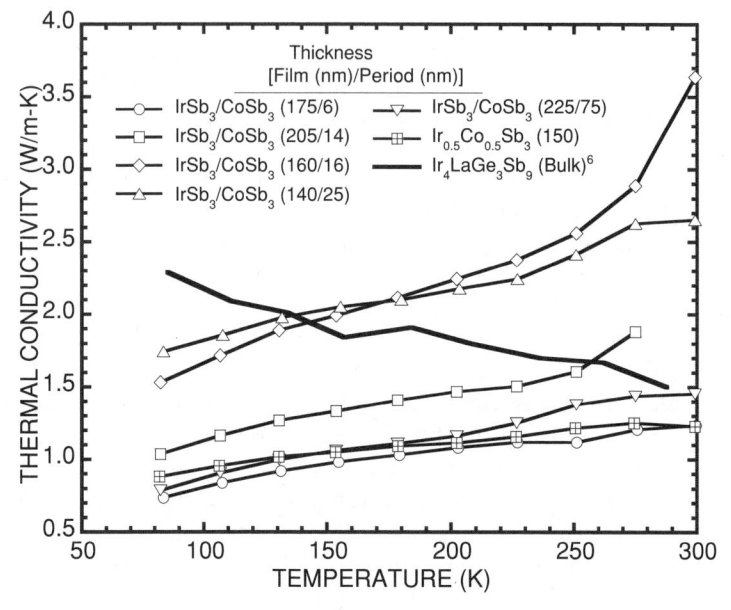

FIGURE 7 Cross-plane thermal conductivity of skutterudite thin films and superlattices.[75] The numbers within the parentheses are the film thickness and the period thickness in nm.

In the cross-plane direction, Baier and Volklein[73] measured the thermal conductivity of $Bi_{0.5}Sb_{1.2}Te_3$ films between 50 and 1000 nm, but no thickness dependence was reported. The thermal conductivity at room temperature is ~0.37 W/m-K, lower than bulk alloys.[74] Song et al.[75] measured the thermal conductivity of polycrystalline $CoSb_3$ and $IrSb_3$ thin films and their alloy films. Figure 7 shows the temperature dependence of the thermal conductivity of these films and a comparison with their bulk counterparts. A significant reduction in thermal conductivity is observed in comparison with bulk counterparts, even for alloy films. More interestingly, the thermal conductivity of the alloy film is comparable to filled skutterudites. The latter has been used to reduce the thermal conductivity of unfilled skutterudites.[76]

5. SEMICONDUCTOR SUPERLATTICES

Superlattice films consist of periodically alternating layers of two different materials stacked upon each other.[77] Modern growth techniques, such as molecular-beam epitaxy (MBE) and metal-organic chemical vapor deposition (MOCVD), have enabled the fabrication of semiconductor superlattices with submonoatomic layer (ML) precision. Thermal transport in semiconductor superlattices has attracted considerable attention due to applications in thermoelectric devices[11,12,78-82] and optoelectronic devices such as quantum well lasers and detectors.[3,4] To increase the efficiency of the thermoelectric devices, these films need to have low thermal conductivity. On the contrary, thin films with high thermal conductivity are desired to dissipate heat in semiconductor lasers and other optoelectronic devices.[10] In this section we emphasize the thermal conductivity of semiconductor superlattices, in which heat is mainly carried by phonons. Metallic superlattices are also of interest for short-wavelength applications such as x-ray and deep ultraviolet lithography, and in magnetic data storage,[83,84] although these studies will not be reviewed in detail here.

The first experiment on the thermal conductivity of semiconductor superlattices was reported by Yao.[85] He found that the in-plane thermal conductivity of GaAs/AlAs superlattices was smaller than the corresponding bulk values (obtained according to the Fourier law with the use of the properties of their bulk constituents). The anisotropy of thermal diffusivity in superlattices was experimentally observed by Chen et al., who found that the cross-plane thermal conductivity was four times smaller than that in the in-plane direction for short-period GaAs/AlAs superlattices used in semiconductor lasers.[19] In recent years, extensive experimental data on the thermal conductivity of various superlattices emerged, including Bi_2Te_3/Sb_2Te_3,[86-88] GaAs/AlAs,[18,89,90] Si/Ge,[91-94] InAs/AlSb,[95] InP/InGaAs,[96] $CoSb_3/IrSb_3$,[75] and PbTe-based superlattices.[97,98] All these experiments confirmed that the thermal conductivities of the superlattices in both directions are significantly lower than the corresponding equivalent values calculated from the Fourier law using the bulk thermal conductivity of their constituent materials. In the cross-plane direction the thermal conductivity values can definitely be reduced below that of their corresponding alloys. In the in-plane direction the reduction is generally above or comparable to that of their equivalent alloys, although a few experimental data indicate that thermal conductivity values lower than these of their corresponding alloys are possible.[97]

The period-thickness dependence of thermal conductivity may be different for

Sec. 5 • SEMICONDUCTOR SUPERLATTICES

different superlattice groups and different growth techniques. For a GaAs/AlAs superlattice the difference in the lattice constants between the adjacent layers is very small. As a consequence, the critical thickness for the formation of misfit dislocations is very large, and the interfaces can have excellent quality, typically consisting of only 1–3 MLs mixing region with some long-range lateral terraces.[99,100] In this case the interface quality can remain almost identical over the considered range of period thickness. Hence, the relative effects of interface scattering or interface resistance on the superlattice thermal conductivity will decrease with increasing period thickness. The experimental data for GaAs/AlAs in the in-plane direction by Yao[85] and in the cross-plane direction by Capinski et al.[18] seem to support this idea. Many other superlattice groups, such as Bi_2Te_3/Sb_2Te_3 and InAs/AlSb, also exhibit behavior similar to GaAs/AlAs superlattices. Although Si and Ge have a relatively large difference in lattice constant, a $Si_xGe_{1-x}/Si_yGe_{1-y}$ superlattice may have a critical thickness much larger than that of superlattice Si/Ge, depending on the composition. Huxtable et al. have shown that for superlattice Si/$Si_{0.7}Ge_{0.3}$ the thermal conductivity scales almost linearly with interface density.[96] An interesting observation made experimentally in Bi_2Te_3/Sb_2Te_3 superlattices[86] (shown in Fig. 8a) and less systematically in GaAs/AlAs superlattices (shown in Fig. 8b)[18] is that in the very thin period limit the thermal conductivity can recover as the period thickness decreases, and thus a minimum exists in the thermal conductivity when plotted as a function of the period thickness. Unlike GaAs/AlAs and Bi_2Te_3/Sb_2Te_3 superlattices, Si/Ge superlattices have a small critical thickness, on the order of 10 nm. The data on Si/Ge superlattices by Lee et al.[91] and by Borca–Tasciuc et al.[92] plotted in Fig. 9, show that the thermal conductivity initially increases with period thickness and then drops as the period thickness is increased beyond 10 nm. This is presumably due to the extension of misfit dislocations for periods larger than the critical thickness.

The temperature dependence of thermal conductivity could be used to dissect the contribution of different scattering mechanisms, as shown in Fig. 10. It is well known that the thermal conductivity of bulk semiconductor materials drops very fast with increasing temperature in the relatively high temperature range due to phonon–phonon scattering.[101,102] In superlattices the presence of interface scatter-

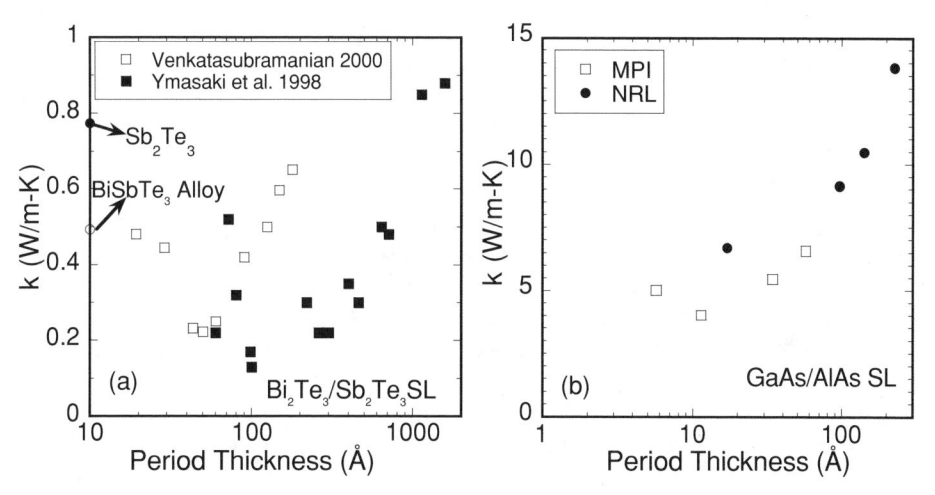

FIGURE 8 Cross-plane thermal conductivity as a function of period thickness for (a) Bi_2Te_3/Sb_2Te_3 SLs and (b) GaAs/AlAs SLs at $T=300$ K.[18,86,87]

180 Chap. 1.7 • THERMAL CONDUCTIVITY OF THIN FILMS AND SUPERLATTICES

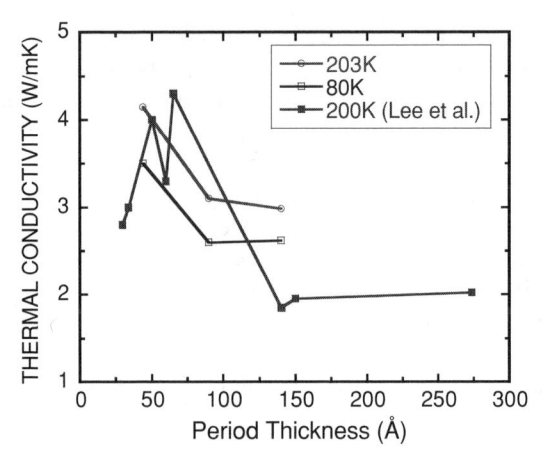

FIGURE 9 Cross-plane thermal conductivity as a function of period thickness in Si/Ge SLs.[91,92]

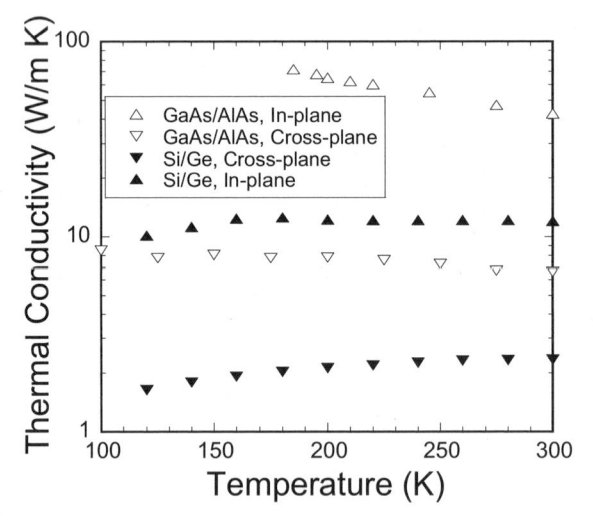

FIGURE 10 In-plane and cross-plane thermal conductivities of the Si/Ge and GaAs/AlAs SLs as a function of temperature. One period of Si/Ge SL consists of Si(80 Å)/Ge(20 Å). In one period of the GaAs/AlAs SLs, the GaAs and AlAs layers have the same thicknes, and the period thicknesses are 140 Å and 55 Å for the GaAs/AlAs SLs used in in-plane and cross-plane directions, respectively.[18,89,94]

ing, as well as dislocation scattering, will significantly modify the temperature dependence of thermal conductivity. The interface roughness and substitution alloying may account for the diffuse interface scattering, which is not sensitive to temperature. For both in-plane and cross-plane directions of a GaAs/AlAs superlattice, the thermal conductivity was found to decrease with increasing temperature at the intermediate temperature range, with a slope much smaller than that of the corresponding bulk materials, especially in the cross-plane direction.[89,90] In the cross-plane direction of Si/Ge and $Si_xGe_{1-x}/Si_yGe_{1-y}$, the opposite trend was observed, i.e. the thermal conductivity increases with increasing temperature.[91-96] In this case, the temperature-dependent scattering mechanisms, such as Umklapp scattering, do not contribute too much to heat conduction, and the interface scattering and dislocation scattering that are less sensitive to temperature dominate the transport. Clearly, the mismatch in acoustic impedance or, more generally, density, specific heat, force, and lattice constants between Si and Ge is more than that of GaAs and AlAs, and a larger interface scattering strength is expected for Si/Ge superlattices. This also contributes to the weaker interface scattering in GaAs/AlAs relative to that in Si/Ge superlattices.

The effects of the growth and annealing temperatures on the thermal conductivity of InAs/AlSb superlattices were studied by Borca-Tasciuc and co-workers and they observed that the thermal conductivity of this superlattice system decreased with increasing annealing and growth temperature.[95] On the other hand, Si/Ge and Ge-quantum-dot superlattices seem to show an opposite trend.[21,80] Although it is known that the growth and annealing temperature will affect the interfaces and defects in superlattice, the detailed mechanisms are still not clear.

Current models on phonon transport in SLs are generally divided into three groups. The first group treats phonons as totally incoherent particles.[22,23,103] The thermal conductivity is usually calculated with the Boltzmann transport equation with boundary conditions involving diffuse interface scattering. These particle models can fit experimental data of several SL systems in the thick-period range. Because the wave features of phonons in SLs are not considered, they fail to explain the thermal conductivity recovery in the short-period limit with period thickness less than ~5 (bulk) unit cells. The second group of models treats phonons as totally coherent waves, and thus phonons in different layers of a SL are coherently correlated, and SL phonon bands can be formed due to the coherent interference of the phonon waves transporting toward and away from the interfaces.[104,105–108] Under this picture thermal conductivity in SLs is usually calculated through the phonon dispersion relation in SLs. The calculated conductivity typically first decreases with increasing period thickness and then approaches a constant with period thickness beyond about 10 MLs. The predictions of the coherent phonon picture at the very thin period limit is similar to some experimental observations, but the thick-period behavior is contrary to the experimental results observed in many SLs, such as GaAs/AlAs and Bi_2Te_3/Sb_2Te_3, which show an increase in thermal conductivity with increasing period thickness.[86,87,89] The third group of traditional models involves lattice dynamics, which usually leads to a temperature-dependent thermal conductivity similar to that of bulk crystalline materials, contrary to most experimental observations.

Apparently, neither coherent wave models nor incoherent particle models alone can explain the period-thickness dependence of thermal conductivity in SLs over the full period-thickness range because each of them deals with an extreme case. Simkin and Mahan proposed a modified lattice dynamics model with a complex wave vector involving the bulk phonon mean free path, and they predicted a minimum of thermal conductivity in the cross-plane direction.[109] However, the calculated thermal conductivity reduction is still lower than the experimental data. Very recently, a partially coherent phonon transport model proposed by Yang and Chen[24] combines the effects of phonon confinement and diffuse interface scattering on the thermal conductivity in superlattices, and is applicable to phonon transport in the partially coherent regime. The period thickness dependence and temperature dependence of the thermal conductivity in the GaAs/AlAs superlattices can be well explained by this model.

6. CONCLUSIONS

In this chapter we have discussed the experimental data of thermal conductivity in metallic, dielectric, semiconductor, and semimetal thin films and semiconductor superlattices. The major points are as follows:

1. For polycrystalline and single-crystalline films, the thermal conductivity is dramatically reduced compared to their bulk counterparts. Phonon and electron scattering at boundaries, grain boundaries, or interfaces plays a crucial role in thermal conductivity reduction. The modified microstructure in thin films also makes significant contribution to the reduction in thermal conductivity.
2. For amorphous films with very short phonon mean free path, the microstructures, particularly mass density and stoichiometry that depend strongly on the processing conditions, play a more significant role in thermal conductivity. Most cross-plane thickness dependence data for films processed under similar conditions can be explained by the interfacial thermal resistance between the film and the boundary rather than size effects arising from the long phonon mean free path.
3. Thermal conductivity in both the in-plane and cross-plane directions of superlattices is significantly reduced compared to the corresponding bulk values. Some experimental data show a minimum in thermal conductivity when the superlattice period thickness is around a few MLs. A partially coherent phonon heat conduction model, combining the effects of phonon confinement and diffuse interface scattering on the thermal conductivity in superlattice, can explain the period-thickness dependence and temperature dependence of the thermal conductivity in the GaAs/AlAs superlattices.

Compared to the large body of work on the electrical properties in thin films, research on thermal transport is still relative scarce. Many challenges should be addressed in future studies, such as measurement techniques, thermal conductivity modeling, and deeper understanding of the phonon scattering mechanisms. The relationship between the microstructures, interface conditions, and thermal conductivity should be emphasized in future experimental research.

7. ACKNOWLEDGMENTS

We would like to acknowledge the support from DoD/ONR MURI on Thermoelectrics, DOE (DE-FG02-02ER45977) on heat transfer in nanostructures, and NSF (CTS-0129088) on nanoscale heat transfer modeling. We also would like to thank contributions of current and former members in G. Chen's group working on thin-film thermophysical properties, particularly Prof. T. Borca-Tasciuc, Diana Borca-Tasciuc, W.L. Liu, and D. Song, and we would like to thank C. Dames for critically reading the manuscript.

8. REFERENCES

1. K. GOODSON AND Y. JU *Heat Conduction in Novel Electronic Films* Annu. Rev. Mater. Sci. **29**, 261–293 (1999).

2. D. G. CAHILL, K. GOODSON, AND A. MAJUMDAR *Thermometry and Thermal Transport in Micro/Nanoscale Solid-State Devices and Structures* J. Heat Trans. **124**, 223–241 (2002).
3. P. BHATTACHARYA, *Semiconductor Optoelectronic Devices* (Prentice Hall, Upper Saddle River, 1997).
4. G. CHEN, *Heat Transfer in Micro- and Nanoscale Photonic Devices* in Ann. Rev. Heat Trans. **7**, 1–57 (1996).
5. T. AMUNDSEN AND T. OLSEN *Size-dependent Thermal Conductivity in Aluminum Films* Phil. Mag. **11**, 561–574 (1965).
6. V. M. ABROSIMOV, B. N. EGOROV, N.S. LIDORENKO, AND I. B. RUBASHOV *Investigation of the Thermal Conductivity of Thin Metallic Films* Soviet Phys. Solid State **11**, 530–532 (1969).
7. K. BANERJEE, A. AMERASEKERA, G. DIXIT, N. CHEUNG, AND C. HU Characterization of Contact and via failure under Short Duration High Pulsed Current Stress *Proc. Int. Reliability Physics Symposium* (1997) pp. 216–220.
8. A. KARIM, S. BJORLIN, J. PIPREK, AND J. E. BOWERS *Long-wavelength Vertical-cavity Lasers and Amplifiers* IEEEJ. Sel. Top. Quantum Electron. **6**, 1244–1253 (2000).
9. E. TOWE, R. F. LEHENY, AND A. YANG *A Historical Perspective of the Development of the Vertical-Cavity Surface-emitting Laser* IEEEJ. Sel. Top. Quantum Electron. **6**, 1458–1464 (2000).
10. L. D. HICKS, T. C. HARMAN, AND M. S. DRESSELHAUS *Experimental Study of the Effect of Quantum-well Structures on the Thermoelectric Figure of Merit* Phys. Rev. B **53**, 10493–10496 (1996).
11. L. D. HICKS AND M. S. DRESSELHAUS *Effect of Quantum-well Structures on the Thermoelectric Figure of Merit* Phys. Rev. B **47**, 16631-16634 (1993).
12. G. CHEN, M. S. DRESSELHAUS, G. DRESSELHAUS, J. P. FLEURIAL, AND T. CAILLAT *Recent developments in thermoelectric materials,* Int. Mat. Rev. 48, 45–66 (2003).
13. H. J. GOLDSMID *Thermoelectric Refrigeration* (Plenum Press, New York, 1964).
14. T. KOGA, X. SUN, S. B. CRONIN, AND M. S. DRESSELHAUS Appl. Phys. Lett. **73**, 2950–2952 (1998).
15. D. G. CAHILL, *Thermal Conductivity Measurement from 30 to 750 K: The 3w method* Rev. Sci. Instrum. **61**, 802–808 (1990).
16. D. G. CAHILL, M. KATIYAR, AND J. R. ABELSON *Thermal Conductivity of a-Si:H Thin Films* Phys. Rev. B **50**, 6077–6081 (1994).
17. W. L. LIU, T. BORCA-TASCIUC, G. CHEN, J. L. LIU, AND K. L. WANG *Anisotropic Thermal Conductivity of Ge Quantum-dot and Symmetrically Strained Si/Ge Superlattices* J. Nanosci. Nanotechnol. **1**, 37–42 (2001).
18. W. S. CAPINSKI, H. J. MARIS, T. RUF, M. CARDONA, K. PLOOG, AND D. S. KATZER *Thermal-conductivity Measurements of GaAs/AlAs Superlattices Using a Picosecond Optical Pump-and-Probe Technique* Phys. Rev. B **59**, 8105–8113 (1999).
19. G. CHEN, C.L. TIEN, X. WU, AND J.S. SMITH *Measurement of Thermal Diffusivity of GaAs/AlGaAs Thin-film Structures* J. Heat Trans. **116**, 325–331 (1994).
20. C. L. TIEN, A. MAJUMDAR, AND F. GERNER eds. *Microscale Energy Transport* (Taylor and Francis, Bristol PA 1998).
21. W. L. LIU *In-plane Thermoelectric Properties of Si/Ge Superlattices* Ph.D. Thesis, University of California at Los Angeles, Department of Mechanical and Aerospace Engineering, 2003.
22. G. CHEN *Size and Interface Effects on Thermal Conductivity of Superlattices and Periodic Thin-Film Structures* J. Heat Trans. **119**, 220–229 (1997).
23. P. HYLDGAARD AND G. D. MAHAN *Phonon Knudson Flow in Superlattices* in *Thermal Conductivity*, Vol. **23**, (Technomic, Lancaster, 1996), pp. 172–181.
24. B. YANG AND G. CHEN *Partially Coherent Phonon Heat Conduction in Superlattices* Phys. Rev. B **67**, 195311–195314 (2003).
25. A. BALANDIN AND K. L. WANG *Significant Decrease of the Lattice Thermal Conductivity due to Phonon Confinement in a Free-standing Semiconductor Quantum Well* Phys. Rev. B **58**, 1544–1549 (1998).
26. P. NATH AND K.L. CHOPRA *Thermal Conductivity of Copper Films* Thin Solid Films **20**, 53–62 (1974).
27. K. L. CHOPRA AND P. NATH *Thermal Conductivity of Ultrathin Metal Films in Multilayer Structures* J. Appl. Phys. **45**, 1923–1925 (1974).
28. E. OGAWA, K. D. LEE, V. BLASCHKE, AND P. HO *Electromigration Reliability Issues in Dual-damascene Cu Interconnections* IEEE Trans. on Reliab. **51**, 403-419 (2002).
29. H. TOYODA, T. KAWANOUE, S. ITO, M. HASUNUMA, AND H. KANEKO *Effects of Aluminum Texture on Electromigration Lifetime* Am. Inst. Phys. Conf. Proc., No.373, (1996), pp. 169-184.
30. C. LEU, H. LIN, C. HU, C. CHIEN, M. J. YANG, M. C. YANG, AND T. HUANG *Effects of Titanium and*

Tantalum Adhesion Layers on the Properties of Sol-Gel Derived $SrBi_2Ta_2O_9$ Thin Films J. Appl. Phys. **92**, 1511–1517 (2002).
31. P. NATH AND K. L. CHOPRA Experimental Determination of the Thermal Conductivity of Thin Films Thin Solid Films **18**, 29–37 (1973).
32. V. WACHTER AND F. VOLKLEIN Method for the Determination of the Thermal Conductivity and the Thermal Diffusivity of Thin Metallic Films Exp. Tech. Phys. 25, 425–431 (1977).
33. K. FUCHS The Conductivity of Thin Metallic Films According to the Electron Theory of Metals Proc. Camb. Phil. Soc. **34**, 100–108 (1938).
34. T. Q. QIU AND C. L. TIEN Size Effects on Nonequilibrium Laser Heating of Metal Films ASME J. Heat Trans. **115**, 842–847 (1993).
35. T. STARZ, U. SCHMIDT, AND F. VOLKLEIN Microsensor for in Situ Thermal Conductivity Measurements of Thin Films Sensors and Materials **7**, 395–403 (1995).
36. S. M. LEE, D. G. CAHILL, AND T. H. ALLEN Thermal Conductivity of Sputtered Oxide Films Phys. Rev. B **52**, 253–257 (1995).
37. M. B. KLEINER, S. A. KUHN, AND W. WEBER Thermal Conductivity Measurements of Thin Silicon Dioxide Films in Integrated Circuits IEEE Trans. Electron Dev. **43**, 1602–1609 (1996).
38. J. H. ORCHARD-WEBB A New Structure for Measuring the Thermal Conductivity of Integrated Circuit Dielectrics in Proc. IEEE Int. Conf. on Microelectronic Test Structures (1991), pp. 41–45.
39. H. A. SCHAFFT, J. S. SUEHLE, AND P. G. A. MIREL Thermal Conductivity Measurements of Thin-Film Silicon Dioxide in Proc. IEEE Int. Conf. on Microelectronic Test Structures (1989), pp. 121–125.
40. K. GOODSON, M. FLIK, L. SU, AND D. ANTONIADIS Annealing-temperature Dependence of the Thermal Conductivity of LPCVD Silicon-dioxide Layers IEEE Electron Device Lett. **14**, 490-492 (1993).
41. R. F. BROTZEN, P. J. LOOS, AND D. P. BRADY Thermal Conductivity of Thin SiO_2 Films Thin Solid Films **207**, 197–201 (1992).
42. D. G. CAHILL, W. K. FORD, K. E. GOODSON, G. D. MAHAN, A. MAJUMDAR, H. J. MARIS, R. MERLIN, AND S. R. PHILLPOT Nanoscale Thermal Transport Appl. Phys. Rev. **93**, 1–30 (2003).
43. J. LAMBROPOULOS, M. JOLLY, C. AMSDEN, S. GILMAN, M. SINICROPI, D. DIAKOMIHALIS, AND S. JACOBS Thermal conductivity of Dielectric Thin Films J. Appl. Phys. **66**, 4230–4242 (1989).
44. D. RISTAU AND J. EBERT Development of a Thermographic Laser Calorimeter Appl. Opt. **25**, 4571–4578 (1986).
45. T. OGDEN, A. RATHSAM, AND J. GILCHRIST Thermal Conductivity of Thick Anodic Oxide Coatings on Aluminum Mater. Lett. **5**, 84–87 (1987).
46. C. HENAGER AND W. PAWLEWICZ, Thermal Conductivities of Thin, Sputtered Optical Films, Appl. Opt. 32, 91-101 (1993).
47. R. TAYLOR, X. WANG, AND X. XU Thermophysical Properties of Thermal Barrier Coatings Surf. and Coat, Techn, **120/121**, 89–95 (1999).
48. R. VASSEN, X. CAO, AND D. STOVER Improvement of New Thermal Barrier Coating Systems using a Layered or Graded Structure Ceram. Eng. Sc. Proc. **22**, 435–442 (2001).
49. D. CLARKE Materials Selection Guidelines for Low Thermal Conductivity Thermal Barrier Coatings Surf. and Coat. Techn. **163/164**, 67-74 (2003).
50. U. SCHULZ, C. LEYENS, K. FRITSCHER, M. PETERS, B. SARUHAN-BRINGS, O. LAVIGNE, J. DORVAUX, R. MEVREL, AND M. CALIEZ Some Recent Trends in Research and Technology of Advanced Thermal Barrier Coatings Aerospace Sc. and Techn. **7**, 73–80 (2003).
51. K. GOODSON, O. KADING, M. ROSLER, AND R. ZACHAI Experimental Investigation of Thermal Conduction Normal to Diamond-silicon Boundaries J. Appl. Phys. **74**, 1385–1392 (1995).
52. H. VERHOEVEN, E. BOETTGER, A. FLOTER, H. REISS, AND R. ZACHAI Thermal Resistance and Electrical Insulation of Thin low-temperature-deposited Diamond Films Diamond Rel. Mater. **6**, 298–302 (1997).
53. S. D. WOLTER, D. A. BORCA-TASCIUC, G. CHEN, N. GOVINDARAJU, R. COLLAZO, F. OKUZUMI, J. T. PRATER, AND Z. SITAR Thermal Conductivity of Epitaxially Textured Diamond Films Diamond Rel. Mater. **12**, 61–64 (2003).
54. M. ASHEGHI, M. N. TOUZELBAEV, K. GOODSON, Y. LEUNG, AND S. WONG Temperature-dependent Thermal Conductivity of Single-crystal Silicon Layers in SOI Substrates J. Heat Trans. **120**, 30–36 (1998).
55. M. ASHEGHI, K. KURABAYASHI, K. GOODSON, R. KASNAVI, AND J. PLUMMER Thermal Conduction in Doped Single-crystal Silicon Films, in Proc. 33rd ASME/AIChE National Heat Transfer Conf., (Albuquerque, NM, 1999).
56. D. SONG, Phonon heat conduction in nano and micro-porous thin films Ph. D. Thesis, University of California at Los Angeles, Department of Mechanical and Aerospace Engineering, 2003.

Sec. 8 · REFERENCES

57. N. SAVVIDES AND H. J. GOLDSMID *The Effect of Boundary Scattering on the High-temperature Thermal Conductivity of Silicon* J. Phys. C: Solid State Phys. **6**, 1701–1708 (1973).
58. Y. C. TAI, C. H. MASTRANGELO, AND R. S. MULLER , *Thermal Conductivity of Heavily Doped Low-pressure Chemical Vapor Deposited Polycrystalline Silicon Films,* J. Appl. Phys. **63**, 1442–7 (1988).
59. F. VOLKLEIN AND H. BATLES *A Microstructure for Measurement of Thermal Conductivity of Polysilicon Thin Films* J. Microelectromech. Syst. **1**, 194–196 (1992).
60. S. UMA, A. MCCONNELL, M. ASHEGHI, K. KURABAYASHI, AND K. GOODSON *Temperature-dependent Thermal Conductivity of Undoped Polycrystalline Silicon Layers* Int. J. Thermophys. **22**, 605–616 (2001).
61. A. MCCONNELL, U. SRINIVASAN, M. ASHEGHI, AND K. GOODSON *Thermal Conductivity of Doped Polysilicon* J. Microelectromech. Syst. **10**, 360-369 (2001).
62. R. NEVILLE, *Solar Energy Conversion: The Solar Cell,* (New York, Elsevier, 1995).
63. H. GOLDSMID, M. KAILA, AND G. PAUL *Thermal conductivity of amorphous silicon* Phys. Status Sol. A **76**, K31–33 (1983).
64. G. POMPE AND E. HEGENBARTH *Thermal Conductivity of Amorphous Si at Low Temperatures* Phys. Status Sol. B **147**, 103–108 (1988).
65. A. ATTAF, M. S. AIDA AND L. HADJERIS *Thermal Conductivity of Hydrogenated Amorphous Silicon* Solid State Commun. **120**, 525–530 (2001).
66. R. ORBACH *Variational Transport in Disordered Systems* Philos. Mag. B **65**, 289–301 (1992).
67. P. B. ALLEN AND J. L. FELDMAN *Thermal Conductivity of Disordered Harmonic Solids* Phys. Rev. B **48**, 12581–12588 (1993).
68. V. M. ABROSIMOV, B. N. YEGOROV, AND N. S. LIDORENKO *An Investigation of the Thermoelectric Figure of Merit of Bismuth Films* Radio Eng. Electro. Phys. **9**, 1578–1579 (1971).
69. F. VOLKLEIN AND E. KESSELER, *Thermal Conductivity and Thermoelectric F and Figure of Merit of $Bi1_{-x}Sb_x$ films with $0 < x \leq 0.3$,* Phys. Stat. Sol. (a) **81**, 585–596 (1984).
70. C. F. GALLO, B. S. CHANDRASEKHAR, AND P. H. SUTTER *Transport Properties of Bismuth Single Crystals* J. Appl. Phys. **34**, 144 (1963).
71. F. VOLKLEIN AND E. KESSLER *Analysis of the Lattice Thermal Conductivity of Thin Films by Means of a Modified Mayadas-Shatzkes Model: The Case of Bismuth Films* Thin Solid Films **142**, 169–181 (1986).
72. F. VOLKLEIN, V. BAIER, U. DILLNER, E. KESSLER *Thermal Conductivity and Diffusivity of a Thin Film SiO_2/Si_3N_4 Sandwich System* Thin Solid Films **188**, 27–33 (1990).
73. V. BAIER AND F. VOLKLEIN *Thermal Conductivity of Thin Films* Phys. Stat. Sol. (a) **118**, K 69 (1990).
74. H. J. GOLDSMID *Thermoelectric Refrigeration* (Plenum Press, New York, 1964).
75. D. W. SONG, W. L. LIU, T. ZENG, T. BORCA-TASCIUC, G. CHEN, C. CAYLOR, AND T. D. SANDS, *Thermal Conductivity of Skutterudite Thin Films and Superlattices,* Appl. Phys. Lett. 77, 3854–3856 (2000).
76. G. NOLAS, G. SLACK, D. MORELLI, T. TRITT, AND A. EHRLICH *The Effect of Rare-earth Flling on the Lattice Thermal Conductivity of Skutterudites* J. Appl. Phys. **79**, 4002–4008 (1996).
77. L. ESAKI AND R. TSU *Superlattice and Negative Differential Conductivity in Semiconductors* IBMJ. Research and Development **14**, 61–61 (1970).
78. G. CHEN *Phonon Heat Conduction in Low-dimensional Structures* Semiconductors and Semimetals **71**, 203–259 (2001).
79. C. WEISBUCH AND B. VINTER, *Quantum Semiconductor Structures* (Academic Press, San Diego, CA, 1991).
80. G. CHEN, B. YANG, AND W. L. LIU *Engineering Nanostructures for Energy Conversion* in Heat Transfer and Fluid Flow in Microscale and Nanoscale Structures Wit Press, Southampton, UK, edited by M. Faghri and B. Sunden, pp. **45–91**, 2003.
81. T. M. TRITT, ed., *Recent Trend in Thermoelectric Materials Research* in Semiconductor and Semimetals **69–71** (Academic Press, San Diego, 2001).
82. R. VENKATASUBRAMANIAN, E. SIIVOLA, T. COLPITTS, AND B. O'QUINN *Thin-film Thermoelectric Devices with High Room-temperature Figures of Merit* Nature **413**, 597–602 (2001).
83. J. INOUE, H. ITOH, S. MAEKAWA, *Transport properties in magnetic superlattices,* J. Phys. Soc. Japan **61**, 1149–1152 (1992).
84. F. TSUI, B. CHEN, J. WELLMAN, C. UHER, AND R. CLARKE *Heat conduction of (111) Co/Cu superlattices* J. Appl. Phys. **81**, 4586–4588 (1997).
85. T. YAO *Thermal properties of AlAs/GaAs superlattices* Appl. Phys. Lett. **51**, 1798–1800 (1987).
86. R. VENKATASUBRAMANIAN *Lattice Thermal Conductivity Reduction and Phonon Localizationlike Behavior in Superlattice Structures,* Phys. Rev. B **61**, 3091–3097 (2000).

87. I. Yamasaki, R. Yamanaka, M. Mikami, H. Sonobe, Y. Mori, and T. Sasaki *Thermoelectric Properties of $Bi_2Te3/Sb2Te3$ Superlattice Structure* in Proc. 17th Int. Conf. on Thermoelectrics, ICT'98 (1998), pp. 210–213.
88. M. N. Touzelbaev, P. Zhou, R. Venkatasubramanian, and K. E. Goodson *Thermal Characterization of $Bi_2Te_3/Sb2Te3$ Superlattices* J. Appl. Phys. **90**, 763–767 (2001).
89. X. Y. Yu, G. Chen, A. Verma, and J. S. Smith *Temperature Dependence of Thermophysical Properties of GaAs/AlAs Periodic Structure* Appl. Phys. Lett. **67**, 3554–3556 (1995).
90. W. S. Capinski and H.J., Maris *Thermal Conductivity of GaAs/AlAs Superlattices* Physica B **219&220**, 699–701 (1996).
91. M. Lee, D. G. Cahill, and R. Venkatasubramanian *Thermal Conductivity of Si-Ge Superlattices* Appl. Phys. Lett. **70**, 2957–2959 (1997).
92. T. Borca-Tasciuc, W. L. Liu, T. Zeng, D. W. Song, C. D. Moore, G. Chen, K. L. Wang, M. S. Goorsky, T. Radetic, R. Gronsky, T. Koga, and M. S. Dresselhaus *Thermal Conductivity of Symmetrically Strained Si/Ge Superlattices* Superlattices and Microstructures **28**, 119–206 (2000).
93. W. L. Liu, T. Borca-Tasciuc, G. Chen, J. L. Liu, and K. L. Wang *Anisotropy Thermal Conductivity of Ge-quantum Dot and Symmetrically Strained Si/Ge Superlattice* J. Nanosc. Nanotechn. **1**, 39–42 (2001).
94. B. Yang, W. L. Liu, J. L. Liu, K. L, Wang, and G. Chen *Measurements of Anisotropic Thermoelectric Properties in Superlattices* Appl. Phys. Lett. **81**, 3588–3590 (2002).
95. T. Borca-Tasciuc, D. Achimov, W. L. Liu, G. Chen, H. Ren, C. H. Lin, and S. S. Pei *Thermal Conductivity of InAs/AlSb Superlattices* Microscale Thermophys. Eng. **5**, 225–231 (2001).
96. S. T. Huxtable, A. Abraham, C. L. Tien, A. Majumdar, C. LaBounty, X. Fan, G. Zeng, J. E. Bowers, A. Shakouri, and E. Croke, *Thermal Conductivity of Si/SiGe and SiGe/SiGe Superlattices* Appl. Phys. Lett. **80**, 1737–1739 (2002).
97. H. Beyer, J. Nurnus, H. Bottner, Roch T Lambrecht, and G. Bauer *Epitaxial Growth and Thermoelectric Properties of Bi_2Te_3 Based Low Dimensional Structures* Appl. Phys. Lett. **80**, 1216–1218 (2000).
98. T. Harman, P. Taylor, M. Walsh, and B. LaForge *Quantum Dot Superlattice Thermoelectric Materials and Devices* Science **297**, 2229–2232 (2002).
99. D. Gammon, B.V. Shanabrook, and D. S. Katzer *Excitons, Phonons, and Interfaces in GaAs/AlAs Quantum-well Structures* Phys. Rev. Lett. **67**, 1547–1550 (1991).
100. T. Ruf, J. Spitzer, V. F. Sapega, V. I. Belitsky, M. Cardona, and K. Ploog *Interface Roughness and Homogeneous Linewidths in Quantum Wells and Superlattices Studied by Resonant Acoustic-phonon Raman Scattering* Phys. Rev. B **50**, 1792–1806 (1994).
101. J. M. Ziman *Electrons and Phonons* (Clarendon, Oxford, 2001).
102. S. Sze, *Physics of Semiconductor Devices* (Wiley, New York, 1981).
103. S. G. Walkauskas, D. A. Broido, K. Kempa, and T. L. Reinicke *Lattice Thermal Conductivity of Wires* J. Appl. Phys. **85**, 2579–2582 (1999).
104. P. Hyldgaard and G. D. Mahan *Phonon Superlattice Transport* Phys. Rev. B **56**, 10754–10757 (1997).
105. S. Tamura, Y. Tanaka, and H. J. Maris *Phonon Group Velocity and Thermal Conduction in Superlattices* Phys. Rev. B **60**, 2627–2630 (1999).
106. W. E. Bies, R. J. Radtke, and H. Ehrenreich *Phonon Dispersion Effects and the Thermal Conductivity Reduction in GaAs/AlAs Superlattices* J. Appl. Phys. **88**, 1498–1503 (2000).
107. A. A. Kiselev, K. W. Kim, and M. A. Stroscio *Thermal Conductivity of Si/Ge Superlattices: A Realistic Model with a Diatomic Unit Cell* Phys. Rev. B **62**, 6896–6999 (2000).
108. B. Yang and G. Chen *Lattice Dynamics Study of Anisotropic Heat Conduction in Superlattices* Microscale Thermophys. Eng. **5**, 107–116 (2001).
109. M. V. Simkin and G. D. Mahan *Minimum Thermal Conductivity of Superlattices* Phys. Rev. Lett. **84**, 927–930 (2000).

Chapter 2.1

MEASUREMENT TECHNIQUES AND CONSIDERATIONS FOR DETERMINING THERMAL CONDUCTIVITY OF BULK MATERIALS

Terry M. Tritt

Department of Physics and Astronomy, Clemson University, Clemson, SC, USA

David Weston

Department of Physics and Astronomy, Clemson University, Clemson, SC, USA
Michelin Americas Research and Development Corporation, Greenville, SC, USA

1. INTRODUCTION

The accurate measurement and characterization of the thermal conductivity of bulk materials can pose many challenges. For instance, loss terms of the heat input intended to flow through the sample usually exist and can be most difficult to quantify. This chapter will provide an overview of the more typical measurement and characterization techniques used to determine the thermal conductivity of bulk materials. Some of the potential systematic errors that can arise and the corrections that need to be considered will be presented. This overview is not intended to serve as a complete description of all the available measurement techniques for bulk materials, of which there are many. However, it should provide an introduction and summary of the characterization and measurement techniques of thermal conductivity of bulk materials and give an extensive reference set for more in-depth dialogue of the concepts and techniques. In addition, this overview should serve as an indication of the care that must be taken in performing these measurements.

Many methods exist for the determination or measurement of thermal conductivity of a material, such as the steady-state (absolute or comparative) technique,[1] the 3ω technique,[2] and the thermal diffusivity measurement.[3] Each of these techniques has its own advantages as well as its inherent limitations, with some techniques more appropriate to specific sample geometry, such as the 3ω technique for thin films. The 3ω technique and methods for the measurement of thermal conductivity of thin films is discussed in detail in a later chapter.[4] Therefore, this chapter will focus on the measurement techniques that are more appropriate for "bulk-like" solid-state materials. The techniques presented will include: the more common steady-state method, the comparative technique, the radial flow method, the laser flash diffusivity method (for high temperatures), the "pulsed power or Maldonado technique,"[5] and the parallel thermal conductance (PTC) technique[6] (for single-crystal needle-like samples). Thermal conductivity measurements are difficult to make with relatively high accuracy, certainly better than within 5%. Many excellent texts and techniques discuss in detail many of the corrections and potential errors one must consider, and the reader is referred to these references.[7–10]

A word of caution is extended to those who are just beginning to perform thermal conductivity measurements. Only thorough understanding of the issues related to these measurements coupled with careful experimental design will yield the desired goal of highly reliable thermal conductivity measurements. Extensive efforts were expended in the late 1950s and 1960s in relation to the measurement and characterization of the thermal conductivity of solid-state materials. These efforts were made by a generation of scientists, who for the most part are no longer active, and this expertise would be lost to us unless we are aware of the great strides they made during their time. They did not have the advantage of computer data acquisition and had to painstakingly acquire the data through careful and considerate measurement techniques. The same issues are prevalent today and the same care has to be employed, yet we have many advantages, such as high-speed, high-density data acquisition. Hopefully, this chapter will serve not only as a testament to those researchers of past generations whose great care in experimental design and thought still stands today but as an extensive resource for the next-generation researchers. Hopefully, the reader will gain a deeper appreciation of these points in the following pages. Readers are encouraged to delve into the papers of authors such as G. White, R. Berman, G. Slack, and R. Pohl, as well as many other superb experimentalists of their age. An excellent and extensive reference for thermal conductivity measurements can be found in Vol. 1 of Ref. 7.

2. STEADY-STATE METHOD (ABSOLUTE METHOD)

Determination of the thermal conductance of a sample is a solid-state transport property measurement in which a temperature difference (ΔT) across a sample is measured in response to an applied amount of heating power. This is essentially a measure of the heat flow through the sample. The thermal conductivity (κ) is given by the slope of a power versus ΔT sweep at a fixed base temperature with the dimensions of the specific sample taken into account. The thermal conductivity as derived from a typical "steady-state" measurement method is

$$\kappa_{TOT} = P_{SAM}L_S/A\Delta T, \qquad (1)$$

where κ_{TOT} is the total thermal conductivity, P_{SAM} is the power flowing through the sample, L_S is the length between thermocouples, ΔT is the temperature difference measured, and A is the cross-sectional area of the sample through which the power flows.

2.1. Overview of Heat Loss and Thermal Contact Issues

However, determination of P_{SAM} is not such a simple task. Typically, an input power, P_{IN}, is applied to one end of the sample, and the power through the sample is

$$P_{SAM} = P_{IN} - P_{LOSS}, \qquad (2)$$

where P_{LOSS} is the power lost to radiation, heat conduction through gases or through the connection leads, or losses due to heat convection currents. The losses can be substantial unless sufficient care is taken in the design of the measurement apparatus and setup. These losses cannot be completely eliminated, but as stated an appropriate experimental design would include design considerations to minimize or sufficiently account for each loss term. For example, the thermocouple wires should be small diameter (0.001 inch) and possess low thermal conductivity, such as

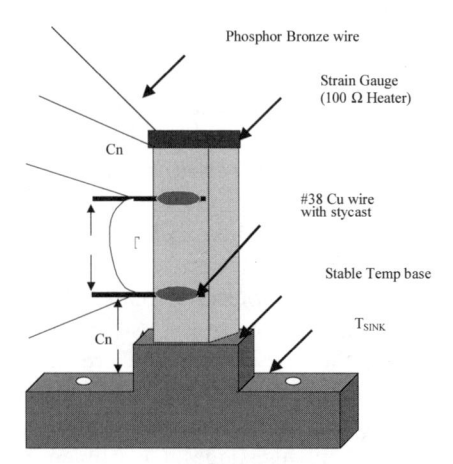

FIGURE 1 Diagram of the steady-state thermal conductivity method used in one of our laboratories at Clemson University (TMT). Small copper wires (flags) are attached to the samples on which the Cn–Cr thermocouples (0.001 inch diameter) are attached. A thin resistance heater (100 Ω strain gauge) is attached to the top of the sample for the power source, I^2R. Phosphor bronze current leads [# 38 (0.004')] are attached to the strain gauge resistor heater. The typical sample size is 2–3 mm for the width and/or thickness, and the total sample length is approximately 6–10 mm. (We acknowledge Prof. C. Uher, Univ. of Michigan, for mounting suggestions).

chromel or constantan wire. Also, we recommend using phosphor bronze (0.004 inch) lead wires for the input heater power. Phosphor bronze yields a relatively low thermal conductivity material, yet one that also has reasonable values of electrical conductivity. Heat loss through the connection wires of the input heater or the thermocouple leads can be calculated and an estimate obtained for this correction factor, which is typically less than 1–2% in a good experimental design. Correction issues related to convection or radiation losses can be more difficult to quantify. An illustration of a sample mounted for steady-state thermal conductivity measurements is shown in Fig. 1. The heater power is supplied by I^2R Joule heating of a 100-Ω strain gauge resistance heater attached to the top of the sample. The complete description of the sample mounting and measurement techniques as well as a thorough description of the apparatus has been given previously.[11] One advantage of this specific apparatus is that it is mounted on a removable puck system attached to a closed-cycle refrigerator system for varying the temperature. This allows the sample to be mounted outside the apparatus and then essentially "plugged" into the measurement system, meaning little "downtime" of the apparatus. The measurement sequence, temperature control and data acquisition are obtained by a high-density computer-controlled experimental data acquisition program that uses Labview software.

The steady-state thermal conductivity technique requires uniform heat flow through the sample; therefore, excellent heat sinking of the sample to the stable temperature base as well as the heater and thermocouples to the sample is an absolute necessity. Kopp and Slack discuss the subject of thermal anchoring and the associated errors quite thoroughly.[12] It is strongly suggested that each researcher (and especially students) definitely test any technique (and their mounting skills) by measuring known standards and comparing values. Insufficient thermal anchoring can be difficult to detect except by experienced researchers. There are, however, a few systematic "checks" that can be performed to rule out errors due to poor thermal anchoring. For example, thermal conductivity measurements are very important for evaluating the potential of thermoelectric materials.[13] Two papers give excellent reviews of the issues related to accurate measurements of the electrical and thermal transport properties of thermoelectric materials, and readers are encouraged to examine them.[14,15] The checks previously mentioned typically relate to simultaneously measuring the thermopower (α) where $\Delta V_{SAM} = \alpha \Delta T$, using separate sample voltage leads attached to the sample. If a significant time lag exists between changes in sample voltage and changes in the temperature gradient (measured by the thermocouples), then this suggests a problem with thermal anchoring. Since the sample is its own best thermometer, one can compare changes in the thermocouple voltage to that of the sample voltage as ΔT is varied, by using an oscilloscope to measure the time differences in the response of these two voltages to a change in ΔT. A poor thermal contact between the thermocouple and the sample can also be revealed by a change in pressure in the sample space. As the gas or air is pumped out, the thermal link between the sample and the thermocouple can change (deteriorate) and the measured thermopower would then change, due to an erroneous temperature difference measurement. A large difference between measurements taken under vacuum and in a gas atmosphere usually indicates poor thermal contact between sample and thermocouple. However, if only thermal conductivity is measured, heat loss due to convection (or conduction via the gas medium) could be causing the differences. It is suggested that an individual's mounting techniques be developed for thermopower measurements to more fully understand thermal contact issues before proceeding with thermal conductivity measurements.

The authors prefer using a differential thermocouple for thermal conductivity measurements; thus only two lead wires are coming off of the sample to minimize conduction heat loss. Each thermocouple leg must be a low-thermal-conductivity material (Cu leads should be avoided). For faster sample throughput, it is often desirable to use a permanently mounted thermocouple and thermally anchor the sample and the thermocouple through a common medium with varnish or some other thin contacting adhesive. This has advantages but is more likely to cause a thermal anchoring problem resulting in an erroneous ΔT reading. At low temperatures, calibration and subtraction of the lead contributions can also be a source of error; since ΔT is smaller, as is the signal-to-noise ratio of the thermocouple voltage.

2.2. Heat Loss Terms

Heat loss due to radiation effects between the surroundings and the sample, as well as convection currents or conduction through any lead wires can be substantial. The radiation loss (which can be quite large near room temperature) is given by

$$Q = \varepsilon \sigma_{S-B} A (T_{SAM}^4 - T_{SURR}^4), \qquad (3)$$

where T_{SAM} and T_{SURR} are the temperature of the sample and the surroundings, respectively, σ_{S-B} is the Stephan–Boltzmann constant $\sigma_{S-B} = 5.7 \times 10^{-8}$ W/m^2-K^4, and ε ($0 < \varepsilon < 1$) is the emissivity. Proper thermal shielding and thermal anchoring are essential, and a thermal heat shield that is thermally anchored to the sample heat sink is the first step. A heat shield and sample probe / base are described in detail for our apparatus at Clemson University in Ref. 11. One might also attach heaters at various points along the heat shield to allow the possibility of being able to match the gradient along the sample. Using a feedback circuit coupled with computer data acquisition and control can allow this to be automated. It is also suggested to metal (e.g., Au) plate the inside of the heat shield in order to aid its reflectivity. Usually the upper temperature limit for measuring thermal conductivity of a sample is restricted by radiation losses. A more complete description of the issues related to high-temperature thermal conductivity measurements can be found in an extensive chapter by M. Laubitz.[16]

Thus, one of the disadvantages in using a standard steady-state method is that for temperatures above $T \approx 150$ K, radiation loss can become a relatively serious problem and the question is how to effectively deal with these losses. For relatively large samples (short and fat), radiation effects are typically negligible below 200 K. In order to correct for the radiation loss at higher temperatures, one must either measure the radiation or determine an estimate for the losses. One way in which to account for the radiation correction is as follows. Once the total thermal conductivity (κ_T) is measured, the Wiedemann–Franz Law ($\kappa_E = L_0 \sigma T$, where L_0 is the Lorentz number, σ is the electrical conductivity, and T is the temperature) is used to extract the electronic contribution (κ_E) to the thermal conductivity from the previously measured electrical conductivity. From this, the lattice contribution (κ_L) to the thermal conductivity is then determined, where $\kappa_L = \kappa_T - \kappa_E$. Upon inspection of the resulting plot of the lattice thermal conductivity (κ_L) versus temperature, the curve most likely will exhibit a characteristic temperature dependence above the low-temperature phonon peak, the so-called crossover temperature.[17] Typically for a crystalline material, κ_L will exhibit an inverse temperature dependence at high temperatures, $\kappa_L \approx 1/T$, due to phonon–phonon scattering. Other types of material may demonstrate other temperature dependence of κ_L, such as $1/T^{0.5}$, or it may be independent of temperature.[18]

If the heat conduction and radiation losses have been minimized with the appropriate measurement system design, then radiation losses should be very small below $T \approx 150$ K. Therefore, κ_L versus T is mathematically fit above the low-temperature phonon peak (if one exists) to $T \approx 150$ K. This fit is applied and the lattice thermal conductivity is extrapolated to $T \approx 300$ K. Then the extrapolated κ_L and the measured κ_L are compared to each other, and their difference is designated by $\Delta\kappa_L$. The sample temperature is given by $T + \Delta T$, where T is the base temperature, or the temperature of the surroundings, T_{SURR} (i.e., $T = T_{SURR}$). After a Taylor expansion of $(T_{SAM})^4 - (T_{SURR})^4$ the radiation loss is found to be proportional to T^3 as a function of the overall temperature. Thus, if $\Delta\kappa_L$ displays a temperature dependence proportional to T^3, one can be relatively confident that $\Delta\kappa_L$ is due to radiation losses. For example, the quasicrystal sample, shown in Fig. 2, displays a very characteristic shape; in this case the curve goes as $1/T$ from 80 K to 150 K. Above 150 K the curve deviates from this fit. If a curve of $1/T$ is plotted in conjunction with the calculated lattice and the difference between the two is called $\Delta\kappa_L$, it is observed that, near room temperature, the difference in the calculated

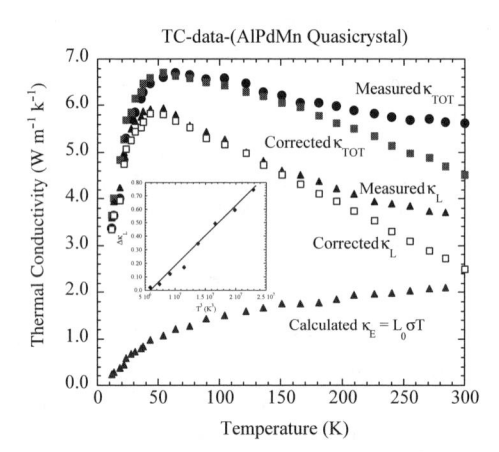

FIGURE 2 Plot of the thermal conductivity as a function of temperature for an AlPdMn quasicrystal. The total, lattice, and electronic contributions to the thermal conductivity are shown as closed circles, closed diamonds, and closed triangles, respectively. The $\kappa_E = L_0\sigma T$ is calculated from the Wiedemann-Franz relationship discussed in the text. The total, lattice, and electronic thermal conductivities are labeled, κ_{TOT}, κ_L, and κ_E, respectively. The difference between the extrapolated or corrected and measured lattice thermal conductivity, $\Delta\kappa_L$, is plotted as a function of T^3 in the inset, illustrating losses due to radiation effects.

and measured lattice thermal conductivity is on the order of 1 W m^{-1} K^{-1}. An inset in Fig. 2 shows $\Delta\kappa_L$ plotted as a function of T^3, and a very linear curve is observed. Thus, $\Delta\kappa_L$ is assumed to be due to radiation loss that is proportional to T^3 and can be corrected for in the original measurements. The total thermal conductivity corrected for radiation losses can be determined by adding the calculated electronic thermal conductivity and the corrected lattice thermal conductivity together. With this iterative process a first level of corrections for radiative losses is hopefully achieved. These corrections should certainly be less than 15–20% at room temperature, otherwise better experimental design or a different sample size should be employed. It also may be necessary to change the emissivity of the sample by applying a thin coating, but this can contribute to other loss effects.

Heat losses can also be due to convection or circulating gas flow around the sample. Heat convection is a more subtle effect and less readily able to quantify as conduction or radiation losses. The best way to minimize these convection losses is to operate the measurement with the sample in a moderate vacuum (10^{-4}–10^{-5} torr). If convection is still a problem, then possibly baffles can be integrated to disturb the convection currents.

The other substantial heat loss mechanism is due to heat conduction. Thermal resistance of leads, heaters, etc., as well as interface anchoring between the sample, the heater, and the heat sink is important. A high vacuum will certainly reduce the heat loss due to conduction through any gaseous medium around the sample. Another source of heat conduction loss is via the thermocouple lead wires or other measurement leads attached to the sample for temperature measurement. Long lead lengths of small diameter (small A) with sufficient thermal anchoring (so essentially no ΔT arises between the sample and shield) are important for minimizing this effect. One must accurately determine the power through the sample by considering the various loss mechanisms.

Even if one minimizes or effectively measures many of these losses, the accurate determination of the sample length (L_S) and cross-sectional area (A) can be a challenge to achieving high precision within a 5–10% uncertainty. Again, measuring known standards and thoroughly calibrating the apparatus are essential to understanding its resolution. Several standards with different values for their thermal conductivity are suggested. Pyrex and Pyroceram are suggested as low-thermal-conductivity standards ($\kappa < 4$–5 W m^{-1}K^{-1}), calibrated 304 stainless steel for intermediate values ($\kappa < 20$ W m^{-1}K^{-1}), and HOPG graphite for higher values ($20 < \kappa < 100$ W m^{-1}K^{-1}). The latter two can both be obtained from the National Institute of Standards and Technology.[19]

3. THE COMPARATIVE TECHNIQUE

Other techniques may also be considered and are just as valid as the previously described absolute steady-state method. In the comparative technique a known standard is put in series between the heater and the sample. This technique, also a steady-state heat flow technique, achieves the best results when the thermal conductivity of the standard is comparable to that of the sample. Indeed, the same types of errors and corrections must be considered as for the absolute steady-state technique. In addition, there are more sources of potential error due to thermal contact effects in the comparative technique than in the absolute technique, since more thermal contact points are involved.

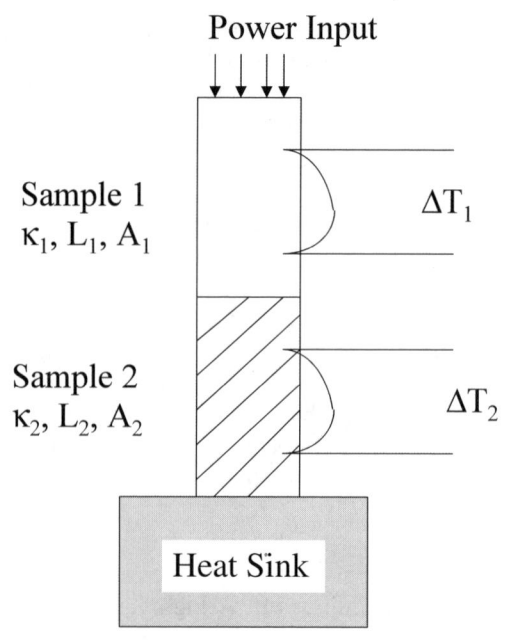

FIGURE 3 Configuration for measuring of the thermal conductivity using a comparative technique, where the sample is in series with a known standard.

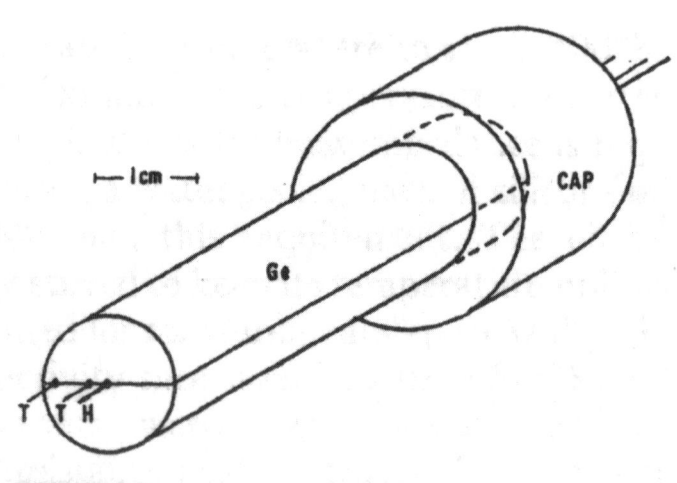

FIGURE 4 Configuration for measuring thermal conductivity using a radial flow technique. From Slack and Glassbrenner, *Thermal Conductivity of Germanium from 3 K to 1020 K*, Phys. Rev. **120**, No 3, (1960). This requires rather large samples but radiation losses are minimized, thus making it appropriate for high temperature.

The power through the standard (κ_1) is equal to the power through the sample (κ_2), and if the thermal conductivity of the standards κ_1 is known, then the thermal conductivity of the sample, κ_2, is

$$\kappa_2 = \kappa_1 \{A_1 \Delta T_1 L_1 / A_2 \Delta T_2 L_2\}. \tag{4}$$

An illustration of a sample mounted for a comparative technique measurement is shown in Fig. 3. One can also add another standard on the other side of the sample (i.e., the sample is sandwiched between two known or standard materials), known as the double comparative technique. But again, more thermal contact points are added which can be potential sources of error.

4. THE RADIAL FLOW METHOD

The conventional longitudinal heat flow method can be satisfactory at low temperatures, but serious errors can occur at high temperatures due to radiative losses directly from the heater and from the sample surface. In the radial heat flow method, heat is applied internally to the sample, generally minimizing radiative losses from the heat source. As presented by Tye,[20] radial flow methods have been applied to solids having a wide range of thermal conductivities. Since radial flow methods are relatively more difficult to apply than linear flow methods, they are not commonly employed below room temperature. An illustration diagram of the radial flow method is shown in Fig. 4.[21]

Internal sample heating has been accomplished in a variety of sample geometries, including imbedding in the sample, at the center of a hollow sample, and by direct electrical heating of the sample itself. The symmetry of the sample geometry must correspond to the geometry of the heater and permit inclusion of the heater. Chapter 4 of Tye[20] gives a comprehensive review of five classes of apparatus in radial methods.

Class 1. The simplest to employ is a cylindrical geometry with a central source or sink of power and assumed 'infinite' length, therefore without end guards.

Class 2. Cylindrical geometry frequently consisting of stacked disks and a central source or sink but of finite length, therefore having end guards.

Class 3. Spherical and ellipsoidal geometry with a completely enclosed heat source, having some sample preparation difficulties.

Class 4. Concentric cylindrical sample geometry consisting of known and unknown thermal conductivities with a central heat source or sink and analyzed by comparative methods.

Class 5. Electrically self-heating samples, having cylindrical geometry where the radial temperature distribution is analyzed.

In the frequently employed cylindrical symmetry, heat is generated along the axis of a cylinder. At steady-state conditions the radial temperature field is measured at two different radii. For heat flow in a cylinder between radii r_1 and r_2, assuming no significant longitudinal heat loss, thermal conductivity κ is found by solving

$$P = -\kappa [T_{r_1} - T_{r_2}] / \int_{r_2}^{r_1} \frac{dr}{2\pi r} \tag{5}$$

for κ:

$$\kappa = \frac{P\ln(r_2/r_1)}{2\pi L \Delta T},\qquad(6)$$

where P is heat energy input per unit time, L is sample length, ΔT is the temperature difference between the thermocouples, and r_1 and r_2 are the radial positions of the inner and outer thermocouples, respectively.

Slack and Glassbrenner employed a combination of linear and radial methods to measure the thermal conductivity of germanium from 3 to 1020 K.[22] This was accomplished by using different samples in two different apparatus, and then relating the two measurements in overlapping temperature ranges by correcting the low-temperature results to match the high-temperature curve. A conventional linear method was used below 300 K and a radial method above 300 K to limit error due to radiation losses. Figure 4, taken from reference 22, illustrates the sample configuration for the radial flow method. The sample size was relatively large, a 6-cm-long cylinder with a 1.3-cm radius (but this is generally considered small for a radial flow sample). The cylinder was sectioned and longitudinal grooves were cut for the central heater and thermocouples at r_1 and r_2, which were cemented in place. In cylindrical geometry, according to Carslaw and Jaeger, as long as the length-to-diameter ratio is greater than 4 to 1, the end-loss error is less than 0.5%.[23] Since in Eq. (5), κ depends on $\ln(r_2/r_1)$, the measurement of r_1 and r_2 is less critical than the measurement of thermocouple separation in the linear geometry as shown in Fig. 1. Even so, Glassbrenner and Slack reported an improvement in the radial thermocouple location error from 20% in 1960 to 5% in 1964 by the use of alumina tubing in holes in the sample.[24]

More recently (1995), thermal diffusivity was measured at high temperatures (800–1800 K) by Khedari *et al.* by employing a cylindrical geometry radial heat flow apparatus and a periodic stationary method.[25] In the periodic stationary method the heat source is sine modulated, creating a radial thermal sine wave through the sample. Thermal diffusivity is calculated from the phase ratio and amplitude ratio coming from the thermocouple signals at r_1 and r_2. The thermocouples were placed in axial holes in the sample. The time constants were adjusted by adjusting the experimental dimensions.

The uniformity of heat flow, the influence of axial heat flow (the infinite cylinder assumption), the thermocouple position measurement error, and the disturbance of the thermal field around the thermocouples were all experimentally evaluated by Khedari *et al.*[25] They found a dependency of the measurements on the frequency and a gap between the diffusivity calculated from the phase ratio and the diffusivity calculated from the amplitude ratio.

Benigni and Rogez[26] continued the work of Khedari *et al.*, and in 1997 they reported experimental data and a new model showing three factors acting in parallel that influenced the total thermocouple contact conductance, correcting much of the error reported by Khedari *et al.* These three factors, solid conduction at contact points, gas conduction, and radiation within the drilled hole containing a thermocouple, were found to be the main problems to be corrected to increase accuracy.

Radial methods are not often used for low-temperature measurement of thermal conductivity since relatively easier longitudinal methods usually offer satisfactory results. In addition, the radial flow method typically requires much larger samples than the longitudinal or linear methods, and this can be difficult to achieve with high-grade research polycrystalline or single-crystal materials. The high-temperature advantage due to minimum radiative losses from the heat source is the primary reason to choose a radial method.

The next several techniques will just be summarized, and it is suggested that the reader refer to the references for a more complete description. Many are not steady-state measurements but are more like quasi-linear or pulsed power heat flow methods. Some are incorporated into commercial systems, and one should contact the companies for specific details of the measurement systems. A technique that is becoming popular for thermoelectric materials, as well as for many nonconducting low-thermal-conductivity systems, is the "3-ω technique."[2,27] The 3-ω technique was originally developed for measuring the thermal conductivity of glasses and other amorphous solids and is appropriate for measuring the thermal conductivity of very thin samples or thin films. It is discussed in this book.

5. LASER-FLASH DIFFUSIVITY

Another technique for measuring the thermal properties of thin-film and bulk samples is the laser-flash thermal diffusivity method.[28] In this technique one face of a sample is irradiated by a short ($\leq$ 1 ms) laser pulse. An IR detector monitors the temperature rise of the opposite side of the sample. The thermal diffusivity is calculated from the temperature rise versus time profile. Algorithms exist for correcting various losses typically present in this measurement. The thermal conductivity is related to the thermal diffusivity, $D = \kappa/\rho_d C_p$, where ρ_d is the density, and C_p is the heat capacity. At high temperatures where the heat capacity (C_P) is a constant, the thermal diffusivity measurement essentially yields the thermal conductivity. Therefore, in principle, this technique can be used to measure thermal conductivity. However, the utility of this method requires fairly stringent sample preparation requirements. In order to prevent "flash-throughs" to the IR detector, there is very little flexibility in the required sample geometry (typically thin disks or plates). In addition, the sample surfaces must be highly emissive to maximize the amount of thermal energy transmitted from the front surface and to maximize the signal observed by the IR detector. Usually this requires the application of a thin coating of graphite to the sample surfaces. If good adhesion is not achieved, this coating procedure can potentially be a source of significant error.

Commercial units are available that allow measurement of thermal diffusivity at temperatures from 77 K to $\sim$2300 K.[29] These units are typically automated and reasonably easy to use. The thermal diffusivity is related to thermal conductivity through the specific heat and sample density; i.e., the laser-flash method is sometimes used to determine thermal conductivity indirectly when the specific heat and density have been measured in separate experiments. However, these systems require a relatively large sample size, a 2-inch diameter disk for some systems. This can be difficult to obtain for a "research sample."

6. THE PULSE-POWER METHOD ("MALDONADO" TECHNIQUE)

Traditional methods for measuring thermal conductivity typically require relatively long waiting times between measurements to enable the sample to reach steady-

198   Chap. 2.1 · MEASUREMENT TECHNIQUES FOR DETERMINING THERMAL CONDUCTIVITY

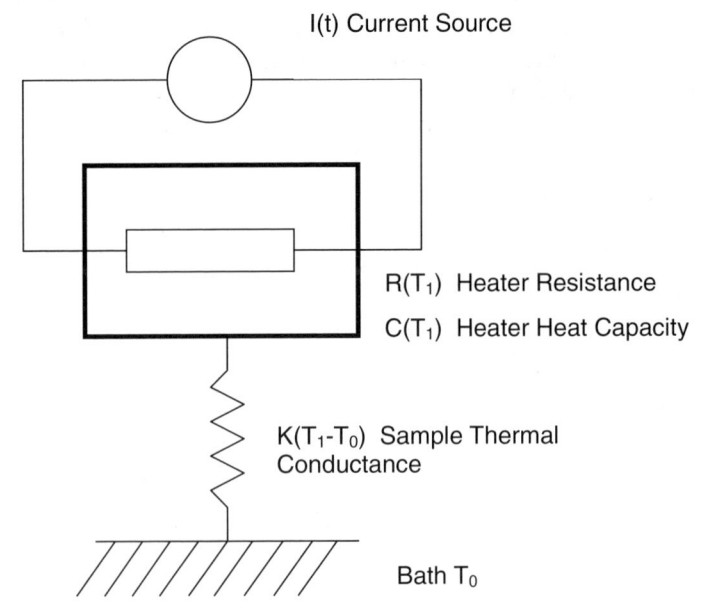

FIGURE 5 Schematic for "Maldonado" technique as described in Ref. 5.

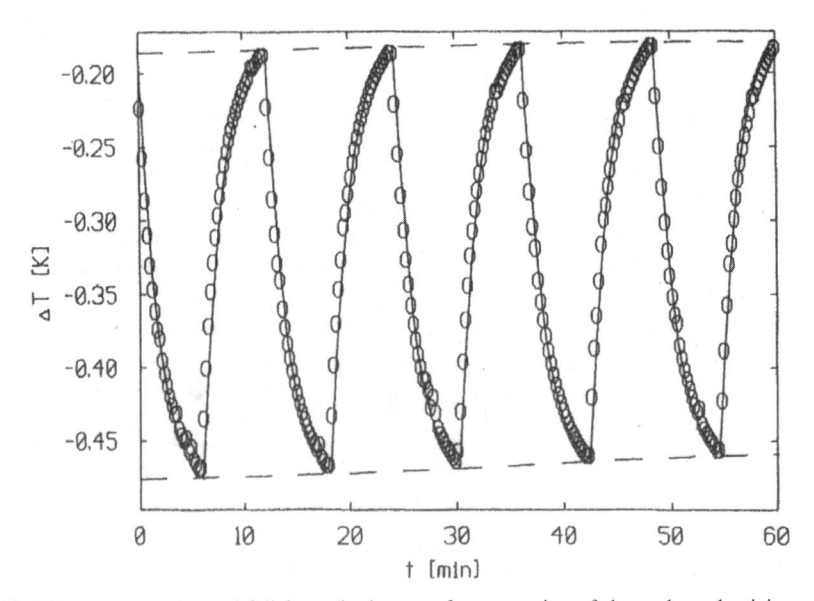

FIGURE 6 Temperature rise and fall for pulsed power for measuring of thermal conductivity using the "Maldonado" technique. From Ref. 5. The time dependence of the temperature difference across the sample is showing where (———) represents a simulation and (o) represents experimental data.

Sec. 7 • THE PULSE-POWER METHOD ("MALDONADO" TECHNIQUE)

state considerations. These time delays may allow various offset drifts to influence the measurement. With the pulse-power method described here, the bath temperature is slowly drifted, while the heater current that generates the thermal gradient is pulsed with a square wave.[5] Maldonado applied this technique in 1992 for the simultaneous measurement of heat conductivity and thermoelectric power.[5] A diagram of the sample setup is shown in Fig. 5. Here we describe only the application to the measurement of thermal conductivity. Since thermal equilibrium is never established, time between measurements can be significantly reduced, facilitating higher-measurement resolution. The experimental setup is basically the same as in steady-state measurements except that the heating current is pulsed with a square wave of constant current, thereby creating small thermal gradients. Figure 6 shows the response of the sample to the pulsed power with the "Maldonado" method. The experimental setup is basically the same as in steady-state measurements except that the heating current is pulsed with a square wave of constant-amplitude current, thereby creating small thermal gradients. No steady-state thermal gradient is established or measured.

The heat balance equation for the heater in Fig. 5 is written as the sum of the current dissipated in the heater and the heat conducted by the sample:

$$\frac{dQ}{dt} = C(T_1)\frac{dT_1}{dt} = R(T_1)I^2(t) - K(T_1 - T_0). \qquad (7)$$

In Fig. 5, dQ/dT is the time rate of change of heat in the heater, T_1 is the heater temperature, C is the heater heat capacity, R is the heater resistance, and K is the sample thermal conductance.

Since K is a function of temperature, the temperature difference, $T_1 - T_0$ is kept small with respect to the mean sample temperature so that K is considered as a function of mean sample temperature. The bath temperature T_0 is allowed to drift slowly and a periodic square-wave current with period 2τ is applied through the resistance heater, which causes T_1 to vary with period 2τ. Maldonado arrives at a solution of Eq. (1) by including several simplifications. Since, $C(T)$, $R(T)$, and $K(T)$ are smooth functions of T, then T_0 is used instead of T_1 as the argument of C, R, and K. In addition, an adiabatic approximation is employed by considering T_0 as nearly constant, since the temperature drift is slow compared to the periodic current. The solution has a sawtooth form as shown in Maldonado's figure (Fig. 6). The difference between the smooth curves through the maxima and minima yields a relation for thermal conductance:

$$K = \frac{RI_0^2}{\Delta T_{pp}}\tanh\left(\frac{K\tau}{2C}\right). \qquad (8)$$

The overall accuracy is reported by Maldonado to be better than 5%, with the principal error sources being ΔT measurement and, of course, sample geometry measurement for calculation of thermal conductivity from the thermal conductance. An advantage of this method is that the sample temperature is slowly slewed while the measurement is performed and can save time in the measurement sequence since achieving a steady-state is not necessary. This technique has recently been employed in a commercial device sold by Quantum Design Corporation.[30]

7. PARALLEL THERMAL CONDUCTANCE TECHNIQUE

Thermal conductivity measurements on relatively small samples are an additional, yet very important, challenge for the scientific community. For several reasons, including size dependence effects or, perhaps, natural size limitations, it is important to be able to measure these smaller samples. In order to measure the thermal conductivity of small needlelike single-crystal samples ($2.0 \times 0.05 \times 0.1$ mm^3), such as pentatellurides and single-carbon fibers, a new technique called parallel thermal conductance (PTC) was recently developed.[6] A thermal potentiometer measurement method had been developed previous to the PTC technique by Piraux, Issi, and Coopmans to evaluate the thermal conductivity of thin carbon fibers.[31] However, the measurement was quite laborious and tedious.

In the more typical steady-state method for measuring thermal conductivity, thermocouples are attached to the sample to measure the temperature gradient and a heater is included to supply the gradient. However, attaching thermocouples and heaters to small samples essentially minimizes the necessary sensitivity needed to perform these measurements. Thus, the measurement of the thermal conductivity of small samples and thin films has been a formidable challenge, with few successes, due to heat loss and radiation effects. It is also difficult for the small samples to support the heaters and thermocouples without causing damage to the sample.

In essence, the PTC technique is a variation of a steady-state technique (P vs. ΔT), which is adapted to measure the thermal conductivity of these types of samples as described in the following. Due to the inability of a sample of this size to support a heater and a thermocouple, a sample holder or stage had to been developed. Figure 7 illustrates the mounting and setup for the PTC technique. This technique requires that the thermal conductance of the sample holder itself must be measured first, which determines the base line or background thermal conduction and losses associated with the sample stage. The second step consists of attach-

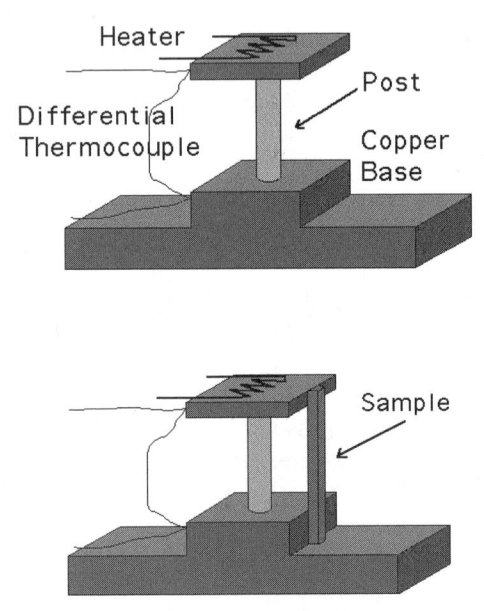

FIGURE 7 Sample configuration and mount for measuring thermal conductivity using the PTC technique.

ing the sample and measuring the new thermal conductance of the system. By subtraction, PTC is calculated. This conductance is due to the sample, thermal contacts and blackbody radiation from the sample. The contributions of heater and thermocouple lead wires and radiation effects of the holder have been subtracted. Of course, one of the necessary components of this technique is the reproducibility of the thermal conductance of the holder as a function of time and temperature, as well as being able to achieve a very low overall thermal conductance of the holder. The results from this technique enabled the measurement of several small samples that could not be measured by other techniques.[32] However, accurate determination of sample dimensions can be a limiting factor in determining absolute thermal conductivity values.

8. Z-METERS OR HARMAN TECHNIQUE

Another widely used technique for characterizing thermoelectric materials is the "Harman Technique" or Z-meter.[33-36] This is a direct method for obtaining the figure of merit, ZT, of a material or a device. Consider the voltage as a function of time for a thermoelectric material under various conditions. If there is no ΔT and $I = 0$, then $V_S = 0$, where V_S is the sample voltage. A current, I, is applied, and the voltage increases by IR_S, where R_S is the sample resistance, to the value V_{IR}. Recall that when a current is applied to a thermoelectric material a temperature gradient, ΔT, arises from the Peltier effect ($Q_P = \alpha IT$) and a voltage, V_{TE}, will add to the IR_S voltage. Under steady-state or adiabatic conditions, the heat pumped by the Peltier effect will be equal to heat carried by the thermal conduction;

$$\alpha IT = \kappa A \Delta T / L. \tag{9}$$

One can derive a relationship between ZT and the adiabatic voltage ($V_A = V_{IR} + V_{TE}$) and the IR sample voltage, V_{IR}[51,52]

$$ZT = V_A/V_{IR} - 1 = \alpha^2 T/\rho\kappa, \tag{10}$$

where ρ is the measured electrical resistivity of the sample. This relationship is a reasonable approximation to ZT but assumes ideal conditions unless corrections are accounted for, such as contacts, radiation effects, and losses. The first criterion is that the sample typically needs to possess a $ZT \geq 0.1$. Also, contact effects, sample heating from the contacts and the sample resistance, should be negligible, and ΔT effects from contact resistance differences can also be negligible. The thermal conductivity can be estimated from the Harman technique in two ways: first measure R and α, then ZT from Eq. (8), and the thermal conductivity can then be determined. Another way is to use Eq. (9) in the form

$$I = \kappa(A/\alpha TL)\Delta T. \tag{11}$$

Then at a constant temperature, the following relationship holds, where $I = \kappa(A/\alpha TL) \Delta T = \kappa C_0 \Delta T$, where C_0 is a constant at a given T. Thus, the linear part of the slope of an $I - \Delta T$ plot will yield the thermal conductivity. The ZT determined from the Harman method is essentially an "effective ZT," or, in other words, the operating figure of merit of the device. This technique requires that essentially no contact resistance effects exist (recall $I^2\{R_{C1} - R_{C2}\} = \Delta P_C \approx \Delta T$ across the sample from $I^2 R$ heating). Information from this technique should be compared to the measurements of the individual parameters that go into ZT. It should not be a substitute for knowing the parameters.

9. SUMMARY

In summary, several methods have been presented and discussed for the experimental determination of the thermal conductivity of a bulk solid-state material. Much care must be taken in the experimental design of the apparatus and the evaluation of the resulting data. Issues relating to understanding loss terms such as radiation or heat conduction were discussed along with issues related to appropriate heat sinking of the sample and corresponding measurement lead wires. Several techniques were discussed and one must then determine which technique is most appropriate for the specific sample geometry and the specific measurement equipment and apparatus available. However, the determination of the thermal conductivity of a material to within less than 5% uncertainty may be quite formidable, and most often a serious limiting factor may be the accurate determination of the overall sample dimensions.

10. REFERENCES

1. See, for example, *Thermal Conductivity of Solids*, J. E. PARROTT and A. D. STUCKES (Pion Limited Press, 1975), T. C. HARMAN and J. M. HONIG, *Thermoelectric and Thermomagnetic Effects and Applications* (McGraw-Hill, New York, 1967).
2. D. G. CAHILL, *Thermal Conductivity Measurement from 30 to 750 K: The 3ω Method* Rev. Sci. Instrum. **61**, 802 (2001).
3. See, for example, S. E. GUSTAFSSON and E. KARAWACKI, *Transient hot-strip probe for measuring thermal properties of insulating solids and liquids* Rev. Sci. Instrum. **54**, 744 (1983) and references therein.
4. T. BORCA-TASCIUC and GANG CHEN *Experimental Techniques for Thin Film Thermal Conductivity Characterization* Chapter 2.2 of this book.
5. O. MALDONADO *Pulse Method for Simultaneous Measurement of Electric Thermopower and Heat Conductivity at Low Temperatures* Cryogenics **32**, 908 (1992).
6. B. M. ZAWILSKI, R. T. LITTLETON IV, and TERRY M. TRITT *Description of the Parallel Thermal Conductance Technique for the Measurement of the Thermal Conductivity of Small Diameter Samples* Rev. Sci. Instrum. **72**, 1770 (2001).
7. G. A. SLACK, *Solid State Physics* (Academic Press, New York, 1979).
8. R. P. TYE, ed. *Thermal Conductivity Vol. I and Vol. II* Academic Press, New York, 1969.
9. R. BERMAN, *Thermal Conduction in Solids* Clarendon Press, Oxford, 1976.
10. *Methods of Experimental Physics: Solid State Physics*, Vol. 6, editor: L. Marton, Academic Press, New York, 1959.
11. AMY L. POPE, B. M. ZAWILSKI, and TERRY M. TRITT *Thermal Conductivity Measurements on Removable Sample Mounts* Cryogenics **41**, 725 (2001)
12. J. KOPP and G. A. SLACK *Thermal Contact Problems in Low Temperature Thermocouple Thermometry* Cryogenics, p 22, Feb. 1971.
13. See for example (a.) *Thermoelectric Materials: Structure, Properties and Applications*. TERRY M. TRITT *Encyclopedia of Materials: Science and Technology*, Volume **10**, pp 1-11, K. H. J. BUSCHOW, R. W. CAHN, M. C. FLEMINGS, B. ILSCHNER, E. J. KRAMER, S. MAHAJAN, (eds) Elsevier Press LTD, Oxford, Major Reference Works, London, UK (b) *Thermoelectrics*: Basic Principles and New Materials Developments, G. S. NOLAS, J. SHARP and H. J. GOLDSMID, Springer Series in Materials Science Volume **45**, 2002.
14. TERRY M. TRITT *Measurement and Characterization of Thermoelectric Materials* T. M. TRITT, M. KANATZIDIS, G. MAHAN and H. B. LYONS JR. (Editors), *1997 Materials Research Society Symposium Proceedings Volume 478: Thermoelectric Materials, New Directions and Approaches* p 25, 1997.

15. C. UHER *Thermoelectric Property Measurements* Naval Research Reviews, *Thermoelectric Materials,* Vol. XLVIII, p 44, 1996.
16. M. J. LAUBITZ in *Thermal Conductivity* edited by R. P. TYE, vol. **1**, Chap. 3 (Academic Press, London; 1969), p. 111.
17. *The crossover temperature of the low temperature phonon peak is the peak or "hump" in the thermal conductivity where there is a gradual change in the dominant scattering effect, such as from phonons-phonon interactions at high temperatures to boundary scattering at low temperatures.* See for example: Figure 3.1, R. BERMAN, *Thermal Conduction in Solids*, Oxford University Press, (1976) New York
18. P. G. KLEMENS Chapter 1, p.1 *Thermal Conductivity Vol.1,* edited by R. P. TYE, Academic Press, New York, 1969.
19. NIST website: http://www.nist.gov
20. Chapter 4 (1969) *Thermal Conductivity Vol. I and Vol. II*, edited by R. P. TYE Academic Press, New York, 1969.
21. Chapter 1, pg. 186, Fig. 1, 1969 *Thermal Conductivity Vol. I and Vol. II* edited by R. P. TYE, Academic Press, New York, 1969.
22. GLEN A. SLACK, and C. GLASSBRENNER *Thermal Conductivity of Germanium from 3K to 1020K*, Physical Review, Volume 120, No 3, Nov 1, 1960.
23. H. S. CARSLAW and J. C. JAEGER *Conduction of Heat in Solids* Oxford University Press, Oxford, 1959.
24. C. J. GLASSBRENNER and GLEN A. SLACK *Thermal Conductivity of Silicon and Germanium from 3°K to the Melting Point* Physical Review, V 134, N 4A, 18 May 1964.
25. J. KHEDARI, P. BENIGNI, J. ROGEZ, and J. C. MATHIEU *New Apparatus for Thermal Diffusivity Measurements of Refractory Solid Materials by the Periodic Stationary Method* Rev. Sci. Instrum. 66 (1), January, 1995.
26. P. BENIGNI and J. ROGEZ *High Temperature Thermal Diffusivity Measurement by the Periodic Cylindrical Method: The problem of Contact Thermocouple Thermometry*, Rev. Sci. Instrum. **68** (3), July, 1997.
27. DAVID CAHILL, HENRY E. FISCHER, TOM KLITSNER, E. T. SWARTZ, and R. O. POHL *Thermal Conductivity of Thin Films: Measurements and Understanding* J. Vac. Sci. Technol. A, 7, 1260, 1989.
28. F. I. CHU, R. E. TAYLOR and A. B. DONALDSON, *Thermal diffusivity measurements at high temperatures by a flash method* Jour. of Appl. Phys. **51**, 336, 1980.
29. See for example: (a) A. B. DONALDSON and R. E. TAYLOR *Thermal diffusivity measurement by a radial heat flow method* J. Appl. Phys. **46**, 4584, 1975. J. W. VANDERSANDE and R. O. POHL *Simple Apparatus for the Measurement of Thermal Diffusivity between 80-500 K using the Modified Angstrom Method,* Rev. Sci. Instrum. **51**, 1694, 1980. J. GEMBAROVIC and R. E. TAYLOR *A New Data Reduction Procedure in the Flash Method of Measuring Thermal Diffusivity* Rev. Sci. Instrum **65**, 3535, 1994.
30. Thermal Transport Option for the Physical property Measurement system, Quantum Design Corp, San Diego CA. *See also Physical Properties Measurement System: Hardware and Operation Manual.*
31. L. PIRAUX, J.-P. ISSI, and P. COOPMANS Measurement **52**, (1987).
32. B. M. ZAWILSKI, R. T. LITTLETON IV, and TERRY T. TRITT *Investigation of the Thermal* Conductivity of the Mixed Pentatellurides $Hf_{1-X}Zr_XTe_5$, Appl. Phys. Lett. **77**, 2319 (2000).
33. T. C. HARMAN *Special Techniques for Measurement of Thermoelectric Properties.* J. Appl. Phys. **30**, 1373, 1959.
34. T. C. HARMAN, J. H. CAHN and M. J. LOGAN *Measurement of Thermal Conductivity by Utilization of the Peltier Effect* J. Appl. Phys. **30**, 1351, 1959.
35. A. W. PENN *The Corrections Used in the Adiabatic Measurement of Thermal Conductivity using the Peltier Effect.* J. Sci. Instrum. **41**, 626, 1964.
36. A. E. BOWLEY, L. E. J. COWLES, G. J. WILLIAMS, and H. J. GOLDSMID *Measurementof the Figure of Merit of a Thermoelectric Material* J. Sci. Instrum. **38**, 433, 1961.

Chapter 2.2

EXPERIMENTAL TECHNIQUES FOR THIN-FILM THERMAL CONDUCTIVITY CHARACTERIZATION

T. Borca-Tasciuc

Department of Mechanical, Aerospace and Nuclear Engineering Rensselaer Polytechnic Institute Troy, NY, USA

G. Chen

Mechanical Engineering Department, Massachusetts Institute of Technology, Cambridge, MA, USA

1. INTRODUCTION

Knowledge of the thermal conductivity of thin films and multilayer thin-film structures is critical for a wide range of applications in microelectronics, photonics, microelectromechanical systems, and thermoelectrics.[1-4] The last 20 years have seen significant developments in thin-film thermal conductivity measurement techniques.[5-8] Despite these advances, the characterization of the thermal conductivity of thin films remains a challenging task. Direct measurements of the thermal conductivity, for example, typically require the determination of the heat flux and the temperature drop between two points of the sample. Figure 1 shows a typical thin-film sample configuration and the experimental challenges associated with the thermal transport characterization of a thin-film-on-substrate system. Often, the thermal conductivity of thin films is anisotropic, as with polycrystalline thin films with columnar grains[9] or superlattices made of periodic alternating thin layers. The determination of the thermal conductivity in different directions (e.g., the cross-plane direction, which is perpendicular to the film plane, or the in-plane direction which is parallel to the film plane) encounters different challenges. For obtaining

Sample configuration and characterization challenges

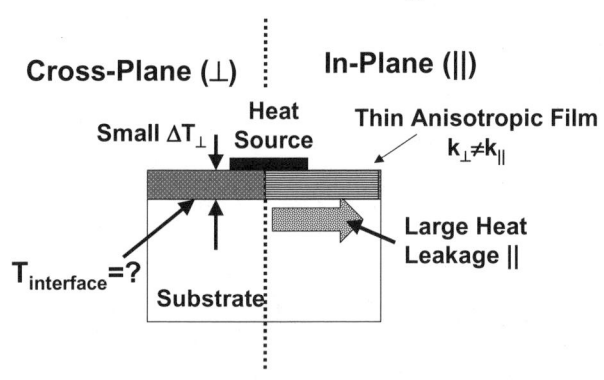

FIGURE 1 Typical sample configuration and challenges raised by the thermal conductivity characterization of thin films. Because the film typically has anisotropic thermal properties, measurements must be carried out in both the cross-plane and in-plane directions. For the cross-plane direction the difficulties consist of creating a reasonable temperature drop across the film without creating a large temperature rise in the substrate and experimentally determining the temperature drop across the film. The in-plane thermal conductivity measurement is affected by heat leakage through the substrate, which makes it difficult to determine the actual heat flow in the plane of the film.

the cross-plane thermal conductivity, the experiment typically requires finding out the temperature drop across the film thickness, which ranges from nanometers to microns. The difficulties consist of (1) creating a reasonable temperature drop across the film without creating a large temperature rise in the substrate, and (2) experimentally measuring the temperature drop across the film. On the other hand, the in-plane thermal conductivity measurement may look easier because temperature sensors can be placed along different locations on the film surface. However, heat leakage through the substrate makes it difficult to determine the actual heat flow in the plane of the film.

To overcome these difficulties, different strategies have been developed to measure the thin-film thermal conductivity, or thermal diffusivity, in different directions. These strategies are shown schematically in Fig. 2. In the cross-plane direction, the general strategies involve (1) creating a large heat flux through the film while minimizing the heat flux in the substrate and (2) measuring the surface temperature rather than the temperature drop. These can be realized by, for example, using microfabricated heaters directly deposited on the film. Furthermore, by using a small heater width, the heat flux going through the film is large, while the heat spreading inside the substrate significantly reduces the temperature drop in the substrate. Another technique is to use transient or modulated heating that limits the heat-affected region to small volumes within the film or its immediate surroundings. As for the in-plane direction, one often-used strategy is to remove the substrate, such that the heat flux through the film can be uniquely determined, or to deposit the films on thin, low-thermal-conductivity substrates, which minimizes heat leakage. Other methods have also been developed that take advantage of lateral heat spreading surrounding small heat sources.

In the following sections, different methods for thin-film thermophysical property measurements are presented. The techniques are categorized based on different heating and temperature sensing methods: electrical heating and sensing based

Sec. 1 · INTRODUCTION

Thermal characterization strategies

Cross-Plane (⊥)

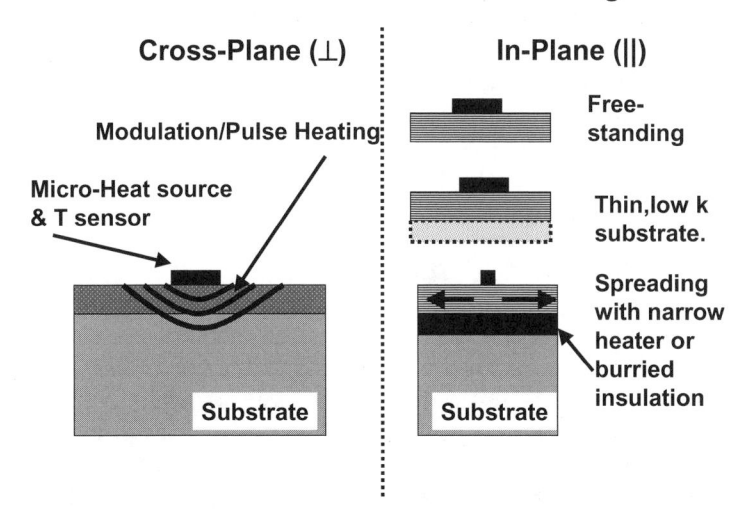

Modulation/Pulse Heating
Micro-Heat source & T sensor
Substrate

In-Plane (∥)

Free-standing
Thin, low k substrate.
Spreading with narrow heater or burried insulation
Substrate

FIGURE 2 Strategies developed to measure the thin-film thermal conductivity, or thermal diffusivity, in different directions. In the cross-plane direction, the general strategies are creating a large heat flux through the film while minimizing the heat flux in the substrate and measuring the surface temperature rather than the temperature drop. For the in-plane direction, often-used strategies are to remove the substrate such that the heat flux through the film can be uniquely determined or to deposit the films on thin, low-thermal-conductivity substrates, which minimizes heat leakages. Other methods take advantage of the lateral heat spreading that surrounds small heat sources.

TABLE 1. Summary of Thin-Film Thermophysical Properties Characterization Techniques

Thin-Film Thermophysical Characterization Techniques		
Electrical	**Optical heating**	**Hybrid**
3ω ⊥	**Time domain**	AC calorimetry ∥ ⊥
2 strips 3ω ∥ ⊥	-photoreflectance ⊥	
	-transient grating ∥	
2 sensor steady ∥	**Frequency domain**	Photo-thermoelectric ∥ ⊥
	-photoreflectance ∥ ⊥	
Membrane ∥	-emission ⊥	
	-displacement ⊥	Electro-reflectance ⊥
Bridge ∥	-deflection ∥ ⊥	
Spreader ∥	Photo-acoustic ⊥	Electro-emission ⊥

The techniques are categorized based on different heating and temperature sensing methods: electrical heating and sensing, optical heating, and combined electrical/optical methods.

The symbols on the right of each method indicate the direction [in-plane (∥), cross-plane (⊥), or both] along which the thermophysical properties characterization is performed.

on microheaters and sensors, optical heating and sensing methods, and combined electrical/optical methods. The methods presented in this chapter and their range of applications are summarized in Table 1. Representative methods in each category are selected for a more detailed discussion. Complementary information on thermophysical property characterization methods can also be found in several other review papers.[5–8]

2. ELECTRICAL HEATING AND SENSING

The thin-film thermal conductivity characterization methods presented in this section use electrical heating and sensing techniques. Some of the methods to be discussed employ the heaters also as temperature sensors, while in other techniques separate heaters and temperature sensors are used. Moreover, some of the techniques use the film itself as a heater and a temperature sensor. The advantages of electrical heating and sensing methods, compared to optical heating and sensing techniques later discussed, are that the amount of heat transfer into the sample can be controlled precisely and the temperature rise accurately determined. Such information makes it easier to obtain the thermal conductivity directly. Many of the techniques discussed employ microheaters and microsensors because high heat fluxes can be created and temperature rises can be pinpointed at micron scales. In the following discussion, the content is further divided into the cross-plane and the in-plane thermal conductivity measurements.

2.1. Cross-Plane Thermal Conductivity Measurements of Thin Films

Experimentally measuring the temperature drop across a film thickness that may be as small as a few nanometers makes the cross-plane thermal conductivity characterization of thin films a challenging task. Direct probing of the temperature at the film–substrate interface by electrical sensing is typically hard to achieve without the removal of the substrate, unless the sensors are fabricated before the deposition of the film. If the latter approach is taken, one must ensure that the films deposited on the temperature sensors have similar microstructures to the ones used in reality and that the sensors used are compatible to the fabrication processes. Some measurements on spin-on polymers adopt this approach,[10] but for most applications embedding temperature sensors underneath the film is not practical. Consequently, several methods have been developed to infer the temperature drop across the film. These techniques will be presented in the remainder of this section.

2.1.1. The 3ω method

A widely implemented approach for measuring the cross-plane thermal conductivity of thin films is the 3ω method.[11] Although initially developed for measuring the thermal conductivity of bulk materials, the method was later extended to the thermal characterization of thin films down to 20 nm thick.[11–19] Moreover, the 3ω technique was recently adapted for measurement of the in-plane and cross-plane thermal conductivity of anisotropic films and freestanding membranes.[20–22] This technique is currently one of the most popular methods for thin-film cross-plane thermal conductivity characterization; therefore it is discussed in more detail.

In the 3ω method a thin metallic strip is deposited onto the sample surface to act

Sec. 2 · ELECTRICAL HEATING AND SENSING

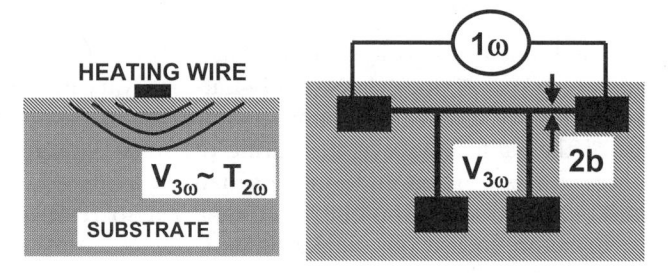

FIGURE 3. (a) cross-sectional view and (b) top view of the heater/temperature sensor deposited on the film on substrate sample in the 3ω method.

as both a heater and a temperature sensor as shown in Fig. 3. An AC current,

$$I(t) = I_0 \cos(\omega t), \quad (1)$$

with angular modulation frequency (ω) and amplitude I_0 passing through the strip, generates a heating source with power

$$P(t) = I_0^2 R_h \cos^2(\omega t) = \left(\frac{I_0^2 R_h}{2}\right)_{DC} + \left(\frac{I_0^2 R_h \cos(2\omega t)}{2}\right)_{2\omega}, \quad (2)$$

where R_h is the resistance of the strip under the experimental conditions. Therefore, the corresponding temperature rise in the sample is also a superposition of a DC component and a 2ω modulated AC component:

$$T(t) = T_{DC} + T_{2\omega} \cos(2\omega t + \varphi), \quad (3)$$

where $T_{2\omega}$ is the amplitude of the AC temperature rise and φ is the phase shift induced by the thermal mass of the system. If the resistance of the heater depends linearly on temperature, there is also a 2ω variation in the resistance of the heater:

$$R_h(t) = R_0\{1 + C_{rt}[T_{DC} + T_{2\omega} \cos(2\omega t + \varphi)]\}$$

$$= R_0(1 + C_{rt}TDC)_{DC} + (R_0 C_{rt} T_{2\omega} \cos(2\omega t + \varphi))_{2\omega}, \quad (4)$$

where C_{rt} is the temperature coefficient of resistance (TCR) for the metallic heater and R_0 is the heater resistance under no heating conditions. The voltage drop across the strip can be calculated by multiplying the current [Eq. (1)] and the resistance [Eq. (4)]:

$$V(t) = I(t)R_h(t) = [I_0 R_0 (1 + C_{rt} T_{DC}) \cos(\omega t)]_{\text{power\_source}}$$

$$+ \left(\frac{I_0 R_0 C_{rt} T_{2\omega}}{2} \cos(3\omega t + \varphi)\right)_{3\omega\_\text{mod}} + \left(\frac{I_0 R_0 C_{rt} T_{2\omega}}{2} \cos(\omega t + \varphi)\right)_{1\omega\_\text{mod}}. \quad (5)$$

This expression contains the voltage drop at the 1ω frequency based on the DC resistance of the heater and two new components proportional to the amplitude of the temperature rise in the heater, modulated respectively at 1ω and 3ω frequencies. The 3ω voltage component is detectable by a lock-in amplifier and is used to measure the temperature amplitude of the heater:

$$T_{2\omega} = \frac{2V_{3\omega}}{I_0 R_0 C_{rt}} \simeq \frac{2V_{3\omega}}{C_{rt} V_{1\omega}}, \quad (6)$$

where $V_{1\omega}$ is the amplitude of the voltage applied across the heater. The frequency-dependent temperature rise of the heater is obtained by varying the modulation frequency of the current at a constant applied voltage $V_{1\omega}$.

One challenge of the 3ω technique is measuring the small 3ω signals, which are usually about three orders of magnitude smaller than the amplitude of the applied voltage. Typically, cancellation techniques are employed to remove the large 1ω voltage component before measuring the 3ω signal.[11]

The heat conduction model used to obtain the substrate and the thin-film thermal conductivities is of crucial importance to the accuracy of the thermal conductivity determination. A simplified model using a line source approximation for the heater, one-dimensional frequency-independent heat conduction across the thin film, and two-dimensional heat transport in a semi-infinite substrate, is often used in the data analysis.[12] Under these assumptions the temperature amplitude of the heater can be written as

$$T_{S+F} = T_S + \frac{p\,d_F}{2\,b\,\text{w}\,k_F}, \tag{7}$$

where b is the half-width of the heater, p/w is the power amplitude dissipated per unit length of the heater, d_F is the film thickness, k_F is the cross-plane thermal conductivity of the film, and T_S is the temperature rise at the film–substrate interface.

The heater measures only the temperature rise on the top surface of the sample (T_{S+F}) which includes the temperature drops across the film and substrate. To determine the thermal conductivity of the film, one must know the temperature at the interface between the film and the substrate (T_S). This can be determined by either of two approaches. One approach is to calculate the temperature rise at the film–substrate interface. From the line source approximation the temperature amplitude of the heater on a bare semi-infinite substrate is[12,23]

$$T_S = \frac{p}{\pi\,\text{w}\,k_S}\left[0.5\ln\left\{\frac{\alpha_S}{b^2}\right\} - 0.5\ln(2\omega) + \eta\right] - i\left(\frac{p}{4\,\text{w}\,k_S}\right) = \frac{p}{\pi\,\text{w}\,k_S} f_{\text{linear}}(\ln\omega), \tag{8}$$

where α_S is the thermal diffusivity of the substrate, η is a constant ~ 1,[12] and f_{linear} is a linear function of $\ln \omega$. Combining Eq., (7) and (8), it becomes clear that the substrate thermal conductivity can be determined from the slope of the real part of the experimental signal as a function of $\ln(\omega)$. We will call this method the slope method.

A different way to estimate the temperature drop across a thin film is to infer it experimentally. The differential technique[23] measures the difference of the top surface temperature rise between the sample and a reference without the film or an identically structured film of lesser thickness. Figure 4 shows the method schematic and examples[24] of the temperature rise measured by 2-μm- and 30-μm-width heaters deposited onto the specimen (T_F) and reference sample (T_R). The sample is a 1.075-μm Ge quantum-dot superlattice film deposited on the Si substrate with an intermediate 50 nm Si buffer layer. The electrical insulation between the heater and the film is provided by an ~ 100 nm Si_xN_y layer.

Although Eq. (7) and Eq. (8) are often used in data analysis, one must bear in mind that these are approximations and have limitations. A more detailed, two-dimensional heat conduction model has been developed and used to analyze the

Sec. 2 • ELECTRICAL HEATING AND SENSING

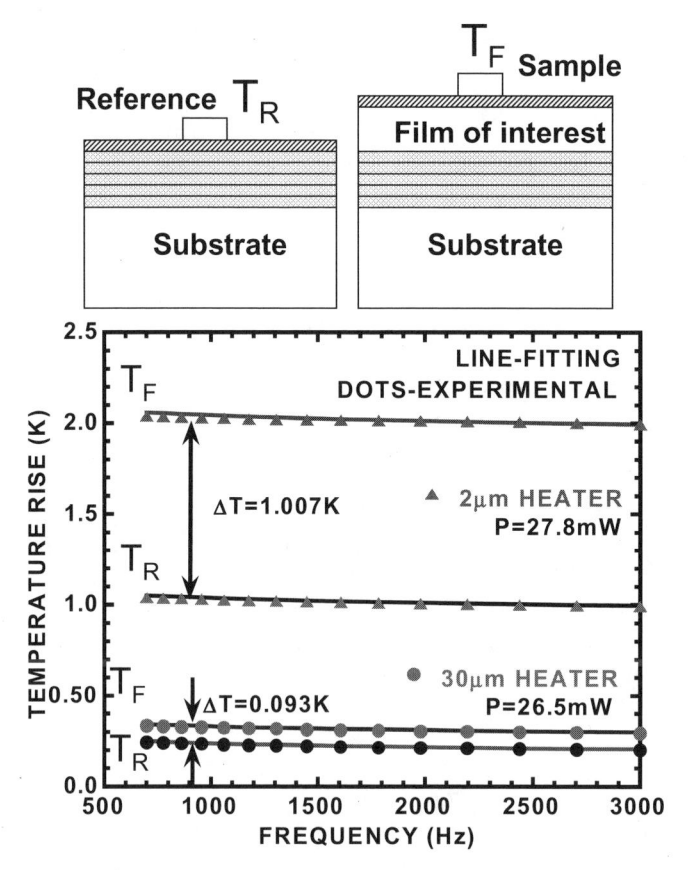

FIGURE 4. Example of an experimentally determined temperature drop across a Ge quantum-dot superlattice film using a differential 3ω method. In this method similar heaters are deposited on the sample and a reference without the film of interest. The temperature drop across the film is given by the difference in the temperature rise of similar heaters under similar power dissipation conditions. If the experiment is performed using a pair of large-width or narrow-width heaters, the anisotropic properties of the film may be determined. The solid lines are calculated by using Eq. (9) and the fitted thermophysical properties of the sample.

validity of these simplifications.[23] Some of the findings are discussed later in the context of an anisotropic film deposited on an isotropic substrate. The model indicates that for the line source approximation [Eq. (8)] to be valid (within 1% error), the thermal penetration depth in the substrate $\sqrt{\frac{\alpha_S}{2\omega}}$ must be at least 5 times larger than the heater half-width. Also, for the semi-infinite substrate approximation required by Eq. (8) to be valid (within 1% error), the thermal penetration depth is at least 5 times smaller than the substrate thickness. Since these conditions put opposite requirements on the frequency range of the measurement, it implies that if the ratio between the substrate thickness and heater half-width is less than 25, it is not possible to simultaneously minimize the substrate effects and still satisfy the line source approximation within 1% error. In addition to these considerations, the specific heat of the heater and the film, the thermal resistance of the film, and the thermal boundary resistance between the film and the substrate also affect the accuracy of Eq. (8) and, consequently, the substrate thermal conductivity determination using the slope method, which is explained in a later section.

212 Chap. 2.2 · EXPERIMENTAL TECHNIQUES FOR THIN-FILM THERMAL CONDUCTIVITY CHARACTERIZATION

The approximation of steady-state one-dimensional heat conduction across the thin film [i.e., Eq. (7)] requires that the substrate thermal conductivity be much higher than that of the film. This is easily understood from the limiting case of a semi-infinite substrate deposited with a film of identical material as the substrate. In this situation the temperature sensor on the surface measures the same temperature as without the film, and, correspondingly, the temperature drop across the film is zero (neglecting thermal boundary resistance). Furthermore, Eq. (7) is based on one-dimensional heat conduction across the thin film and requires minimization of the heat spreading effects inside the film, which depends on the film thermal conductivity anisotropy K_{Fxy} (ratio between the in-plane and cross-plane thermal conductivities) and the aspect ratio between heater width 2.6 and film thickness d_F. For example, if the cross-plane film thermal conductivity is much smaller than the substrate thermal conductivity and $\sqrt{k_{Fxy}}\frac{d_F}{b} < 10$, the heat spreading effect could be accounted for by the one-dimensional heat conduction model if the width of the heater is replaced by a "corrected" width of $2b + 0.76 d_F \sqrt{k_{Fxy}}$.

In situations where such conditions do not hold, the general expression for the heater temperature rise on a multilayer film–on–finite substrate system with anisotropic thermophysical properties must be used. Neglecting the contributions from the heat capacity of the heater and thermal boundary resistances, the complex temperature amplitude of a heater dissipating p/w W/m peak electrical power per unit length is[23,25]

$$\Delta T = \frac{-p}{\pi w k_{y_1}} \int_0^\infty \frac{1}{A_1 B_1} \frac{\sin^2(b\lambda)}{b^2 \lambda^2} d\lambda, \qquad (9)$$

where

$$A_{i-1} = \frac{A_i \frac{k_{y_i} B_i}{k_{y_{i-1}} B_{i-1}} - \tanh(\varphi_{i-1})}{1 - A_i \frac{k_{y_i} B_i}{k_{y_{i-1}} B_{i-1}} \tanh(\varphi_{i-1})}, \; i = \text{from 2 to 12} \qquad (10)$$

$$B_i = \left(k_{xyi} \lambda^2 + \frac{i 2\omega}{\alpha_{yi}} \right)^{1/2}, \qquad (11)$$

$$\varphi_i = B_i d_i \quad k_{xy} = k_x/k_y. \qquad (12)$$

In these expressions, n is the total number of layers including the substrate, subscript i corresponds to the i_{th} layer starting from the top, subscript y corresponds to the direction perpendicular to the film/substrate interface (cross-plane), b is the heater half-width, k_y and k_x are respectively the cross-plane and the in-plane thermal conductivity of the layer, ω is the angular modulation frequency of the electrical current, d is the layer thickness, and α is the thermal diffusivity. If the substrate layer ($i = n$), is semi-infinite, $A_n = -1$. When the substrate has a finite thickness, the value of A_n depends on the boundary condition at the bottom surface of the substrate: $A_n = -\tanh(B_n d_n)$ for an adiabatic boundary condition or $A_n = -1/\tanh(B_n d_n)$ if the isothermal boundary condition is more appropriate.

The above analytical expressions do not include the effects of the thermal boundary resistance between the heater and the film and the heat capacity of the heater. For a heater of thickness d_h and heat capacity $(\rho c)_h$, with thermal boundary resistance R_{th} between the heater and the first film, and neglecting heat conduction effects inside the heater, the complex temperature rise of the heater becomes[23]

$$T_h = \frac{\Delta T + R_{th}p/2bw}{1 + (\rho c)_h d_h i2\omega(R_{th} + \Delta T 2bw/p)}, \quad (13)$$

where ΔT is the average complex temperature rise determined from Eq. (9) for a multilayer-film-on-substrate structure under the same heating power with zero heater heat capacity and zero thermal boundary resistance. The thermal boundary resistance, and equivalently, the film heat capacity and thermal resistance, impacts the determination of the substrate thermal conductivity when using the slope method [based on the simple expression in Eq. (7) and Eq. (8)]. Indeed, experimental results of SiO_2 films on silicon substrates show that the substrate thermal conductivities determined from the slope method are smaller than standard values,[15] and this trend is attributed to the effect of the additional thermal resistance of the SiO_2 film.[23]

The 3ω technique employs modulation of the heat source, which brings several advantages. One is that the AC temperature field can be controlled by the modulation frequency. Choosing a reasonably high modulation frequency range, the AC temperature field is confined to the region close to the heater such that the substrate can be treated as semi-infinite. This approach avoids the influence of the boundary condition at the substrate side. Another advantage of the modulation technique is that the AC signal is less sensitive to the radiation heat loss compared to DC measurement methods. The AC modulation also leads to the possibility of determining the substrate properties in addition to the film properties.

Moreover, by performing the measurements over a wide frequency range it is possible to determine both the thermal conductivity and volumetric heat capacity of thin films.[26] The thermal conductivity measurement is performed at relatively low frequencies, where the temperature drop across the film is frequency independent. On the other hand, at adequately high frequencies, the heat-affected region can effectively be confined into the film, in which case the temperature drop across the film becomes sensitive also to the volumetric heat capacity of the film.

2.1.2. Steady-State Method

A steady-state method has also been employed to determine the cross-plane thermal conductivity of dielectric films and the thermal boundary resistance between a microfabricated heater/thermometer and the substrate.[27,28] In the 3ω technique just one metallic strip acts as both the heater and the temperature sensor, whereas in the steady-state method at least two, and sometimes three, strips are deposited onto the film. One of the strips has a large width and serves as the heater and the thermometer for the temperature rise of the film surface. A second thermometer is situated in the vicinity of the heating strip and provides the temperature rise of the substrate at a known distance away from the heater. One can use a two-dimensional heat conduction model and a known thermal conductivity of the substrate, to infer the temperature rise of the substrate at the heater location from the temperature rise of the second thermometer. If the substrate thermal conductivity is unknown, a third thermometer situated at a different position away from the heater can be used to determine the substrate thermal conductivity. Typically the difference between the temperature of the second thermometer and the temperature of the substrate underneath the first thermometer is small because of the short separation of the thermometers and because of the large thermal conductivity of the substrate. The temperature drop across the film is then determined by taking the difference between the experimental temperature rise of the heating strip and the predicted temperature rise of the substrate at the heater location. The heat spreading into the film is typically neglected for low-thermal-conductivity films, and a one-dimensional heat conduction model is used to determine the thermal conductivity of the film. For high-thermal-conductivity films such as silicon, heat spreading may allow the in-plane thermal conductivity for special sample configurations to be determined, as discussed later. The thermal boundary resistance between the heater and the substrate could also be determined by performing the experiment with the thermometer array deposited directly onto the substrate.[27] Because the measurement is done at steady state, the boundary condition at the backside of the substrate is important for determining the temperature underneath the heater.

2.2. In-Plane Thermal Conductivity Measurements

As shown in Fig. 1, the main challenge in determining the in-plane thermal conductivity of a thin film is estimating the heat transfer rate along the film because the heat leakage to the substrate can easily overwhelm the heat flow along the film. To address this issue, several strategies have been developed, such as (1) depositing the films on thin substrates of low thermal conductivity, (2) making freestanding film structures by removing the substrate, (3) using microheaters and temperature sensors and/or special sample configurations to sense the lateral heat spreading in the film.

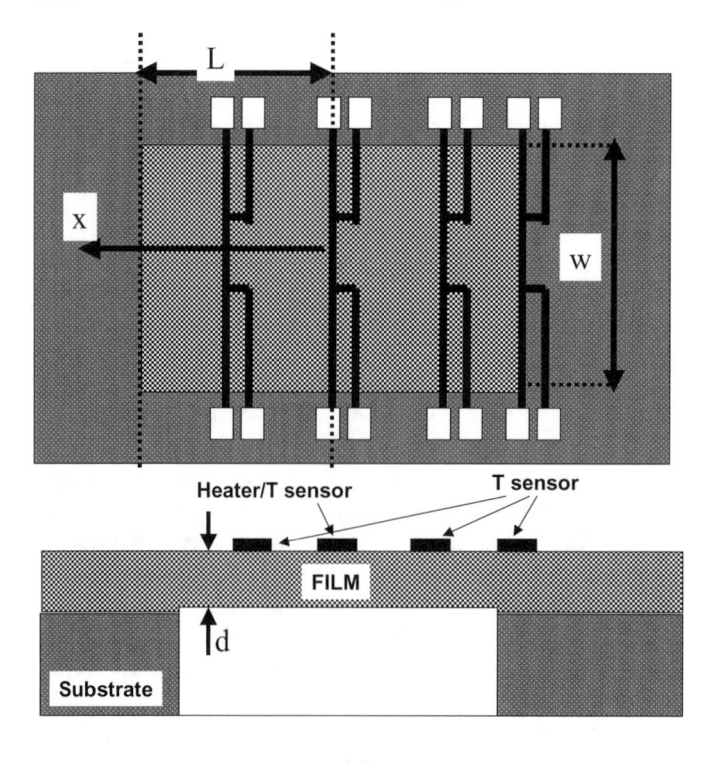

(a)

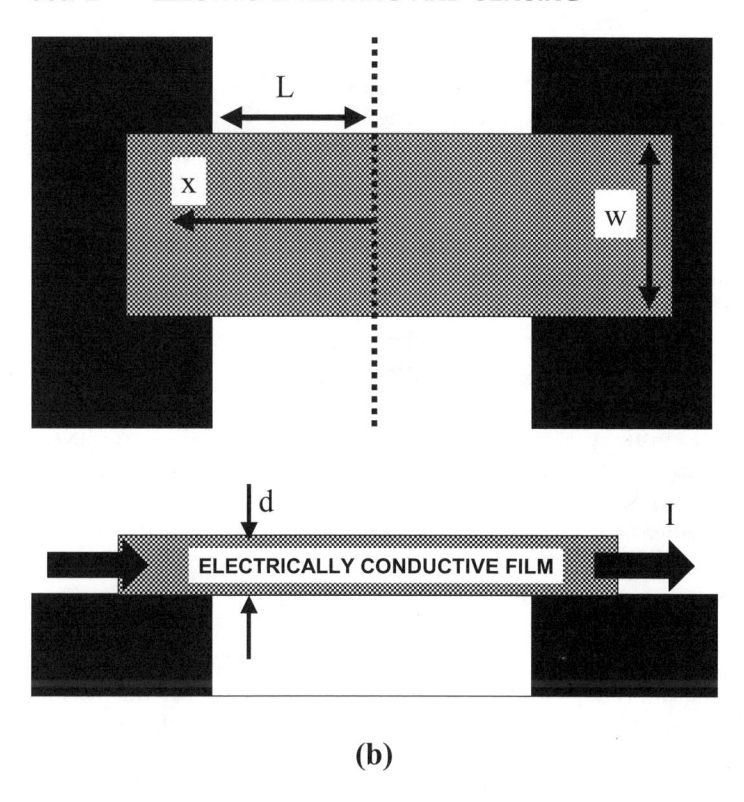

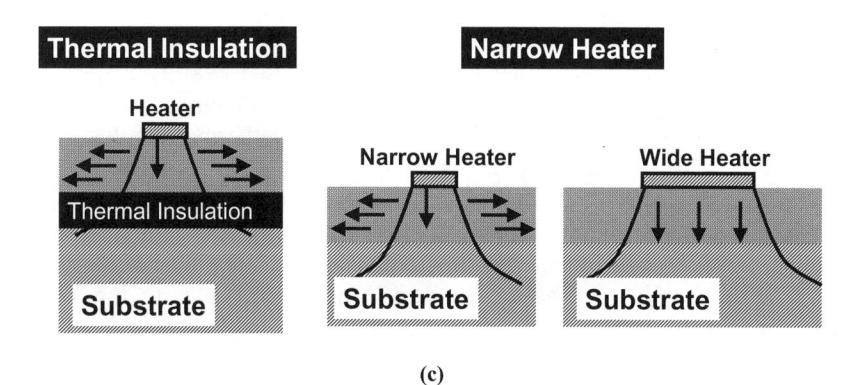

FIGURE 5 Major experimental methods for in-plane thermal conductivity characterization: (a) configuration of the membrane method; (b) configuration of the bridge method; (c) in-plane heat spreading methods using buried thermal barriers and narrow-width heaters. A qualitative comparison between the spreading effects of narrow and wide heaters is shown.

Several in-plane thermal conductivity measurements have been developed with the first two strategies by making suspended structures.[29–40] Representative concepts are shown in Fig. 5. In one configuration the heater and temperature sensors are fabricated on a large membrane (Fig. 5a).[30–39] The membrane can be a thin-film-on-thin-substrate structure suspended between massive heat sinks;[29,30] or the substrate underneath the film may be removed over an area, making the film a free-standing structure with the frame playing the role of a heat sink.[32–38] In both situations, heat generated from the heater spreads from the middle of the membrane toward its edges and the temperature profile is detected by one[32] (the heater itself), two,[34–38] or more thermometers[39] situated at various locations of the membrane. Alternatively, the membrane can be shaped as a cantilever beam and the heater can be suspended at one edge,[33,35,40] in which case the heat will be conducted toward the heat-sink edge of the film. On the other hand, if the membrane is conducting or semiconducting, another configuration shown in Fig. 5b is to shape the membrane into a bridge and to pass current directly through the film.[29,31] In this case heat spreads out along the bridge axis direction and the average temperature of the bridge is measured and correlated to the thermal conductivity.

The in-plane thermal conductivity of film-on-thick-substrate systems can be measured by exploring the heat spreading effect (Fig. 5c) into the film by using small heaters[10,23] or buried thermally insulating layers, which provide additional thermal resistance between the film and the substrate, to force more heat flow along the film plane direction.[41,42]

The foregoing strategies and their applications to various in-plane thermal conductivity methods will be discussed in greater detail.

2.2.1. Membrane Method

The principle of the membrane method for the in-plane thermal conductivity characterization is shown in Fig. 5a. The membrane film is supported around the edges by a relatively massive frame, usually the substrate onto which the film is grown, which also plays the role of heat sink. In the middle of the membrane, parallel with the membrane width, a thin, narrow strip of electrically conducting material acts as a heater and temperature sensor when electrical current passes through it. The temperature rise of the heater is determined typically by measuring the change in its electrical resistance. The experiments are performed in a vacuum ambient to minimize the effect of convection heat transfer. The temperature response of the strip under steady-state, pulsed, and modulation heating, coupled with appropriate heat transport modeling, can be used to determine the thermal conductivity, thermal diffusivity, and heat capacity of the membrane.

Steady-State Methods. The steady-state heating method for in-plane thermal conductivity measurement is discussed first, followed by transient heating techniques. In the simplest case, assuming one-dimensional steady-state heat conduction in the plane of the film and along the direction perpendicular to the axis of the heating strip, the thermal conductivity of the film can be determined as

$$k = \frac{pL}{wd(T_h - T_s)}, \qquad (14)$$

where p/w is power dissipated in the heater per unit length, L is the distance from

Sec. 2 · ELECTRICAL HEATING AND SENSING

the heater to the heat sink, T_h is the heater temperature rise, and T_s is the temperature of the sink.

If the membrane is made of more than one film, and/or the substrate is not an ideal heat sink, a more complex thermal resistance network should be employed instead. Figure 6 shows an example of a one-dimensional thermal network used to describe the heat conduction in a Si membrane.[43] In this experimental configuration the heater at the center of the membrane, which also acts as a resistive thermometer, is electrically insulated from the Si membrane by a Si_3N_4 layer. The thermal resistance network thus includes the thermal resistance of the Si_3N_4 film directly underneath the heater, the silicon membrane resistance, the spreading thermal resistance of the substrate, and contact resistance between the silicon substrate and a large heat sink. The spreading resistance in the substrate and thermal contact resistance can be important if the measured membrane has a relatively high thermal conductivity. In order to determine the substrate spreading resistance, a temperature sensor should be deposited on the substrate at the place where the membrane meets the substrate. Alternatively, temperature sensors can also be deposited on the membrane. Such an arrangement, however, can lead to complications due to the (usually) high thermal conductivity of the sensor material, which can create additional heat leakage along the temperature sensors. Figure 7 shows the various thermal resistances determined experimentally as a function of temperature. The thermal resistance of the Si_3N_4 layer is determined from experimentally measured thermal conductivity of the Si_3N_4 layer by using the 3ω method. The in-plane thermal conductivity of the silicon membranes is calculated by isolating the thermal resistance of the membrane, $R_{Membrane}$, from the total thermal resistance, R_{heater}, after subtracting the thermal resistance of the Si_3N_4 layer underneath the heater. The combined spreading and the contact resistance in the substrate and between the substrate and sample holder are determined by the edge temperature sensor. Figure 7b shows an example of the measured thermal conductivity of a 4.67-μm-thick silicon membrane, as determined by the aforementioned method, together with bulk experimental data of silicon.

Several conditions must be met in order to fulfill the one-dimensional approximation: (1) the designed geometry of the membrane must force heat conduction in the film along the direction perpendicular to the heater length; (2) the heat con-

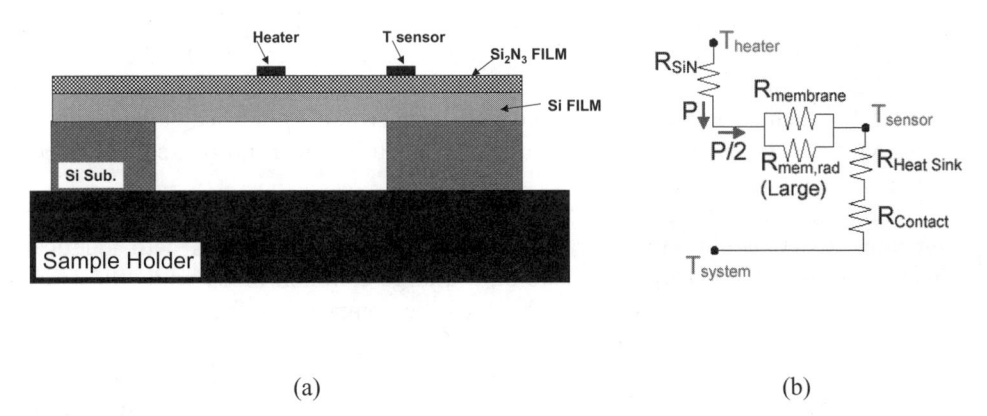

FIGURE 6. (a) Sketch of the sample and (b) thermal resistance network for a membrane on a nonideal heat sink.

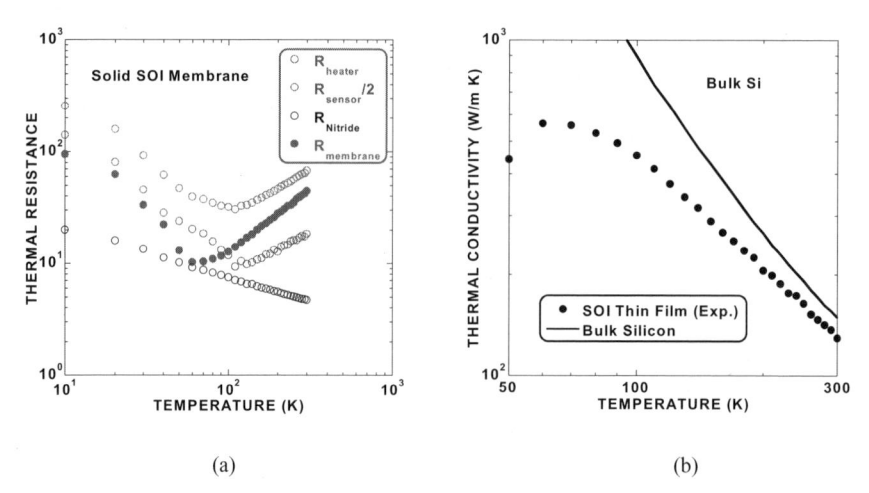

FIGURE 7 (a) Experimentally determined thermal resistance, including the total heater thermal resistance, the Si_3N_4 thermal resistance, and the heat sink thermal resistance for sample configuration as shown in Fig. 6(a); and (b) thermal conductivity of a 4.67-μm single-crystal silicon membrane together with that of bulk silicon crystal.

duction loss along the heater and temperature sensors must be reduced to a minimum; (3) radiation loss must be minimized.

One way to address the first requirement is to shape the membrane as a cantilever beam connected with just one side to the substrate while the heater is suspended at the opposite side,[33,35] as opposed to anchoring the membrane with all sides to the substrate. To reduce heat conduction loss along the heater and temperature sensors, one can use IC fabrication methods to minimize the heater/sensor cross-sectional area. Finally, the radiation heat loss can be reduced by performing the experiments under small temperature rise or by coating the high-temperature-rise areas with low-emissivity materials. However, if the aforementioned conditions cannot be met, more sophisticated heat transport models must be applied to determine the in-plane thermal conductivity of the film. Alternatively, AC or transient-based heating and measurements can reduce the effects of heat loss through the metal heater and by thermal radiation.[34]

The following paragraphs describe several implementations of in-plane thermal conductivity characterization by steady-state membrane methods. In the first example bulk heaters and temperature sensors are employed, while in the second example the membrane is instrumented with a microfabricated heater and microfabricated temperature sensors. Additional reports using microfabricated test structures are briefly discussed.

In one of the earliest methods[40] to determine the in-plane thermal conductivity of thin films, pairs of film-on-substrate and bare substrate specimens with identical dimensions are mounted with one side onto a large common block while the other sides are suspended. On the side opposite to the common block, a small piece of lead sheet serving as a radiative heat sink is attached to each membrane. The experiment is carried out in vacuum by heating the common block above the ambient temperature, which is kept constant. The method does not use microfabricated thermometers. Instead, fine thermocouples are used to monitor the temperature of the common block and each suspended heat sink. The heat transfer rate through each sample is calculated from the radiative heat loss to the ambient of the suspended heat sink. A one-dimensional heat conduction model is used to relate the temperature drop along the samples to the heat transfer rate and to infer the thermal conductivity of the film from the difference between the effective thermal conductivities of the film-on-substrate and bare substrate samples. The method relies on one's capability of making geometrically identical sample and substrate specimens, with identical heat sinks attached to each sample, and is limited to film-on-substrate systems on which the in-plane heat conduction along the film is significant compared to the total heat flow through the film-on-substrate system, such as copper films on thin mica substrates where the reported contribution of the films was ~30–35%. Moreover, careful consideration must be given to radiation loss through the membrane itself if the sample has a large emmissivity or if the experiment is performed at high temperature.

In this technique the film thermal conductivity was extracted from the thermal conductivity of the film-on-substrate system. For sufficient accuracy the product of thermal conductivity and film thickness must be comparable to the product of thermal conductivity and substrate thickness. One strategy is to deposit the films on thin substrates of extremely low thermal conductivity. In this method,[30] two opposite sides of the membrane are anchored between two relatively massive copper blocks, which serve as heat sinks. Through microfabrication techniques, a heater/temperature sensor is deposited in the middle of the freestanding section of the membrane, with its axis parallel to the membrane sides that are anchored to the copper blocks. The heat transfer model assumes the heat flow within the foil is essentially one dimensional and takes into account radiation losses and the heat conduction loss at the edges of the heater. To determine simultaneously the unknown thermal conductivity and emmissivity of the membrane, one performs measurements on at least two identical foils of the same thickness but different lengths. The properties of the insulating foils are first measured. Then thin films, either electrically conducting or insulating, can be deposited onto the free side (no heater) of the foils, and the experiment is repeated. The film properties are extracted by subtracting the foil properties from the effective thermal properties of the film-on-substrate system. The method allows for *in situ* thermal conductivity measurement of thin films during deposition.

In this example the heating and sensing strips were patterned on the membrane through microfabrication technology.[44] Due to its intrinsic advantages, such as ease of miniaturization, integration, and batch fabrication, microfabrication is employed to make better test structures for the in-plane thermal conductivity measurements. For example, the technology may allows us to selectively remove the substrate from underneath a well-defined area of the film. Film-only freestanding test structures fabricated in this way make it possible to determine the in-plane thermal conductivity of low-thermal-conductivity thin films deposited on thick and high-thermal-conductivity substrates. This approach has been used, for instance, to determine the thermal conductivity of silicon nitride,[34] SiO_2–Si_3N_4 sandwich films,[32] thermal CMOS MEMS films,[35] doped polysilicon,[36] and single-crystal silicon films,[38] which were all initially deposited on silicon substrates. Moreover, microfabrication facilitates deposition of thermometer arrays for measuring the temperature profile of the membrane at variable distances from the heater.[34–39] With a small constant current passing through the sensor strip, the voltage change across the sensor reflects the temperature changes of the film at the sensor location. This strategy reduces the uncertainty due to the unknown thermal boundary resistance between the heater and the film underneath. However, when multiple-sensor arrays are used, the heat losses introduced by the sensors must be minimized.[39]

Transient Heating Methods. This section addresses thin-film in-plane thermophysical properties characterization methods based on transient or periodic electrical heating and the corresponding temperature sensing. The heating source is obtained by passing a nonsteady electric current through the heating strip. Relative to the time dependence of the heating source, the transient techniques can be categorized as pulse and modulation techniques. Some techniques are used to determine both thermal conductivity and thermal diffusivity of the film, since under transient heating the temperature response of the film depends also on its thermal mass. In the following, the pulsed heating technique is presented first, followed by the modulation heating technique.

The heat pulse method employs a heat pulse induced by passing an electric current pulse through the heating strip. For a rectangular heat pulse of duration t_0 and for a simple one-dimensional transient heat conduction model, the analytical solution for the temperature response $T(x,t)$ of a thin film membrane is[34]

$$T(x,t) - T_0 = \frac{2A}{k}(L-x) + \sum_{n=0}^{\infty} B_n \cos(\lambda_n x) \exp(-\alpha \lambda_n^2 t) \qquad \text{for } 0 < t < t_0,$$

(15)

$$T(x,t) - T_0 = \sum_{n=0}^{\infty} B_n \cos(\lambda_n x) \exp(-\alpha \lambda_n^2 t)\left[1 - \exp(\alpha \lambda_n^2 t_0)\right] \qquad \text{for } t > t_0,$$

where α is the thermal diffusivity along the membrane, $2A$ is the magnitude of the heat flux along x, and

$$B_n = -\frac{16AL}{k(2n+1)^2\pi^2}, \qquad \lambda_n = \frac{(2n+1)\pi}{2L}.$$

(16)

The experiments measure the transient temperature profile at the sensor position during and after the heat pulse. If the specific heat and density of the sample are known, the thermal diffusivity or the thermal conductivity can be found by fitting the experimental temperature profile with the help of Eq. (15). Some examples using the electrical pulse heating technique include thermal diffusivity measurement of freestanding silicon nitride films,[34] thermal diffusivity and specific heat capacity of SiO_2–Si_3N_4 sandwich films,[34] and heat capacity of thin organic foils.[30] A one-dimensional theoretical model which includes radiative heat transfer is presented in Ref. 30.

Modulation heating techniques employ an AC current modulated at angular frequency ω passing through the heating strip, which generates a DC and an AC heating component modulated at 2ω. This is similar to electrical heating in the 3ω method. The solution for the temperature rise in the membrane is the superposition between a DC temperature component and an AC temperature (T_{ac}) modulated at 2ω angular frequency. For modulation frequencies where the thermal penetration depth is much larger than the thickness, the temperature gradients across the membrane thickness can be neglected and heat transfer assumed to take place just in the plane of the membrane. Using complex representation, one can write the temperature rise at a position x away from the heater as

Sec. 2 · ELECTRICAL HEATING AND SENSING

$$T_{ac} = \theta(x)e^{i2\omega t}, \tag{17}$$

where θ is the complex amplitude of the ac temperature rise. In a purely one-dimensional heat conduction model with no convection and radiation losses, the solution for the complex temperature rise in the semi-infinite membrane is

$$\theta(x) = \frac{q}{km}\frac{1}{1+e^{2mL}}\left[e^{mx} - e^{m(2L-x)}\right], \tag{18}$$

where q is the amplitude of the heat flux generated by the heater and $m = \sqrt{\frac{i2\omega}{\alpha}}$. Due to the membrane symmetry, half of the generated heat flux in the heater contributes to the temperature rise. Equation (15) contains both the amplitude and the phase of the AC temperature and can be used to determine the thermophysical properties of the film by fitting the experimental temperature signals collected at different locations and modulation frequencies. Furthermore, if $L \rightarrow \infty$ (the heater is far away from the heat sink), Eq. (15) becomes

$$\theta(x) = \frac{q}{km}e^{-mx}, \tag{19}$$

after taking the logarithm of the temperature amplitude, we obtain for Eq. (16):

$$\ln\{\text{Amp}[\theta(x)]\} = \ln\left(\frac{q}{k}\sqrt{0.5\alpha}\right) - 0.5\ln(\omega) - \sqrt{\frac{\omega}{\alpha}}x. \tag{20}$$

The phase of the temperature signal is

$$\text{phase} = \frac{\pi}{4} + \sqrt{\frac{\omega}{\alpha}}x. \tag{21}$$

Equations (17) and (18) suggest several ways to back out the thermal diffusivity of the film. If the thermal signal is collected at the same frequency for different locations (using an array of thermometers), the thermal diffusivity of the membrane can be inferred from the slope$_x$ of the phase (in radians) and/or the slope of the ln(Amp) plotted as a function of location x:

$$\alpha = \frac{\omega}{\text{slope}_x^2}. \tag{22}$$

Furthermore, if the signal is collected at the same location x_0 but for different modulation frequencies, the thermal diffusivity of the membrane can be inferred from the slope$_{\sqrt{\omega}}$ of the phase and/or the slope of the ln(Amp*$\sqrt{\omega}$) plotted as a function of $\sqrt{\omega}$:

$$\alpha = \frac{x_0^2}{\text{slope}_{\sqrt{\omega}}^2}. \tag{23}$$

In order to determine the thermal conductivity of the membrane, the temperature amplitude for different modulation frequencies is collected at $x = 0$ and plotted as a function of $\frac{1}{\sqrt{2\omega}}$. Equation (17) implies

$$\frac{\sqrt{\alpha}}{k} = \frac{\text{slope }^1/\sqrt{2\omega}}{q}. \tag{24}$$

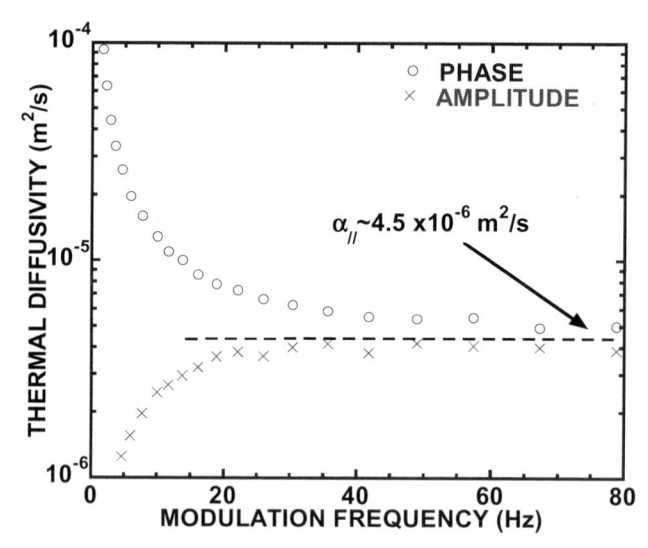

FIGURE 8 Thermal diffusivity of the film calculated based on Eq. (19) as a function of the modulation frequency of the current. At low frequencies, the one-dimensional approximation is not valid, and the model yields frequency-dependent values for thermal diffusivity.

Therefore, the thermal conductivity can be extracted by using the previously determined values of the film diffusivity.

Examples using the modulation heating technique include in-plane thermal diffusivity characterization of silicon nitride films[34] and Si/Ge superlattice structures.[39] For instance, Fig. 8 shows the in-plane thermal diffusivity of a freestanding Si/Ge superlattice membrane,[39] calculated by using Eq. (19) for both amplitude and phase signals and as a function of modulation frequency. At low frequencies the one-dimensional approximation is not valid, and the model yields frequency-dependent values for thermal diffusivity. For frequencies larger than 40 Hz the calculated thermal diffusivity becomes frequency independent, indicating the applicability of the one-dimensional model. In order to include the effect of lateral heat spreading, an analytical two-dimensional heat conduction model was also developed.[39]

2.2.2. Bridge Method

The principle of the bridge method for the in-plane thermal conductivity characterization is shown in Fig. 5b. The method is only applicable if a current can pass through the film itself or another conducting layer with known properties deposited onto the film. The film is patterned as a thin strip, which bridges the gap between two heat sinks. The film itself serves as a heater and temperature sensor when an electric current passes through it. The temperature rise of the heater is determined by measuring the change in its electrical resistance. Similar to the membrane method the experiments are performed in a vacuum ambient to minimize the effect of convection heat transfer. The temperature response of the strip under steady-state, pulsed and modulation heating, coupled with appropriate heat transport modeling, can be used to determine the thermal conductivity, thermal diffusivity, and heat capacity of the film.[29] In the following, applications of steady-state and transient, bridge method will be discussed.

Sec. 2 · ELECTRICAL HEATING AND SENSING

Steady-State Methods. The bridge method constrains the heat flow to one dimension. In this sense it avoids the potential complication of two-dimensional heat conduction effects in the membrane. However, the thermal conductivity depends on temperature profile along the bridge, which further depends on the radiation and convection heat losses (if the measurement is not done in vacuum) and on the thermal resistance at the two ends of the bridge. In a one-dimensional steady-state model, neglecting convection effects, the average temperature of the bridge T_M is[45]

$$T_M - T_0 = \frac{I^2 R(T_0)}{\pi^2 k d w / 2L + 16 \varepsilon \sigma T_0^3 L w}, \tag{25}$$

where σ is the Stefan–Boltzmann constant, T_0 is the ambient and heat sink temperature, ε is emissivity of the bridge surface, and R is the bridge electrical resistance. The unknown thermal conductivity and emissivity can be determined simultaneously by performing measurements on identical films that have different bridge lengths. The technique has been employed, for example, to determine the in-plane thermal conductivity of bismuth films[29] deposited on thin insulating foils, enabling one to measure the thermal conductivity of films as thin as 10 nm and with low thermal conductivities (1 W/m-K). The films are evaporated onto very thin organic foils, as thin as 40 nm and with a thermal conductivity of 0.2–0.25 W/m-K. However, to find the thermal conductivity of the film, the substrate thermal conductivity must be known since the method does not apply for nonconducting materials (such as the organic foils used for substrates). Other in-plane thermal conductivity measurements using the steady-state bridge method have been reported for uniformly doped polysilicon bridges.[31]

A different embodiment of the bridge method is applicable to materials that allow embedding the heater within a spatially well-defined region of the bridge by locally changing the electrical resistivity of the materials. This allows for localized heat generation and temperature sensing. One example of this approach is the thermal conductivity measurement of heavily doped low-pressure chemical-vapor-deposited polycrystalline silicon films.[31] In this method the polycrystalline silicon film is patterned in the shape of a bridge suspended above the silicon substrate on oxide pillars. The bridge has a narrow, lightly doped region at its center and is heavily doped elsewhere. The resistance of the heavily doped region is negligible compared with that of the lightly doped region. This doping profile concentrates heating and temperature sensing at the center of the bridge when a current passes through it. Under a simple one-dimensional heat conduction model, neglecting convection and radiation losses, the temperature profile along the bridge is linear. Moreover, if the bridge supports remain at the ambient temperature during heating, the thermal conductivity of the bridge can be determined from the slope of the power dissipated in the center region as a function of the center temperature of the bridge. However, to infer the temperature of the heater region, the temperature dependence of resistance must be carefully calibrated. One strategy is to submerge the setup in an oil bath with controlled temperature and to measure the current–voltage characteristic of the bridge as a function of the bath temperature. Ideally, small currents should be used during the calibration such that negligible heating is produced into the bridge and the system can be considered isothermal. However, if the probing currents are large, the heating of the bridge center above oil temperature must be considered.[31] The heating of the bridge supports may also need to be taken into account.[31]

Transient Heating Methods. By using heat pulse and modulation heating, one can adapt the bridge method to determine the thermal diffusivity or the specific heat capacitance of the film. Pulse heating is discussed first.

In pulse heating, when an electric current is pulsed through the film the instantaneous experimental temperature rise of the film is related to the time constant of the bridge by[45]

$$T_M - T_0 \sim [1 - \exp(-t/\tau)]. \tag{26}$$

With a one-dimensional transient heat conduction model (which includes radiation loss), the time constant can be related to the specific heat capacity of the bridge[45] by

$$\tau = \frac{\rho c d 2 L w}{\pi^2 k d w/2L + 16\varepsilon\sigma T_0^3 L w}. \tag{27}$$

where ρ is density and c is specific heat of the bridge.

Therefore, in combination with the steady-state method, which yields the thermal conductivity of the film, the transient heat pulse method can be employed to determine the specific heat or thermal diffusivity. To measure the specific heat, one monitors the time dependence of the average film temperature immediately after turning on the heating voltage in a form of a step function, and fits the experimental time constant with Eq. (27). For a film-on-thin-substrate bridge, the specific heat of the substrate foil must be known and cannot be much larger than the film in order to perform the specific heat measurements accurately.

Modulation heating employs an AC current passing through the bridge. Similar to discussions from previous sections, through thermoresistive effects the modulation heating generates a 3ω voltage proportional to the amplitude of the temperature oscillations of the bridge. Thus, a model for heat transport under modulation heating in the bridge setup was developed.[46] At low and high frequencies the exact solutions can be approximated with simpler expressions, facilitating the determination of thermal conductivity and heat capacitance of the film.

At low modulation frequencies the heat capacitance of the bridge has little effect on temperature rise. The thermal conductivity can be determined from the following relation between the root-mean-square values of 3ω voltage ($V_{3\omega}^{\rm rms}$) and current ($I_{\rm rms}$), and the sample's length $2L$, cross section Σ, thermal conductivity k, electrical resistance R, and $dR/dT(R')$:

$$V_{3\omega}^{\rm rms} = \frac{4 I_{\rm rms}^3 R R' 2L}{\pi^4 k \Sigma} \tag{28}$$

At high modulation frequencies the temperature signal is dominated by the heat capacity of the sample, which can be determined from the approximate expression of the 3ω signal:

$$V_{3\omega}^{rms} = \frac{I_{rms}^3 R R'}{4\omega\rho c 2L\Sigma}. \tag{29}$$

Therefore, performing the measurements over a wide frequency range allows both thermal diffusivity and thermal conductivity of the film to be determined with this method.

2.2.3. In-Plane Thermal Conductivity Measurement Without Substrate Removal

The in-plane thermal conductivity methods presented use thin freestanding structures in order to force heat transport in the plane of the film. The techniques require either substrate removal, which implies additional complexity in making the test structures, or depositing the films on thin and low-thermal-conductivity supporting films, which may not be applicable generally, for example, if films must be grown epitaxially. Techniques for in-plane thermal conductivity measurements, without requiring the test structure preparations we have discussed, are presented here and shown schematically in Fig. 5c. Some of the techniques can be employed for simultaneous in-plane and cross-plane thermal conductivity characterization.

In one technique the film to be measured is separated from the substrate by a low-thermal-conductivity layer.[41,42] The heat source can be a heating strip deposited onto the film surface[42] or a low-resistivity region embedded into the film[41] (if the film is semiconducting). When an electric current is passed through the heater, its temperature rise is sensitive to the in-plane thermal conductivity of the film. In one report[41] the steady-state temperature distribution in the film plane is detected by electrical resistance thermometry in metallic strips deposited parallel to the heater at various distances. A heat conduction model was developed to back out the thermal conductivity of the film from the experimental temperature rise and input power. The model considers heat conduction along the film and across the thermal barrier layer to the substrate. In a different example modulation heating and detection in just one strip were used.[42] If a semiconducting heating strip defined by *pn* junctions is also used as a temperature sensor, particular attention must be given to temperature calibration because the temperature coefficient of resistance of semiconductors obtained under nearly isothermal calibration conditions may not be applicable to the experimental condition that has a large temperature gradient across the space-charge region of the *pn* junction.

The heat spreading effect was obtained in the foregoing method with a low-thermal-conductivity layer underneath the film to force more lateral heat spreading. Alternatively, a similar effect can be obtained by using small-width heaters.[20,22,23] No thermal barrier layers are necessary in this case. This method requires the simultaneous determination of the cross-plane thermal conductivity, which can be obtained with a larger-width heater (sensitive to the cross-plane transport). The temperature rise of the heating strip subjected to modulation heating and deposited on a multilayer thin film on substrate system is given by Eq. (9). As discussed in Sect. 2.1.1, the heat spreading effect inside a film is proportional to the film anisotropy and to the ratio between film thickness and heater width. As the film thickness becomes much smaller than the width of the heating strip, the in-plane thermal conductivity determination becomes more difficult. Anisotropic thermal conductivity measurements performed with modulation heating and a pair of different heater widths include dielectric thin films[20] and semiconductor superlattices.[22,24]

3. OPTICAL HEATING METHODS

Optical heating methods use radiation energy as the heat source. Unlike the microfabricated heater and sensor methods discussed in the previous section, optical-heating-based methods typically use the dynamic response of the sample because

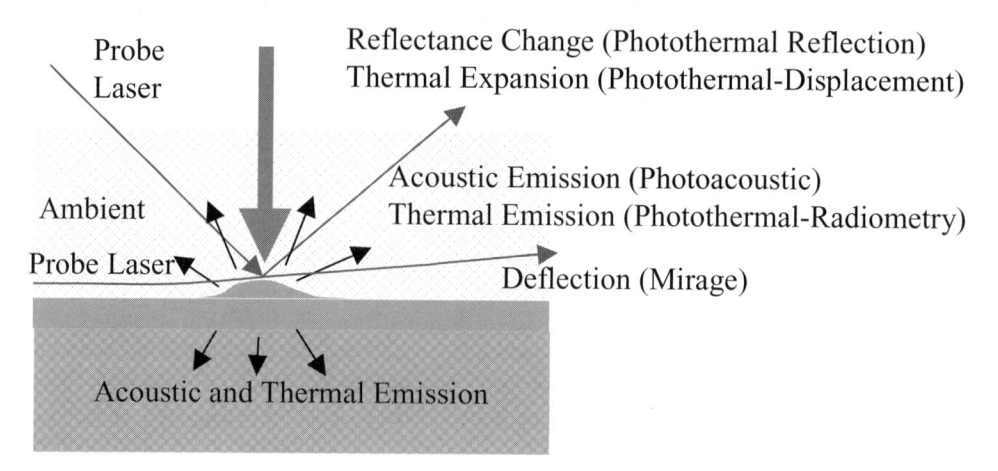

FIGURE 9 Variation of the sample temperature under time-varying heating conditions generates different signatures that can be detected and used for obtaining the thermal diffusivity/effusivity of the samples. Examples of these signatures, as shown, are the thermal emission, refractive index change in both the sample and the surrounding media, and the thermal expansion in both the solids and the surrounding media.

it is usually difficult to determine exactly the amount of heat absorbed in the sample. From the temporal response caused by the sample temperature rise, the thermal diffusivity or thermal effusivity (the product of the thermal conductivity and the volumetric specific heat) of studied samples is often obtained, rather than direct measurements of the thermal conductivity of the sample. Because the density and specific heat of dense thin films normally do not change significantly from their corresponding bulk materials, the bulk density and specific heat can be used to backout the thermal conductivity of thin films.

The variation of the sample temperature under time-varying heating conditions generates different signatures that can be detected and used for obtaining the thermal diffusivity/effusivity of the samples. Examples of these signatures, as shown in Fig. 9, are the thermal emission, refractive index change in both the sample and the surrounding media, and the thermal expansion in the solids and the surrounding media. All these signatures have been explored in the past to determine thin-film thermal diffusivity/effusivity. Conference proceedings and monographs on photothermal and photoacoustic phenomena are excellent resources on these methods.[47,48] We will divide our discussion into time-domain methods and frequency-domain methods. Time-domain methods typically use pulse heating and measure the decay of thermally induced signals, while frequency-domain methods employ modulated heating and measure the amplitude and phase of thermally induced signals. Time-domain signals contain more information because a transient heat pulse includes many different frequency components, but the signals have larger noise compared to frequency-domain signals that can be detected by phase-sensitive lock-in techniques.

3.1. Time-Domain Pump-and-Probe Methods

For thin-film thermophysical properties measurements, it is usually desirable to have the pulse length shorter than the thermal penetration depth of the film. Put in terms of order of magnitude, this can be written

$$t_0 < d^2/3\alpha. \tag{30}$$

For a film thickness of 1 micron and thermal diffusivity of 10^{-5} m^2/s, the thermal signal should be collected in less than 10^{-7} s to avoid the influence of the substrate. This constraint requires that the heating pulse be much shorter than 10^{-7} s. A commercial available laser-flash apparatus typically cannot reach such a temporal resolution. In the past, femtosecond, picosecond,[49,50] and nanosecond lasers[51] were used in the measurement of the thermal diffusivity of thin films.

The temperature response is typically probed with another laser beam through the change of the reflectivity of the samples caused by the temperature dependence of the refractive index. For this reason pulsed laser heating and photothermal reflectance probing are also called the pump-and-probe method. Typically, the reflectivity (r) of the samples varies with temperatures only slightly:

$$\frac{1}{r}\frac{dr}{dT} \sim 10^{-3} - 10^{-5}, \tag{31}$$

with metals typically on the order of 10^{-5} and semiconductors in the range of 10^{-3}–10^{-4}, depending on the laser wavelength. In most cases, to create surface heating and to avoid the complication of electron–hole generation and transport, metal-coated surfaces are preferred. The small temperature dependence of the reflectance means that the thermal signal is small. For a 10°C temperature rise, for example, the thermal signal creates a change of only 10^{-4} in the reflected power.[50] Such a small change can be easily overwhelmed by the noise of the probe laser or the detection circuit.

For nanoscale laser heating the temporal temperature response can be directly measured with fast radiation detectors. The temperature variation generated by a single pulse is captured, but averaging of many pulses is needed to reduce the noise.[51] For femtosecond and picosecond laser heating, photodetectors are not fast enough, and the temporal response is often measured by a time-delayed probe beam. Thermal response to a single pulse is inferred from repetitive measurements

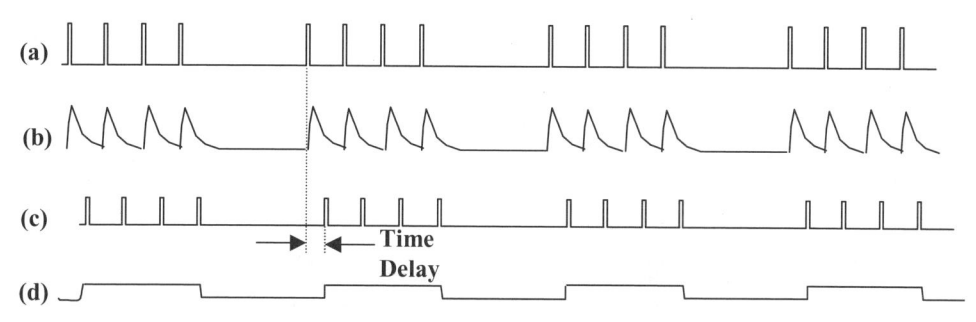

FIGURE 10 Illustration of a time-delayed pump-and-probe detection scheme: (a) heating laser pulse trains are externally modulated; (b) temperature response of the sample to each pulse; (c) reflectance of the probe beam, time-delayed relative to the heating beam, has a small change that depends on transient temperature; and (d) detector measures an average of many reflected pulses within each modulation period; this average signal has a small change that is proportional to the small reflectance change of each pulse and is detected by a lock-in amplifier.

228 Chap. 2.2 • EXPERIMENTAL TECHNIQUES FOR THIN-FILM THERMAL CONDUCTIVITY CHARACTERIZATION

of the probe response to many identical heating pulses. Figure 10 explains the basic idea behind such a time-delayed pump-and-probe method. The laser pulse trains from mode-locked pulse lasers such as a Ti–sapphire laser typically have pulse durations of 10^{-14}–10^{-11} s, and the pulse is repeated about every 100 MHz (Fig.10a). The laser beam is externally modulated by an acousto-optic modulator at a few MHz for phase-locked detection of the small thermal signal. The thermal response of the sample, under the quasi steady-state condition, is shown in Fig. 10b. For detection of the thermal response following heating, a probe laser beam, split from the same laser beam, is directed onto heating area. This probe laser beam has a controllable, longer path length, which creates a time delay relative to the heating (pumping) beam. Due to the temperature rise, there will be a small modulation in the reflected (or transmitted) probe pulses falling in the period when the pump pulses are not blocked by the external acousto-optic modulator (Fig.10c). The photodetector averages over a number of the probe laser pulses (Fig.10d), and the small power variation in the time-averaged reflectance signal of the probe beam is picked up by a lock-in amplifier. The amplitudes of the probe beam intensity variation at different delay times give the thermal response of the sample to the pulse train. This response is often interpreted as that of a single pulse.

The time-delayed pump-and-probe method was originally developed to study the nonequilibrium electron–phonon interactions and was later adapted for a variety of applications, including the thermal diffusivity measurement of thin films and superlattices,[49,50] the study of thermal boundary resistance,[52] film thickness determination, and the study of acoustic phonons.[53] A variety of factors should be considered in the time-delayed pump-and-probe experiment. Some key considerations will be explained later.

The first factor is the time period to be examined. Figure 11 shows a typical probe-response curve to femtosecond laser heating of a metal film on a substrate.[54] In the range of femtoseconds to a few picoseconds, the electrons and, phonons are out of equilibrium, and the probe laser response is mostly from the hot electrons.[55] This region is not of interest if the purpose is to measure the thermal diffusivity and, from which, the thermal conductivity. Also in the figure are acoustic echoes

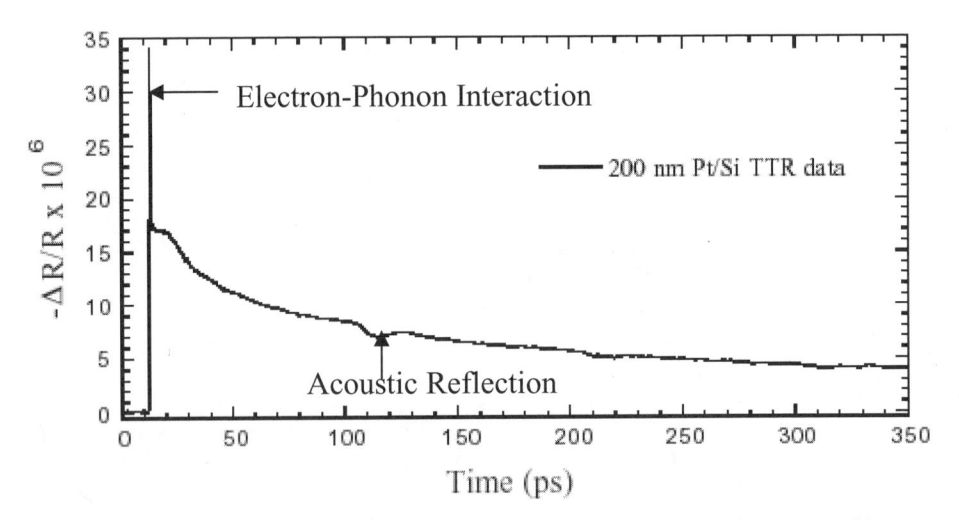

FIGURE 11 Typical behavior of a photothermal reflectance signal. (Courtesy of P. M. Norris)

due to reflection of acoustic waves at the interface between the film and the substrate, which should be ignored in fitting for the thermal diffusivity.[56] The decay of the metallic thin-film temperature after the pulse is governed by (1) thermal boundary resistance between the film and the substrate and (2) thermal diffusion in the layers underneath, including possibly other thin films and the substrate.[57] To use this method to measure the thermal conductivity of the dielectric or semiconducting thin films, one should coat the surface with a metallic thin film to absorb the laser beam. The thickness of the metallic thin film must be thicker than the absorption depth to block the laser beam, yet thin enough so that the temperature of the metallic film can be approximated as uniform during the subsequent diffusion inside the dielectric or semiconducting film.[50] The alternative of including heat conduction inside the metallic thin film is possible but introduces new uncertainties in the thermophysical properties.

On the experimental side, the laser power fluctuations and the probe beam location drift caused by the mechanical delaying stage increase the experimental uncertainties. Methods to minimize these factors have been developed. The utilization of in-phase and out-of-phase components collected by the lock-in amplifier also provides an effective way to minimize the uncertainties in the relative overlapping of the pump and probe beams.[52]

A few other factors should enter the consideration of the data analysis. When a large temperature gradient overlaps with the optical absorption depth, the photothermal reflectance signal may depend on the internal temperature distribution.[58] This effect is important for the short period when the pump and the probe pulses overlap or immediately following the pump pulse. A more severe problem is that the surface temperature may not relax to the uniform temperature state between the pulses and thus the photothermal reflectance signal is not the response to a signal pulse but a quasi-steady-state response to periodic pulses. Proper consideration of the pulse thermal overlap between pulses is important.[50,52] A third question is on the more fundamental side. The thermal diffusion in the substrate is often described by the Fourier heat conduction equation, which is not valid when the time scale of the event is comparable to the phonon relaxation time. Existing experiments show that the thermal conductivity of the substrate inferred from the long time-decay signals is lower than bulk values.[57] Although this may be due to the interface microstructures, the nonlocal phonon transport can be another reason.[59] At this stage no systematic studies exist to clarify this problem.

Because of the short pulse used in the time-domain pump-and-probe method, there is very little lateral heat spreading during the experiment, and the methods are mostly suitable for the measurement of transport properties along the optical axis directions, i.e., in the cross-plane directions of thin films. For the in-plane thermal diffusivity measurement, a technique called transient grating is suitable.[60] This technique splits the original laser beam into three beams. Two of them are used to create a lateral interference pattern on the sample by overlapping the two incident beams at an angle. This interference pattern also creates a transient grating due to the refractive index dependence on the temperature. The third beam, again time delayed, passes through the grating and is diffracted. By measuring the intensity of the diffracted signal as a function of the delay time, the thermal diffusivity along the film-plane direction can be determined. This technique has been used for the characterization of the thermal diffusivity of high-temperature superconducting films.[61]

3.2. Frequency-Domain Photothermal and Photoacoustic Methods

By periodically heating the samples, the signal detection can be accomplished through the use of lock-in amplifiers. Depending on the signature detected, various pump-and-probe methods have been developed and named differently.[47,48,62,63] The thermal penetration depth, determined by the modulation frequency $\sim (\alpha/\pi f)^{1/2}$, is typically longer than the film thickness so that the substrate effect must be taken into account in most of the experiments. This requires that the substrate properties be accurately known or determined by experiment. Some of these methods will be discussed.

3.2.1. Photothermal Reflectance Method

Similar to the time-domain pump-and-probe method, the photothermal reflectance method in the frequency domain is completed by periodically modulating a continuous-wave heating laser, such as an argon laser or a diode laser, and detecting the small periodic change in the intensity of the reflected beam of a continuous-wave probe laser, typically a He–Ne laser or a diode laser. This reflectance modulation is again due to the temperature dependence of the refractive index. Because the modulation is relatively slow, the probe laser is usually a different laser source such that the pump laser can be filtered out before entering the detector. The photothermal reflectance method has been used to measure the thermal diffusivity of thin films and the film thickness.[64] By scanning a focused probe beam around the pumping beam, maps of the amplitude and phase distributions can be obtained for modulation at each frequency. Such information can potentially be used to determine the anisotropic properties of thin films or to obtain the distributions of the thermal diffusivity.[65,66] For such applications, ideally, it is better to scan the probe beam with a fixed pump beam. In reality, however, it is better to fix the probe beam and scan the pumping beam, because the surface reflectance fluctuation will impact more directly on the probe beam. The motion of the heat source during the scanning of the pump laser calls for careful consideration of the scanning speed and modulation frequency such that the final results are equivalent to the scanning of the probe beam relative to a fixed heating spot.[67]

3.2.2. Photothermal Emission Method

Instead of measuring the reflectance change caused by the heating, one can collect the thermal emission from the sample and use it to obtain the thermal diffusivity of thin films. The first use of the photothermal emission signal was based on pulsed laser heating to measure thermal diffusivity of thin films and thermal boundary resistance.[68,69] Modulation-based techniques were also developed for thin-film thermal diffusivity measurements.[70] In a photothermal radiometry experiment, the thermal emission signal from the sample is a convolution of the thermal waves and thermal emission generated in the sample. The thermal emission from the sample can be volumetric, depending on the optical properties of the sample at the detection wavelength. The volumetric emission contributes additional uncertainties to the thermophysical property determination. Thus, if the photothermal radiometry method is used for thermophysical property determination, care should be taken in the sample preparation so that thermal emission is from the surface. On the other hand, the volumetric characteristic of thermal emission (and also absorption) can be utilized to characterize other properties of the samples, such as electron diffusivity and recombination characteristics.[71]

3.2.3. Photothermal Displacement Method

The radiative heating of samples also creates thermal expansion that can be measured by various methods, such as interferometry[72] and the deflection of a probe laser beam. These signals have been used to infer the thermal diffusivity of thin films and the thermal boundary resistance between the film and the substrate.[72,73] The interpretation of the photothermal displacement signal requires solving the thermoelastic equation to determine the profile of the surfaces due to heating and thermal expansion, which in turn needs information on the thermal expansion coefficient and Poisson ratio of the films and the substrate. These additional parameters add to the uncertainty of the thermal diffusivity determination. Despite these disadvantages, in cases when the thermal displacement signal is much larger than photothermal reflectance signals, such as in the measurement of polymer layers that have a large thermal expansion coefficient, the use of the photothermal displacement signal may be preferable.[73]

3.2.4. Photothermal Deflection Method (Mirage Method)

The photothermal deflection method, sometimes also called the Mirage method, explores the deflection of a laser beam when it passes through a temperature gradient caused by the temperature dependence of the refractive index. Usually, the temperature gradient created on the air side is utilized. This technique was originally developed to observe the optical absorption spectroscopy of materials,[74] but was later extended for determining the thermophysical properties of thin films.[75-77] This technique requires the knowledge of the probe beam location relative to the heating beam and relative to the heated surface.

3.2.5. Photoacoustic Method

In the photoacoustic method the acoustic waves generated by the heating of the sample and the subsequent thermal expansion in the gas side can be measured and used to fit the thermophysical properties of the sample. The ambient fluid, typically air, is often used as the medium to couple the acoustic wave.[62,78-80] The detection of the acoustic waves in the gas side often employs commercial microphones. Thermoacoustic spectroscopy has been widely used for characterizing the optical properties of solids.[62] The photoacoustic signal typically depends on the thermophysical properties of the coupling gases that are usually known.[81,82] Compared to the photothermal displacement method, it has fewer unknown parameters. Compared to the photothermal deflection method, it does not need precise determination of the relative location of the probe beam and the sample. In actual determination of the thermophysical properties of thin films, however, the phase delay caused by the acoustic wave propagation must be calibrated and taken into consideration in the final data processing.

In addition to these photothermal techniques, other signal detector methods, such as the through pyroelectric effect,[83] have also been developed and employed in determining thermophysical properties of thin films. In the next section we

4. OPTICAL–ELECTRICAL HYBRID METHODS

Hybrid methods that combine electrical heating and optical detection, or vice versa, are also used. One well-developed method is the AC calorimetry method, in which a laser beam is used to heat the sample and the detection is done by small thermocouples or sensors directly patterned onto the sample, as shown in the insert of Fig. 11.[6,84] The distance between the laser and the sensor is varied. Under appropriate conditions the thermal diffusivity along the film plane can be calculated from phase or amplitude data. Figure 12 gives an example of the phase data for GaAs/AlAs superlattices.[85] For the in-plane property measurement, commercially available thermocouples can satisfy the required frequency response because heating is typically limited to a few Hz up to 100 Hz. For the cross-plane thermal diffusivity, however, the thermocouple response is too slow. Chen *et al.*[86] employed microfabricated resistance thermometers and modulated optical heating to obtain the thermal diffusivity perpendicular to the cross-plane direction of short-period GaAs/AlAs superlattices used in semiconductor lasers.

Although commercial thermocouples have a slow response, the thermoelectric effect is not limited to between two thermocouple strips. A thermocouple can be made by using a conducting film as one leg of the thermocouple and a sharpened metallic strip pressed onto the film as the other leg. Such a thermocouple does not have a junction mass and can have fast thermal response. Such a photothermo-

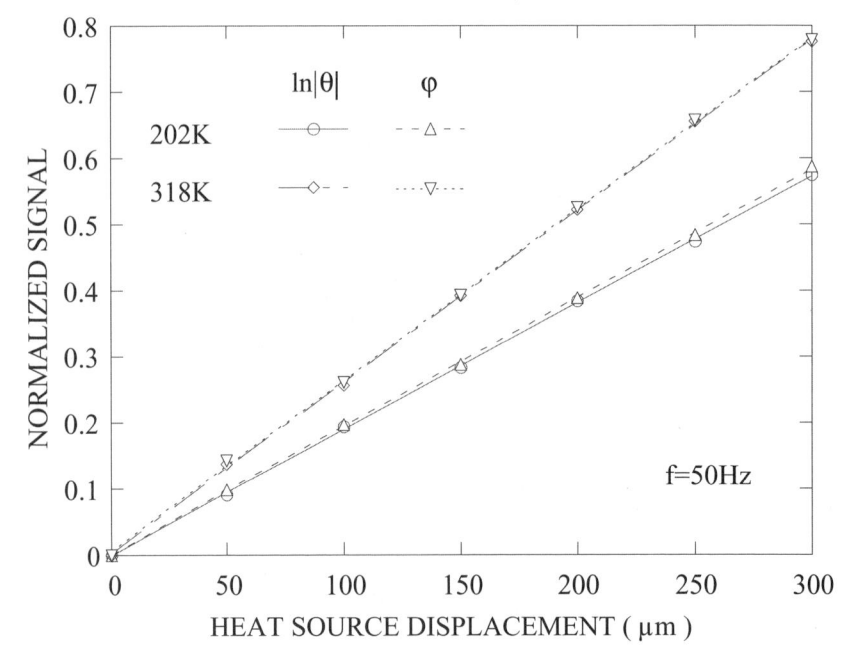

FIGURE 12 Normalized phase and amplitude signals as functions of the relative displacement of a laser beam away from the thermocouple in the ac calorimetry measurements of the in-plane thermal diffusivity of a GaAs/AlAs superlattice.

electric method has been used in the configurations of thermal mapping and single point measurements of thermal diffusivity of thin films[87] and nanowire composite samples.[88] Figure 13 shows an image of the amplitude and phase of a gold-on-glass sample obtained with the photothermoelectric method.

As an alternative to the optical heating method, the sample can be heated with an electrical signal and the thermal signal can be detected by the reflectance change[51] or thermal emission from the sample.[89] As discussed before, one advantage of electrical heating is that power input can be determined precisely. Optical-based detection, however, does not provide this advantage because the detected signals are usually not the absolute temperature rise of the sample.

5. SUMMARY

In this chapter we summarized various methods developed for thin-film thermophysical property measurements. We divide the measurement techniques into three categories: (1) electrically based, (2) optically based, and (3) a hybrid of optical and electrical.

Electrical heating and sensing methods have the advantage that the power input into the sample and the temperature rise can be precisely determined, which allows direct deduction of thermal conductivity. Many recently developed electrical heating and sensing methods rely on microfabrication for heaters and temperature sensors. These methods, although allowing direct determination of thermal conductivity and potentially thermal diffusivity and specific heat, rely heavily on the availability of microfabrication facilities. For conducting samples the external heaters and temperature sensors should be carefully instrumented to minimize the impacts of current leakage into the film.

The optical heating and sensing methods, on the other hand, usually require minimal sample preparation. Since the optical power input and the temperature rise are more difficult to determine, thermal diffusivity is measured, rather than a direct measurement of thermal conductivity, by fitting the normalized time response of sample under transient heating. Such fitting often requires the specific heat and density of the samples as input parameters.

Despite the large number of techniques developed in the past for thin-film thermophysical property measurements, this chapter shows that thin-film thermophy-

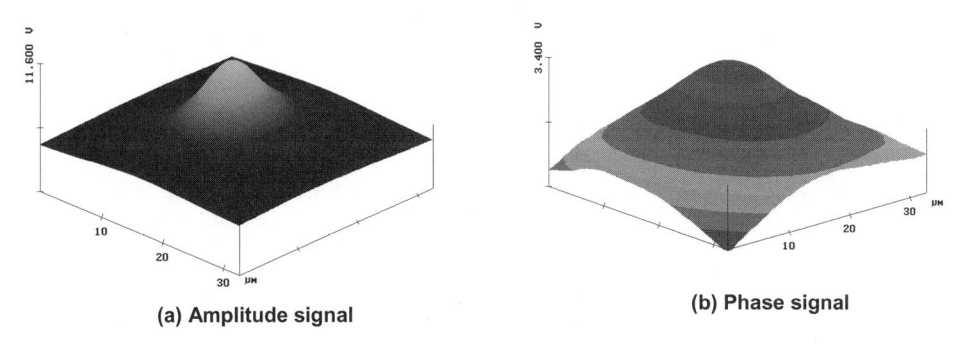

(a) Amplitude signal

(b) Phase signal

FIGURE 13 Two-dimensional normalized phase and amplitude signals as functions of the relative displacement of a laser beam away from the thermocouple junction in the photothermoelectric method.

sical properties measurement is by no means easy. The choice of methods depends on sample constraints and available facilities.

6. ACKNOWLEDGMENTS

T. B. would like to acknowledge the financial support of NSF (DMR-0210587) on nanoscale thermoelectric properties characterization. G. C. would like to acknowledge the contributions from his group on thermophysical property characterization, and including D. Song, W. L. Liu, B. Yang, D. Borca-Tasciuc. G. C. would also like to acknowledge the financial support of DOE (DE-FG02-02ER45977) on heat transfer in nanostructures, NSF (CTS-0129088) on nanoscale heat transfer modeling, and DoD/ONR MURI on thermoelectrics.

7. REFERENCES

1. G. CHEN *Heat Transfer in Micro- and Nanoscale Photonic Devices* Annu. Rev. Heat Trans. **7**, 1–57 (1996).
2. K. GOODSON AND Y. JU *Heat Conduction in Novel Electronic Films* Annu. Rev. Mater. Sci. **29**, 261–293 (1999).
3. G. CHEN, T. BORCA-TASCIUC, B. YANG, D. SONG, W. L. LIU, T. ZENG and D. ACHIMOV *Heat Conduction Mechanisms and Phonon Engineering in Superlattice Structures* Therm. Sci. Eng. **7**, 43–51 (1999).
4. G. CHEN, T. ZENG, T. BORCA-TASCIUC, and D. SONG *Phonon Engineering in Nanostructures for Solid-State Energy Conversion* Mater. Sci. Eng. A **292**, 155–161 (2000).
5. D. G. CAHILL, H. E. FISCHER, T. KLITSNER, E. T. SWARTZ, and R. O. POHL *Thermal Conductivity of Thin Films: Measurement and Understanding* J. Vac., Sci. Technol. A **7**, 1259–1269 (1989).
6. I. HATTA *Thermal Diffusivity Measurement of Thin Films and Multilayered Composites* Int. J. Thermophys. **11**, 293–302 (1990).
7. C. L. TIEN and G. CHEN *Challenges in Microscale Radiative and Conductive Heat Transfer* J. Heat Transf. **116**, 799–807 (1994).
8. K. E. GOODSON and M. I. FLIK *Solid Layer Thermal-Conductivity Measurement Techniques* Appl. Mech. Rev. **47**, 101–112 (1994).
9. H. VERHOEVEN, E. BOETTGER, A. FLOTER, H. REISS, and R. ZACHAI *Thermal Resistance and Electrical Insulation of Thin Low-Temperature-Deposited Diamond Films* Diamond Rel. Mater. **6**, 298–302 (1997).
10. K. KURABAYASHI, M. ASHEGHI, M. TOUZELBAEV, and K. E. GOODSON *Measurement of the Thermal Conductivity Anisotropy in Polyimide Films* J. MEMS **8**, 180–190 (1999).
11. D. G. CAHILL *Thermal Conductivity Measurement from 30 to 750K: the 3ω method* Rev. Sci. Instrum. **61**, 802–808 (1990).
12. S. M. LEE and D. G. CAHILL *Heat Transport in Thin Dielectric Films* J. Appl. Phys. **81**, 2590–2595 (1997).
13. S. M. LEE, D. G. CAHILL, and R. VENKATASUBRAMANIAN *Thermal Conductivity of Si-Ge Superlattices* Appl. Phys. Let. **70**, 2957–2959 (1997).
14. T. BORCA-TASCIUC, J. L. LIU, T. ZENG, W. L. LIU, D. W. SONG, C. D. MOORE, G. CHEN, K. L. WANG, M. S. GOORSKY, T. RADETIC, and R. GRONSKY *Temperature Dependent Thermal Conductivity of Symmetrically Strained Si/Ge Superlattices* ASME HTD **364-3**, 117–123 (1999).
15. J. H. KIM, A. FELDMAN, and D. NOVOTNY *Application of the Three Omega Thermal Conductivity Measurement Method to a Film on a Substrate of Finite Thickness* J. Appl. Phys. **86**, 3959–3963 (1999).
16. R. VENKATASUBRAMANIAN *Lattice Thermal Conductivity Reduction and Phonon Localization Like Behavior in Superlattice Structures* Phys. Rev. B **61**, 3091–3097 (2000).
17. T. BORCA-TASCIUC, W. L. LIU, J. L. LIU, T. ZENG, D. W. SONG, C. D. MOORE, G. CHEN, K. L. WANG, M. S. GOORSKY, T. RADETIC, R. GRONSKY, T. KOGA, and M. S. DRESSELHAUSS *Thermal*

Sec. 7 · REFERENCES

Conductivity of Symmetrically Strained Si/Ge Superlattices Superlattices Microstruct. **28**, 199–206 (2000).

18. S. T. HUXTABLE, A. R. ABRAMSON, C.-L. TIEN, A. MAJUMDAR, C. LABOUNTY, X. FAN, G. ZENG, J. E. BOWERS, A. SHAKOURI, and E. T. CROKE *Thermal Conductivity of Si/SiGe and SiGe/SiGe Superlattices* Appl. Phys. Lett. **80**, 1737–1739 (2002).
19. T. BORCA-TASCIUC, D. W. SONG, J. R. MEYER, I. VURGAFTMAN, M.-J. YANG, B. Z. NOSHO, L. J. WHITMAN, H. LEE, R. U. MARTINELLI, G. W. TURNER, M. J. MANFRA, and G. CHEN *Thermal Conductivity of $AlAs_{0.07}Sb_{0.93}$ and $Al_{0.9}Ga_{0.1}As_{0.07}Sb_{0.93}$ Alloys and $(AlAs)_1/(AlSb)_1$ Digital Alloy Superlattices* J. Appl. Phys. **92**, 4994–4998 (2002).
20. Y. S. JU, K. KURABAYASHI, and K. E. GOODSON *Thermal Characterization of Anisotropic Thin Dielectric Films Using Harmonic Joule Heating* Thin Solid Films **339**, 160–164 (1999).
21. A. R. KUMAR, D.-A. ACHIMOV, T. ZENG, and G. CHEN *Thermal Conductivity of Nanochanneled Alumina* ASME-HTD **366-2**, 393–398 (2000).
22. W. L. LIU, T. BORCA-TASCIUC, G. CHEN, J. L. LIU, and K. L. WANG *Anisotropic Thermal Conductivity of Ge Quantum-Dot and Symmetrically Strained Si/Ge Superlattices* J. Nanosci. Nanotechnol. **1**, 39–42 (2001).
23. T. BORCA-TASCIUC, R. KUMAR, and G. CHEN *Data Reduction in 3ω Method for Thin-Film Thermal Conductivity Determination* Rev. Sci. Instrum. **72**, 2139–2147 (2001).
24. T. BORCA-TASCIUC, W. L. LIU, J. L. LIU, G. CHEN, and K. L. WANG *Anisotropic Thermal Conductivity of a Si/Ge Quantum-Dot Superlattice* ASME-HTD **366-2**, 381–384 (2000).
25. T. BORCA-TASCIUC *Thermal and Thermoelectric Properties of Superlattices* Ph. D. thesis, University of California at Los Angeles, Department of Mechanical and Aerospace Engineering (2000).
26. Y. S. JU AND K. E. GOODSON *Process-Dependent Thermal Transport Properties of Silicon-Dioxide Films Deposited Using Low-Pressure Chemical Vapor Deposition* J. Appl. Phys. **85**, 7130–7134 (1999).
27. D. G. CAHILL, H. E. FISHER, T. KLITSNER, E. T. SWARTZ, and R. O. POHL *Thermal Conductivity of Thin Films: Measurements and Understanding* J. Vac. Sci. Technol. A **7**, 1259–1266 (1989).
28. K. E. GOODSON, M. I. FLIK, L. T. SU, and D. A. ANTONIADIS *Prediction and Measurement of the Thermal Conductivity of Amorphous Dielectric Layers* Trans. ASME J. Heat. Transfer **116**, 317–324 (1994).
29. F. VOLKLEIN and KESSLER E. *A Method for the Measurement of Thermal Conductivity, Thermal Diffusivity, and Other Transport Coefficients of Thin Films* Phys. Stat. Sol. A **81**, 585–596 (1984).
30. F. VOLKLEIN and J. E. KESSLER *Determination of Thermal Conductivity and Thermal Diffusivity of Thin Foils and Films* Exp. Tech. Phys. **33**, 343–_350 (1985).
31. Y. C. TAI, C. H. MASTRANGELO, and R. S. MULLER *Thermal Conductivity of Heavily Doped Low-Pressure Chemical Vapor Deposited Polycrystalline Silicon Films* J. Appl. Phys. **63**, 1442–1447 (1988).
32. F. VOLKLEIN *Thermal Conductivity and Diffusivity of a Thin Film SiO_2-Si_3N_4 Sandwich System* Thin Solids Films **188**, 27–33 (1990).
33. F. VOLKLEIN and H. BALTES *A microstructure for Measurement of Thermal Conductivity of Polysilicon Thin Films* J. MEMS **1**, 193–196 (1992).
34. Z. ZHANG and C. P. GRIGOROPOULOS *Thermal Conductivity of Free-Standing Silicon nitride Thin Films* Rev. Sci. Instrum. **66**, 1115–1120 (1995).
35. M. VON ARX, O. PAUL, and H. BALTES *Process-Dependent Thin-Film Thermal Conductivities for Thermal CMOS MEMS* J. MEMS **9**, 136–145 (2000).
36. A. D. MCCONNELL, S. UMA, and K. E. GOODSON *Thermal Conductivity of Doped Polysilicon Layers* J. MEMS **10**, 360–369 (2001).
37. A. JACQUOT, W. L. LIU, G. CHEN, J.-P. FLEURIAL, A. DAUSCHER, and B. LENOIR *Improvements of On-Membrane Method for Thin-Film Thermal Conductivity and Emissivity Measurements* Proc. ICT2002, 21st Int. Conf. on Thermoelectrics (IEEE, NY, 2002, pp. 353–358).
38. M. ASHEGHI, K. KURABAYASHI, R. KASNAVI, and K. E. GOODSON *Thermal Conduction in Doped Single-Crystal Silicon Films* J. Appl. Phys. **91**, 5079–5088 (2002).
39. BORCA-TASCIUC, T., LIU, W. L., LIU, J. L., WANG, K. L., and G. CHEN *In-Plane Thermoelectric Properties Characterization of a Si/Ge Superlattice Using a Microfabricated Test Structure* in Proc. 2001 National Heat Transfer Conf. (2001).
40. P. NATH, K. L. CHOPRA *Experimental Determination of the Thermal Conductivity of Thin Films* Thin Solid Films **18**, 29–37 (1973).
41. M. ASHEGHI, Y. K. LEUNG, S. S. WONG, and K. E. GOODSON *Phonon-Boundary Scattering in Thin Silicon Layers* Appl. Phys. Lett. **71**, 1798–1800 (1997).
42. Y. S. JU and K. E. GOODSON *Phonon scattering in Silicon Films with Thickness of OrderG of 100nm* Appl. Phys. Lett. **74**, 3005–3007 (1999).

43. D. Song *Phonon Heat Conduction in Nano and Micro-Porous Thin Films* Ph.D. Thesis, University of California at Los Angeles, Department of Mechanical and Aerospace Engineering, 2003.
44. M. Madou *Introduction to microfabrication* (CRC Press, 1997).
45. F. Volklein and T. Starz *Thermal Conductivity of Thin Films – Experimental Methods and Theoretical Interpretation* Proc. 16th Int. Conf. on Thermoelectrics (IEEE, (1997), pp. 711–718.
46. L. Lu, W. Yi, and D. L. Zhang *3ω Method for Specific Heat and Thermal Conductivity Measurements* Rev. Sci. Instrum. **72**, 2996–3003 (2001).
47. A. Mandelis *Principles and Perspectives of Photothermal and Photoacoustic Phenomena* (Elsevier, New York; 1992).
48. F. Scudieri and M. Bertolotti, eds., *Photoacoustic and Photothermal Phenomena* Proc. 10th Int. Conf., Rome, (AIP 1998).
49. C. A. Paddock and G. Eesley *Transient Thermoreflectance from Thin Metal Films* J. Appl. Phys. **60**, 285–290 (1986).
50. W. S. Capinski, H. J. Maris, T. Ruf, M. Cardona, K. Ploog, and D. S. Katzer *Thermal-Conductivity Measurements of GaAs/AlAs Superlattices Using a Picosecond Optical Pump-and-Probe Technique* Phys. Rev. B **59**, 8105–8113 (1999).
51. K. E. Goodson, O. W. Kading, M. Rosler, and R. Zachai *Experimental Investigation of Thermal Conduction Normal to Diamond-Silicon Boundaries* J. Appl. Phys. **77**, 1385–1392 (1995).
52. R. M. Costescu, M. A. Wall, and D.G. Cahill *Thermal Conductance of Epitaxial Interfaces* Phys. Rev. B **67**, 054302 (1–5) (2003).
53. H. Y. Hao and H. J. Maris *Study of Phonon Dispersion in Silicon and Germanium at Long Wavelengths Using Picosecond Ultrasonics* Phys. Rev. Lett. **84**, 5556–5559 (2000).
54. P. M. Norris, A. P. Caffrey, R. Stevens, J. M. Klopf, J. T. McLeskey, and A. N. Smith *Femtosecond Pump-Probe Nondestructive Evaluation of Materials* Rev Sci. Instrum. **74**, 400–406 (2003).
55. G. L. Eesley *Observation of Nonequilibrium Electron Heating in Copper* Phys. Rev. Lett. **51**, 2140–2143 (1983).
56. K. A. Svinarich, W. J. Meng, and G. L. Eesley *Picosecond Acoustic Pulse Reflection from a Metal-Metal Interface* Appl. Phys. Lett. **57**, 1185–1187 (1990).
57. R. J. Stoner and H. J. Maris, Kapitza *Conductance and Heat Flow Between Solids at Temperatures from 50 to 300 K* Phys. Rev. B **48**, 16373–16387 (1993).
58. G. Chen and C. L. Tien *Internal Reflection Effects on Transient Photothermal Reflectance* J. Appl. Phys. **73**, 3461-3466 (1993).
59. G. Chen *Ballistic-Diffusive Heat Conduction Equations* Phys. Rev. Lett. **85**, 2297–2300 (2001).
60. H. J. Eichler *Laser-Induced Dynamic Gratings* (Springer-Verlag, Berlin, 1986).
61. C. D. Marshall, I. M. Fishman, R. C. Dorfman, C. B. Eom, and M. D. Fayer *Thermal Diffusion, Interfacial Thermal Barrier, and Ultrasonic Propagation in Yba2Cu3O7-x Thin Films: Surface-Selective Transient-Grating Experiments* Phys. Rev. B **45**, 10009–10021 (1992).
62. A. C. Tam *Applications of Photoacoustic Sensing Techniques* Rev. Modern Phys. **58**, 381–434 (1986).
63. K. Katayama, H Yui, T. Sawada *Recent Development of Photothermal and Photoacoustic Spectroscopy* Jpn. Soc. Appl. Phys. **70**, 272–276 (2001).
64. A. Rosencwaig, J. Opsal, W. L. Smith, and D. L. Willenborg *Detection of Thermal Waves Through Optical Reflectance* Appl. Phys. Lett. **46**, 1013–1015 (1985).
65. L. Pottier *Micrometer Scale Visualization of Thermal Waves by Photoreflectance Microscopy* Appl. Phys. Lett. **64**, 1618–1619 (1994).
66. G. Langer, J. Hartmann, and M. Reichling *Thermal Conductivity of Thin Metallic Films Measured by Photothermal Profile Analysis* Rev. Sci. Instrum. **68**, 1510–1513 (1997).
67. T. Borca-Tasciuc and G. Chen *Temperature Measurement of Fine Wires by Photothermal Radiometry* Rev. Sci. Instrum. **68**, 8040–8043 (1997).
68. W. P. Leung and A. C. Tam *Thermal Diffusivity in Thin Films Measured by Noncontact Single-Ended Pulsed-Laser Induced Thermal Radiometry* Opt. Lett. **9**, 93–95 (1984).
69. W. P. Leung and A. C. Tam *Thermal Conduction at a Contact Interface by Pulsed Photothermal Radiometry* J. Appl. Phys. **63**, 4505–4510 (1988).
70. J. A. Garcia, A. Mandelis, B. Farahbakhsh, C. Lebowitz and I. Harris *Thermophysical Properties of Thermal Spray Coatings on Carbon Steel Substrates by Photothermal Radiometry* Int. J. Thermophys. **20**, 1587–1602 (1999).
71. A. Mandelis, A. Othonos, and C. Christofides *Non-Contacting Measurements of Photocarrier Lifetimes in Bulk- and Polycrystalline Thin-Film Si Photoconductive Devices by Photothermal Radiometry* J. Appl. Phys. **80**, 5332–5341 (1996).

Sec. 7 · REFERENCES

72. B. S. W. Kuo, J. C. M. Li, and A. W. Schmid *Thermal Conductivity and Interface Thermal Resistance of Si Film on Si Substrate Determined by Photothermal Displacement Interferometry* Appl. Phys. A, Solids Surf. **A55**, 289–296 (1992).
73. C. Hu, E. T. Ogawaa and P. S. Ho *Thermal Diffusivity Measurement of Polymeric Thin Films Using the Photothermal Displacement Technique, II. On-Wafer Measurement* J. Appl. Phys. **86**, 6028–6038 (1999).
74. A. C. Boccara, D. Fournier, and J. Badoz *Thermo-Optical Spectroscopy: Detection by the "Mirage Effect"* Appl. Phys. Lett. **36**, 130–132 (1980).
75. J. P. Roger, F. Lepoutre, D. Fournier, and A. C. Boccara *Thermal Diffusivity Measurement of Micron-Thick Films by Mirage Detection* Thin Solid-Films **155**, 165–174 (1987).
76. P. K. Kuo, M. J. L Lin, C. B. Reyes, L. D. Favro, R. L. Thomas, D. S. Kim, S. Y. Zhang, J. L. Ingelhart, D. Fournier, A. C. Boccara, and N. Yacoubi *Mirage-Effect Measurement of Thermal Diffusivity, Part I: Experiment* Can. J. Phys., **64**, 1165–1167 (1986).
77. P. K. Wong, P. C. Fung, H. L. Tam, and J. Gao *Thermal-Diffusivity Measurements of an Oriented Superconducting Film-Substrate Composite Using Mirage Technique* Phys. Rev. B **51**, 523–533 (1995).
78. M. Akabori, Y. Nagasaka, and A. Nasashima *Measurement of the Thermal Diffusivity of Thin Films on Substrate by the Photoacoustic Method* Int. J. Thermophys. **13**, 499–514 (1992).
79. J. Xiao, M. Qian, and M. A. Wei *Detection of Thermal Diffusivity of Thin Metallic Film Using a Photoacoustic Technique* J. Phys. IV Colloq. **2 (C1)**, 2809–812 (1992).
80. X. Wang, H., Hu, X., Xu *Photo-Acoustic Measurement of Thermal Conductivity of Thin films and Bulk Materials* J. Heat Transfer **123**, 138–144 (2001).
81. A. Rosencwaig and A. Gersho Theory of the Photoacoustic Effect with Solid, J. Appl. Phys. **47**, 64–69 (1976).
82. H. Hu, X. Wang, and X. Xu *Generalized Theory of the Photoacoustic Effect in a Multilayer Material* J Appl. Phys. **86**, 3953–3958 (1999).
83. H. Coufal and P. Hefferle *Thermal Diffusivity Measurements of Thin Films with Pyroelectric Calorimeter* Appl. Phys. A **38**, 213–219 (1985).
84. X. Y. Yu, Z. Liang, and G. Chen *Thermal-Wave Measurement of Thin-Film Thermal Diffusivity with Different Laser Configurations* Rev. Sci. Instrum. **67**, 2312–2316 (1996).
85. X. Y. Yu, G. Chen, A. Verma, and J. S. Smith *Temperature Dependence of Thermophysical Properties of GaAs/AlAs Periodic Structure* Appl. Phys. Lett. **67**, 3554–3556 (1995), **68** 1303 (1996).
86. G. Chen, C. L. Tien, X. Wu, and J. S. Smith *Measurement of Thermal Diffusivity of GaAs/AlGaAs Thin-Film Structures* J. Heat Transfer **116**, 325–331 (1994).
87. T. Borca-Tasciuc and G. Chen *Thermophysical Property Characterization of Thin Films by Scanning Laser Thermoelectric Microscope* Int. J. Thermophys. **19**, 557–567 (1998)
88. D. Borca-Tasciuc, G. Chen, A. Borchevsky, J.-P. Fleurial, M. Ryan, Y.-M. Lin, O. Rabin, and M. S. Dresselhaus *Thermal Characterization of Nanowire Arrays in a-Al$_2$O$_3$*, presented at 2002 MRS Fall Meeting, Nov. 26–30, Symposium V: Nanophase and Nanocomposite Materials, proceeding in press.
89. S. W. Indermuehle and R. B. Peterson *A Phase-Sensitive Technique for the Thermal Characterization of Dielectric Thin Films* J. Heat Transfer **121**, 528–536 (1999).

Chapter 3.1

CERAMICS AND GLASSES

Rong Sun and Mary Anne White

Department of Chemistry and Institute for Research in Materials Dalhousie University, Halifax, Nova Scotia B3H 4J3 Canada

1. INTRODUCTION

Glasses and ceramics are important materials. They have various applications, which require very high thermal conductivity (e.g., as heat sinks) to very low thermal conductivity (e.g., as thermal insulation). Understanding, and hence tailoring, the thermal conductivities of glasses and ceramics can be predicated on Slack's finding[1] that for nonmetallic crystals in which phonons are the dominant heat transport mechanism, the thermal conductivity can be increased by the following factors: (1) low atomic mass; (2) strong interatomic bonding; (3) simple crystal structure (small unit cell); (4) low lattice anharmonicity.

In this chapter we have summarized selected recent findings with regard to thermal conductivities (κ) and, to a lesser extent, thermal diffusivities of ceramics and glasses. Our selection of recent work is meant not to be exhaustive but illustrative of the factors at play in determining κ, with an emphasis on technological importance of the thermal conductivities of the materials.

2. CERAMICS

Ceramics history dates back at least 35,000 years, but new ceramics with new applications are being developed virtually daily. Applications of ceramics are largely based on their high-temperature stability. A recent review of advanced engineering ceramics[2] summarizes many novel ceramics based on monolithic, composite, and cellular architectures. In each application thermal conductivity is an important consideration.

2.1. Traditional Materials with High Thermal Conductivity

Materials with high thermal conductivities are required in many applications, especially those related to microelectronics. For example, a 10 K increase in temperature in a CMOS leads to a twofold increase in failure rate.[3] Furthermore, thermal management in microelectronics is better if heat is uniformly dissipated,[4] making high-thermal-conductivity materials desirable.

2.1.1. Aluminum Nitride (AlN)

Aluminum nitride is not a new material, but its thermal properties have received intense attention during the past decade. Its applications include electronic packaging and heat sinks, because of its high thermal conductivity and the good match of its thermal expansion with that of silicon.[5-9] Furthermore, it has a low dielectric constant and low loss tangent.[10,11] Aluminum nitride has been proposed as an alternative to alumina and beryllia for microelectronic devices.[12,13]

The room-temperature thermal conductivity of pure AlN along the c-axis can be as high as 320 W m^{-1}K^{-1},[14] but the thermal conductivity is usually considerably lower for polycrystalline AlN. By reduction in grain boundaries, the room-temperature thermal conductivity of polycrystalline AlN has been raised from about 40 W m^{-1}K^{-1} to 272 W m^{-1}K^{-1},[15,16] which is close to the intrinsic value. However, pure AlN is difficult to densify, so efforts have been made to include additives without decreasing thermal conductivity.

Both processing and the presence of impurities can affect the thermal conductivity of AlN. The presence of oxygen will reduce the thermal conductivity (the substitution of N sites by O gives concomitant Al vacancies),[17] but other additives can increase the thermal conductivity. Processing that leads to reduction in grain boundaries also can be used to enhance the thermal conductivity.

Additives such as rare-earth lanthanide oxides, generalized as Ln_2O_3, will effectively enhance AlN's thermal conductivity.[18] In one investigation,[18] starting from high-purity AlN powder and lanthanide oxide powders with average particle sizes of 1–5 μm, samples with different oxide additive concentrations were uniaxially pressed at 35 MPa and then isostatically pressed at 200 MPa. The concentration of dopant was equimolar to the initial amount of oxygen in the AlN powder, assuming all the oxygen to be in the form Al_2O_3. Densification was carried out by sintering in a N_2 flow with graphite with a ramp time of 10 K/min at various temperatures above 1000°C and for varying times. Selected results from this study are shown in Table 1.

From these results[18] we see that Y_2O_3 and Sm_2O_3 are the most effective additives, giving the largest increase in thermal conductivity. In all cases, annealing for a longer time also increased thermal conductivity, and, in most cases, a higher anneal temperature had the same effect. Recently, Xu et al.[19] have shown that the thermal conductivity of Sm_2O_3-doped AlN increases with increased sintering time and by not including packing powder in the sintering process; however, their highest room-temperature thermal conductivity was 166 W m^{-1} K^{-1}, somewhat lower than that reported elsewhere,[18] showing again the sensitivity to preparation conditions.

The concentration of the additive also affects thermal conductivity, as shown in Table 2 for added Y_2O_3. Thermal conductivity is maximized in this case with about 8 wt% Y_2O_3. The same conclusion was reached at higher temperatures, up to about 1200°C.[18]

Sec. 2 · CERAMICS

TABLE 1 Room-temperature Thermal Conductivities of AlN–Ln$_2$O$_3$ Ceramics Sintered at Different Temperatures and for Different Durations[18]

Additive	κ (W m^{-1} K^{-1})		
	1850°C for 100 min	1850°C for 1000 min	1950°C for 1000 min
4.91 wt% Y$_2$O$_3$	176	232	246
8.29 wt% Lu$_2$O$_3$	150	203	202
3.42 wt% Sm$_2$O$_3$	184	220	246
3.41 wt% Nd$_2$O$_3$	181	199	229
3.41 wt% Pr$_2$O$_3$	172	194	231

TABLE 2 The effect of added Y$_2$O$_3$ on the room-temperature thermal conductivity of AlN.[18]

Wt % Y$_2$O$_3$	κ (W m^{-1} K^{-1})
0	70
1	104
2	154
4	170
8	176
15	165
20	157
25	142
40	94
75	38

Studies[20] of AlN with added YF$_3$ show that this dopant can be used to reduce the intergranular oxygen content (through sublimation of YOF and formation of Y$_2$O$_3$). In this way, a room-temperature thermal conductivity of AlN of 210 W m^{-1} K^{-1} has been achieved.

The addition of lanthanide oxides and other sintering aids raises the thermal conductivity by acting as "getters" for oxygen impurities.[18] Watari et al.[21] have shown that selected sintering aids can produce high-thermal-conductivity AlN at lowered sintering temperatures, due to the reduction in liquid temperature in this system and the correspondingly lower temperature required to achieve the reaction to form metal oxide. Their results are summarized in Table 3. Li-doped AlN showed higher thermal conductivity and higher strength.

In another study Watari et al. have shown that similar improvements in the room-temperature thermal conductivity of AlN can be made by firing in a N$_2$ (reducing) atmosphere with carbon.[17] The N$_2$ and carbon lead to reduction of the amount of oxygen, and, moreover, the grain boundary phase migrates to the surface of the sample. Both these effects enhance thermal conductivity to as high as 272 W m^{-1}K^{-1} at room temperature for a sample with 8-μm grains,[17] i.e., very close to the value for a single crystal of AlN. However, at lower temperatures the thermal conductivity of the ceramic was much less than that of the single crystal. Analysis of the thermal conductivity of the ceramic leads to a calculated phonon mean free path of 6 μm at T = 100 K, which corresponds closely to the grain size,

TABLE 3 Room-temperature Thermal Conductivities of AlN Samples with Various Sintering Aids, Fired at 1600°C for 6 h[21]

AlN sample	4.0% LiYO$_2$ + 0.5% CaO	0.47% Li$_2$O + 3.53% Y$_2$O$_3$ + 0.5% CaO	3.53% Y$_2$O$_3$ + 0.5% CaO	3.53% Y$_2$O$_3$
κ (W m^{-1}K^{-1})	180	130	100	40

so the predominant resistance is attributed to grain boundaries.[18] This analysis shows that the grain boundaries are not as important at room temperature because, at $T = 300$ K, the phonon mean free path is much smaller than the grains. However, other studies (*vide infra*) have shown that maximizing the grain size is still important for increasing the room-temperature thermal conductivity.

For example, a reheating step in the preparation of AlN has been found to further improve thermal conductivity.[22] In this study three heating procedures were followed for Dy$_2$O$_3$/Li$_2$O/CaO-doped AlN: (A) sintering at 1600°C for 5 h; (B) sintering at 1600°C for 1 h, followed by reheating at 1600C for 15 h; or (C) sintering at 1600°C for 1 h, followed by reheating for 45 h. The results are presented in Table 4.

A detailed study of the annealing time and the relationship to microstructure and thermal conductivity of Y$_2$O$_3$-doped AlN has been made by Pezzotti *et al.*[23] Their results are presented in Table 5. In general, they found that longer annealing times lead to increased AlN grain sizes, larger aggregates of Y$_2$O$_3$ at the junctions of AlN grains (rather than continuous wetting on isolated grains), and, hence, enhanced thermal conductivity. Similar recent studies of AlN with CaF$_2$ and Y$_2$O$_3$ confirm that during sintering at 1650°C the inhomogeneities move from the grain boundaries to form discrete pockets, thereby enhancing the densification and increasing thermal conductivity.[24]

In summary, the thermal conductivity of AlN can be maximized by reduction of oxide impurities, through the use of dopants or N$_2$ atmosphere treatments, increase in grain size (through long sintering or annealing times), and removal of impurities from grain boundaries.

TABLE 4 Thermal Conductivities at Room Temperature for Doped AlN Following Different Sintering Treatment.[22]

κ (W m^{-1}K^{-1}) Dopants in AlN	1600°C for 5 h	1600°C for 1 h, followed by 1600°C for 15 h	1600°C for 1 h, followed by 1600°C for 45 h
4% Li$_2$O + 2% CaO	54	64	112
0.5% Li$_2$O + 0.5% CaO + 7% Dy$_2$O$_3$	89	95	163
0.5% Li$_2$O + 0.5% CaO + 9% Dy$_2$O$_3$	84	90	150

Sec. 2 · CERAMICS

TABLE 5 Influence of the Anneal Time at 1800°C on the Room-Temperature Thermal Conductivity, Grain Size, Grain Boundary Thickness, and the Size of the Y_2O_3-Based Phase Trapped at Triple Junctions of the AlN Matrix of 5 wt% Y_2O_3-Doped AlN[23]

Annealing time (h)	0	5	10	20	30	50
κ (W m^{-1}K^{-1})	174	207	221	224	223	224
AlN grain size (μm)	2.5	6.3	8.3	10.0	10.7	11.2
Grain boundary thickness (μm)	0.45	0.25	0.08	0.01	$\sim$0.005	$\sim$0.005
Y_2O_3 triple-grain junction (μm)	0.75	2.3	2.9	3.3	3.4	3.4

2.1.2. Silicon Nitride (Si$_3$N$_4$)

Silicon nitride (Si$_3$N$_4$) can have a room-temperature thermal conductivity as high as 200 to 320 W m^{-1}K^{-1}.[25] Furthermore, its mechanical properties are superior to those of AlN, making Si$_3$N$_4$ very useful as an electrical substrate. However, relatively low thermal conductivities have been reported for Si$_3$N$_4$ ceramics, ranging from 20 to 70 W m^{-1}K^{-1}.[26]

Hot pressing[27,28,29] and hot isostatic pressing (HIP) or a combination of these two methods[30] can be used to develop high-thermal-conductivity Si$_3$N$_4$. For example, Watari et al.[30] developed a HIPed Si$_3$N$_4$ with a room-temperature thermal conductivity of 102 W m^{-1}K^{-1} perpendicular to the hot-pressing axis, compared with 93 W m^{-1}K^{-1} for the HIP sample. In this case the process included mixing of high-purity 0.8 μm α-Si$_3$N$_4$ raw powder and 3.5 wt% Y_2O_3, followed by sieving, then hot pressing at 1800°C in flowing N_2, then HIP sintering at 2400°C for 2 h under flowing N_2. The HIP treatment resulted in large β-Si$_3$N$_4$ grains, dimensions 10–50 μm, and the elimination of smaller grains. The grain size is much larger than the room-temperature phonon mean free path (ca. 20 nm), showing that the predominant thermal resistance mechanism is point defects and dislocations within the grains.[30]

Other researchers have succeeded in improving the thermal conductivity of Si$_3$N$_4$ by adding β-Si$_3$N$_4$ particles as seeds.[31–33] Although seeding leads to larger grains, the mechanical properties deteriorate. Furthermore, above a certain grain size, the room-temperature thermal conductivity no longer increases,[26] for reasons described earlier.

As for AlN, the thermal conductivity of Si$_3$N$_4$ can be improved by the addition of dopants and control of the sintering time and temperature. For example, Si$_3$N$_4$ sintered with 5 wt% Y_2O_3 and 2–4 wt% MgO can have room-temperature thermal conductivities as high as about 80 W m^{-1} K^{-1}, if sintered at a temperature sufficiently high to remove Mg$_5$Y$_6$Si$_5$O$_{24}$.[34]

The ceramic Si$_3$N$_4$ with the highest room-temperature thermal conductivity, 110–150 W m^{-1} K^{-1}, has been produced by chemical vapor deposition, which allowed careful control of the oxygen content.[35] However, because of the millimeter-sized grains, this led to a material with poor mechanical properties. As we now know, such large grains are not required for high thermal conductivity, so there is scope for development of methods to produce small-grain, high-purity Si_3N_4, which could have both high thermal conductivity and good mechanical properties.

2.1.3. Alumina (Al_2O_3)

Alumina (Al_2O_3,) is a useful ceramic because of its great abundance (alumina-containing materials make up about 15% of the earth's crust). Although not as high in thermal conductivity as AlN or Si_3N_4, it still has relatively high room-temperature thermal conductivity (ca. 30 W m^{-1} K^{-1}), low thermal expansion, and high compressive strength, making it useful where good thermal shock resistance is required. Furthermore, alumina is resistant to most chemicals, and is a good electrical insulator with high wear resistance. Therefore, alumina is extensively used for mechanical and electronic devices in the ceramic industry.[36]

Nunes Dos Santos et al.[37] have systematically studied Nb_2O_5-doped Al_2O_3 with variation in Nb_2O_5 concentration and sintering temperature. Nb_2O_5 is useful because of the Nb_2O_5–Al_2O_3 eutectic at 1698 K, allowing densification and, hence, enhanced mechanical properties, via formation of a transient liquid phase.[38] After sintering at a relatively high temperature (1723 K), the porosity can be reduced to 3% for a sample with 6% Nb_2O_5. However, the addition Nb_2O_5 has detrimental effects on thermal conductivity, reducing the room-temperature value from ca. 15 to 4 W m^{-1} K^{-1}, when 6% Nb_2O_5 is added. The plausible reason is that Nb acts as an impurity in the Al_2O_3 lattice and thus increases the phonon scattering.[37] It would appear that other means to improve the mechanical properties of alumina must be sought if high thermal conductivity is required.

2.2. Novel Materials with Various Applications

2.2.1. Ceramic Composites

Besides traditional high-thermal-conductivity ceramics, a number of novel ceramic composite materials have been developed in recent years.

2.2.1.1. Diamond Film on Aluminum Nitride. Diamond film (up to 12 μm) can be synthesized on AlN ceramic substrates (giving DF/AlN) by hot-filament chemical vapor deposition (HFCVD). With a high electrical resistance and low dielectric coefficient, DF/AlN has potential applications in electronic devices. The room-temperature thermal conductivity has been investigated as a function of film thickness; results are given in Table 6. The thermal conductivity increased with film thickness, giving a maximum thermal conductivity of 205 W m^{-1}K^{-1} at room temperature, 73% greater than the AlN substrate.[39] Furthermore, the adhesion of the film to the AlN is very good due to the formation of aluminum carbide.

2.2.1.2. Silicon Carbide Fiber-Reinforced Ceramic Matrix Composite (SiC-CMC). Silicon carbide fiber-reinforced ceramic matrix composites are being developed for thermostructural materials, but most cannot be used at high temperature in air due to oxidation resistance and/or heat resistance of the fiber and interphase.[40,41] However, Ishikawa et al.[42] have reported a SiC-CMC, which maintains its strength (over 600 MPa) and stability in air up to 1600°C, with thermal conductivities as high as 35 W m^{-1} K^{-1} at 1000°C. The fibers were made from close-

Sec. 2 · CERAMICS

TABLE 6 Room-temperature Thermal Conductivity of DF/AlN Composites with Varying Thicknesses of Diamond Film[39]

Thickness of diamond film (μm)	2	6	9	12
κ (W m^{-1}K^{-1})	132	183	197	206

packed, fine hexagonal columns of sintered β-silicon carbide crystals, with no apparent second phase at the grain boundaries.

2.2.1.3. Carbon Fiber-Incorporated Alumina Ceramics. In a continuing search for high-thermal-conductivity materials for microelectronics applications, Ma and Hng[43] recently developed an Al$_2$O$_3$-layered substrate in which carbon fibers had been incorporated. Different thicknesses of the alumina layers were used to investigate the effect of the alumina/carbon fiber thickness ratio. The results, presented in Table 7, show dramatic improvement of the thermal conductivity as the carbon content increases. A further favorable feature for microelectronics applications is that the dielectric constant also decreases as the carbon content increases.

2.2.1.4. Ceramic Fibers. Due to their low thermal conductivities, ceramic fibers have important applications related to high-temperature thermal isolation. Maqsood *et al.* have investigated the thermal conductivities of several commercial ceramic fibers as functions of temperature and load pressure,[44] as summarized in Tables 8 and 9. The variation in thermal conductivity with temperature is rather small. In addition, they determined the variation of the room-temperature thermal conductivity with pressure up to 9.7 kN m^{-2}. At most, the thermal conductivity increased by 12% (for Nextel) on increasing the load pressure. Under all these conditions Nextel/VK-80 was found to have the best thermal insulating ability.

2.2.1.5. Glass-Ceramic Superconductor. The thermal conductivities of superconductors are known to be exceptionally low below T_c because the Cooper pairs do not interact with the thermal phonons; this can allow superconductors to be used as thermal switches in the vicinity of the critical temperature.[45] Recent investigations of a glass-ceramic superconductor, (Bi$_{2-\delta-\gamma}$Ga$_\delta$Tl$_\gamma$)Sr$_2$Ca$_2$Cu$_3$O$_{10+x}$, pre-

TABLE 7 Room-Temperature Thermal Conductivity and Dielectric Constant at 7.5 MHz for Carbon-Fiber-Incorporated Alumina Substrates, as a Function of the Ratio of the Thickness of the Carbon layer to the Aluminum Layer.[43]

Thickness ratio	0	0.037	0.062	0.130	0.213	0.588
κ (W m^{-1}K^{-1})	8.0	12.7	15.2	22.5	30.6	52.1
Dielectric constant	6.02	3.65	3.16	3.04	2.61	1.24

TABLE 8 Room-Temperature Thermal Conductivities of Ceramic Fibers at Ambient Pressure[44]

Trademark (Composition)	Temperature (K)	κ (W m^{-1}K^{-1})
VK (60Al$_2$O$_3$/40SiO$_2$)	293	0.0376
ABK (70Al$_2$O$_3$/28SiO$_2$/2B$_2$O$_3$)	293	0.0345
Nexel/VK-80 (80Al$_2$O$_3$/20SiO$_2$)	294	0.0305

TABLE 9 Room-Temperature Thermal Conductivities of Ceramic Fibers under a Load of 1.7 kN m^{-2} [44]

Trademark	Temperature (K)	κ (W m^{-1}K^{-1})
VK-60	298	0.0566
	473	0.0793
	673	0.1074
	873	0.1320
	1073	0.1700
ABK	298	0.0531
	473	0.0657
	673	0.0721
	873	0.1037
	1073	0.1300
Nextel/VK-80	298	0.0465
	473	0.0501
	673	0.0562
	873	0.0795
	1073	0.1100

pared by melt quenching, show[46] the expected decrease in thermal conductivity far below T_c, but a small peak in thermal conductivity, with κ of the order of 0.5 W m^{-1} K^{-1} (with the exact value depending on the composition), near T_c. The peak has been attributed to electron–phonon coupling.[46]

2.2.2. Other Ceramics

2.2.2.1. Rare-Earth Based Ceramics. Rare-earth-based ceramics have applications in the nuclear industry, particularly in neutron absorption, nuclear control rods, and radioactive waste containment. Quantification of their thermal properties, especially high-temperature thermal conductivity, is an important part of their assessment. The high-temperature thermal conductivities of LaAlO$_3$ and La$_2$Zr$_2$O$_7$ (high-temperature container materials, solid electrolytes and potential host materials for fixation of radioactive waste), SmZr$_2$O$_7$, Eu$_2$Zr$_2$O$_7$, Gd$_2$Zr$_2$O$_7$, and GdAlO$_3$ (neutron absorption and control rod materials) are given in Table 10.[47,48]

In all these materials the thermal conductivity drops with increasing temperature up to about 1000 K, indicating that the dominant heat transport mechanism is phononic. As the temperature increases, the phonon mean free path decreases until it reaches its minimum value, of the order of interatomic distances. At higher

Sec. 2 · CERAMICS

TABLE 10 Thermal Conductivities of Several Rare-Earth-Based Ceramics Used in the Nuclear Industry[47,48]

κ (W m^{-1}K^{-1})	\multicolumn{8}{c}{Temperature (K)}
	673	773	873	973	1073	1173	1273	1373
LaAlO$_3$	2.63	2.22	2.00	1.92	1.90	1.87	1.88	1.87
SmZr$_2$O$_7$	1.73	1.67	1.58	1.49	1.51	1.50	1.50	1.55
Eu$_2$Zr$_2$O$_7$	1.66	1.63	1.56	1.57	1.58	1.58	1.60	1.65
GdAlO$_3$	6.58	4.98	4.04	3.56	3.38	3.38	3.42	3.69
La$_2$Zr$_2$O$_7$	2.27	1.82	1.64	1.47	1.42	1.33	1.29	1.29
Gd$_2$Zr$_2$O$_7$	1.78	1.33	1.24	1.16	1.02	0.98	1.02	0.98

temperatures the thermal conductivity increases slightly, especially for SmZr$_2$O$_7$, Eu$_2$Zr$_2$O$_7$ and GdAlO$_3$, indicating a contribution from radiative heat transfer.[47]

2.2.2.2. Magnesium Silicon Nitride (MgSiN$_2$).

Magnesium silicon nitride (MgSiN$_2$), has been proposed as an alternative heat sink material for integrated circuits.[1,49] Nonoptimized electrical and mechanical properties are comparable with those of Al$_2$O$_3$ and AlN,[49,50] and it had been suggested that the optimized room-temperature thermal conductivity could be 30 to 50 W m^{-1} K^{-1}.[51] Bruls et al.[52] carried out a systematic study of MgSiN$_2$ ceramics with and without sintering additives, and optimized the production of low-oxygen (< 1 wt%), fully dense, large-grain ceramics with no grain boundary phases and secondary phases as separate grains. From these samples, which gave a maximum room-temperature thermal conductivity of 21–25 W m^{-1} K^{-1}, they concluded that the intrinsic thermal conductivity of MgSiN$_2$ would not exceed 35 W m^{-1} K^{-1}.[52]

2.2.2.3. Thermoelectric Ceramics.

The development of novel materials for thermoelectric applications is an active field, and ceramic materials are being investigated in this context. Oxides are especially important because of their resistance to oxidation. The main aim of thermoelectrics is to develop materials with high values of the figure of merit, ZT, where $Z = S^2\sigma/\kappa$ (S is the Seebeck coefficient, σ is the electrical conductivity), and T is the temperature (in K).

Bi$_{2-x}$Pb$_x$Sr$_{3-y}$Y$_y$Co$_2$O$_{9-\delta}$ is a polycrystalline ceramic with high potential as a good thermoelectric material because of its high-temperature stability. A recent study[53] varied x and y and found that $x = y = 0.5$ gave the best thermoelectric properties. The thermal conductivity (see Table 11) decreased as the temperature increased, indicating dominance of thermal phonon–phonon resistance, showing that ZT would be improved as temperature is increased. The value of ZT was 0.006 at $T = 300$ K, rising to 0.052 at $T = 800$ K.

Katsuyama et al.[54] reported studies of another thermoelectric ceramic, (Zn$_{1-y}$Mg$_y$)$_{1-x}$Al$_x$O, as functions of both x and y. The thermal conductivity was found to decrease as the temperature increased, but without a systematic dependence on x when $y = 0$. However, the value of ZT for $y = 0$ was found to increase with decreasing x and increasing T, to a maximum ZT for Zn$_{0.9975}$Al$_{0.0025}$O of 0.074 at 1073 K. When $x = 0.0025$ and y was varied, ZT was optimized at $y = 0.10$,

TABLE 11 Thermal Conductivity of $Bi_{2-x}Pb_xSr_{3-y}Y_yCo_2O_{9-\delta}$, for $x = y = 0.5$[53]

Temperature (K)	300	400	600	800
κ (W m^{-1}K^{-1})	1.0	0.9	0.8	0.8

corresponding to $(Zn_{0.90}Mg_{0.10})_{0.9975}Al_{0.0025}O$, with a maximum ZT of 0.10 at 1073 K.

3. GLASSES

3.1. Introduction

The origins of glass are thought to be in Mesopotamia, more than 5000 years ago. Although this likely was soda lime glass, we know now that many more materials are glass formers. Technically important glasses range from polymers and other soft materials to metallic alloys. Here we concentrate on the thermal conductivity and related properties of one glass family, chalcogenide glasses, i.e., those containing S, O, Se, and Te, as illustrative of several of the factors governing thermal conductivity of glasses.

3.2. Chalcogenide Glasses

Chalcogenide glasses have been extensively studied for a long time. They are normally p-type semiconductors[55] and can be used as switching and memory devices.[56] One of the features of amorphous solids is that the thermal conductivity increases with increasing temperature and approaches a nearly-temperature-independent value near the softening temperature.[57]

Velinov and Gateshiki[58] studied the thermal diffusivities of $Ge_xAs_{40-x}(S/Se)_{60}$. The thermal diffusivity was found to vary with the average coordination number, Z, and this was explained on the basis of changes in the chemical ordering and in the network topology. Their results are summarized in Table 12.

Hegab et al.[59] studied another chalcogenide glass, $Te_{82.2}Ge_{13.22}Si_{4.58}$. They investigated the thermal conductivity and other properties as a function of temperature and film thickness; the results for the bulk sample are summarized in Table 13. They show an increase in thermal conductivity with increasing temperature, typical of an amorphous sample. The electrical response shows memory-type switching.

Philip et al.[60] reported the thermal conductivity and heat capacity of $Pb_{20}Ge_xSe_{80-x}$ and $Pb_yGe_{42-y}Se_{58}$ as functions of the Ge and Pb concentrations. The thermal conductivity increases with increasing x or y up to $x = 21$ and $y = 8$ and then decreases sharply, as summarized in Tables 14 and 15. These materials exhibit changeover from p-type to n-type semiconductors at specific compositions, and the p to n transition reduces the phonon mean free path and, consequently, reduces the thermal conductivity for $x > 21$ and $y > 8$.[60]

Srinivasan et al.[61] studied the thermal diffusivity, optical band gap, and mean coordination number, $\langle r \rangle$, of $Ge_xSb_5Se_{90-x}$ and $Ge_xSb_{10}Se_{90-x}$ glasses. As shown in Tables 16 and 17, the thermal diffusivity reaches a peak at $\langle r \rangle = 2.6$, indicating that

Sec. 3 · GLASSES

TABLE 12 Values of Average Coordination Number, Z, and Room Temperature Thermal Diffusivity of $Ge_xAs_{40-x}S_{60}$ and $Ge_xAs_{40-x}S_{60}$ [58]

x	Z	Thermal diffusivity of $Ge_xAs_{40-x}S_{60}$ (10^{-3} cm^2 s^{-1})	Thermal diffusivity of $Ge_xAs_{40-x}Se_{60}$ (10^{-3} cm^2 s^{-1})
0	2.40	2.40	2.16
0.10	2.50	1.89	1.67
0.15	2.55	1.51	1.60
0.22	2.62	1.70	1.89
0.25	2.65	1.92	1.90
0.27	2.67	1.98	1.95
0.32	2.72	2.52	2.03
0.36	2.76	2.72	2.34

TABLE 13 Thermal Conductivity as a Function of Temperature for Bulk $Te_{82.2}Ge_{13.22}Si_{4.58}$ [59]

T (K)	300	304	307	316	320	327	329	335	341	345	350
κ (W m^{-1} K^{-1})	2.6	4.0	4.7	6.4	8.4	9.8	10.3	11.3	13.5	15.0	15.6

the network is most ordered at this composition and then decreases with further increase in $\langle r \rangle$.

The room-temperature thermal diffusivity of chalcogenide glasses of composition Ge_xTe_{1-x} has revealed a relationship between thermal properties and network ordering.[62] The thermal diffusivity exhibited peaks at $x = 0.2$ and $x = 0.5$ (see Table 18). The first peak was explained by the chemically ordered network model,[63,64,65,66] in which $x = 0.2$ represents the crossover from a one-dimensional network to a three-dimensional network. The second peak corresponds to the formation of GeTe.[62]

TABLE 14 Composition Dependence of Room-Temperature Thermal Conductivity of $Pb_{20}Ge_xSe_{80-x}$ [60]

x	17	19	21	22	24
κ (10^{-3} W cm^{-1} K^{-1})	3.97	3.99	4.01	3.72	3.41

TABLE 15 Composition Dependence of Room-Temperature Thermal Conductivity of $Pb_yGe_{42-y}Se_{58}$ [60]

y	0	4	6	8	10	14	20
κ (10^{-3} W cm^{-1} K^{-1})	3.73	3.79	3.81	3.83	3.62	3.62	3.64

TABLE 16 Values of Mean Coordination Number, $\langle r \rangle$, and Room-Temperature Thermal Diffusivity of $Ge_xSb_5Se_{90-x}$.[61]

x	$\langle r \rangle$	Thermal Diffusivity (10^{-2} cm^2 s^{-1})
12.5	2.30	0.81
15.0	2.35	0.85
17.5	2.40	0.88
20.0	2.45	0.95
22.5	2.50	1.06
25.0	2.55	1.26
27.5	2.60	1.69
30.0	2.65	1.55
32.5	2.70	1.28
35.0	2.75	1.15

TABLE 17 Values of Mean Coordination Number, $\langle r \rangle$, and Room-Temperature Thermal Diffusivity of $Ge_xSb_{10}Se_{90-x}$.[61]

x	$\langle r \rangle$	Thermal Diffusivity (10^{-2} cm^2 s^{-1})
10.0	2.30	0.93
12.5	2.35	0.99
15.0	2.40	1.01
17.5	2.45	1.08
20.0	2.50	1.21
22.5	2.55	1.58
25.0	2.60	1.94
27.5	2.65	1.85
30.0	2.70	1.81
32.5	2.75	1.71

TABLE 18 Room-Temperature Thermal Diffusivity of Ge_xTe_{1-x} with Varying Concentration of Ge[62]

x	0.00	0.10	0.20	0.25	0.30	0.35	0.40	0.50	0.65	0.80
Thermal diffusivity (cm^2 s^{-1})	0.013	0.0090	0.027	0.003	0.003	0.003	0.004	0.015	0.006	0.008

3.3. Other Glasses

As one example of other glass systems, we present the results of thermal diffusivity measurements of vacuum-melted, low-silica, calcium aluminosilicate glass with various concentration of Nd_2O_3 in Table 19.[67] The thermal diffusivity decreases as the concentration of Nd_2O_3 increases, indicating that Nd^{3+} acts as a network modifier.[67]

TABLE 19 Room-Temperature Values of Thermal Diffusivity and Thermal Conductivity of Calcium Aluminosilicate Glass of Composition 47.4 wt% CaO, (41.5 - x) wt% Al_2O_3, 7.0 wt% SiO_2, 4.1 wt% MgO, x wt% Nd_2O_3 [67]

x	Thermal Diffusivity (10^{-3} cm^2 s^{-1})	κ (W m^{-1}K^{-1})
0.0	5.69±0.05	1.548±0.031
0.5	5.63±0.05	1.525±0.030
1.0	5.67±0.11	1.559±0.043
1.5	5.65±0.09	1.545±0.038
2.0	5.55±0.04	1.534±0.031
2.5	5.44±0.04	1.494±0.030
3.0	5.55±0.04	1.541±0.031
3.5	5.44±0.03	1.495±0.030
4.0	5.39±0.03	1.490±0.030
4.5	5.36±0.03	1.477±0.029
5.0	5.22±0.03	1.433±0.029

4. CONCLUSIONS

In many applications of ceramic and glass materials, appropriate thermal conductivity is intrinsically linked to their applications. Through selected examples, we have illustrated how recent advances in thermal conductivity allow us to better understand thermal conductivities of ceramics and glasses, thereby advancing uses of these important materials.

5. REFERENCES

1. G. A. SLACK *Nonmetallic Crystals with High Thermal Conductivity* J. Phys. Chem. Solids **34**, 321–335 (1973).
2. P. GREIL *Advanced Engineering Ceramics* Adv. Eng. Mater. **4**, 247–254 (2002).
3. D. R. FEAR and S. THOMAS *Emerging Materials Challenges in Microelectronics Packaging* MRS Bulletin **28**(1), 68–74 (2003).
4. V. P. ATLURI, R. V. MAHAJAN, P. R. PATEL, D. MALLIK, J. TANG, V. S. WAKHARKAR, G. M. CHRYSLER, C.-P. CHIU, G. N. CHOKSI, and R. S. VISWANATH *Critical Aspects of High-Performance Microprocessor Packaging* MRS Bull. **28**(1), 21–34 (2003).
5. W. WERDECKER and F. ALDINGER *Aluminum Nitride – An Alternative Ceramic Substrate for High Power Applications in Microcircuits* IEEE Trans. Comp., Hybrids, Manuf. Technol. **7**, 399–404 (1984).
6. Y. KUROKAWA, K UTSUMI, H. TAKAMIZAWA, T. KAMATA, and S. NOGUCHI *AlN Subtrates with High Thermal Conductivity* IEEE Trans. Comp., Hybrids, Manuf. Technol. **8**, 247–252 (1985).
7. D. G. BRUNNER and K. H. WIENAND *Metallized Aluminium Nitride Ceramics— Potential, Properties, Applications* Interceram. **37**(4), 29–32 (1988).
8. N. ICHINOSE *Aluminium Nitride Ceramics for Substrates* Mater. Sci. Forum **34-36**, 663–667 (1988).
9. F. MIYASHIRO, N. IWASE, A. TSUGE, F. UENO, M. NAKAHASHI, and T. TAKAHASHI *High Thermal Conductivity Aluminium Nitride Ceramic Substrates and Packages* IEEE Trans. Comp., Hybrids, Manuf. Technol. **13**, 313–319 (1990).
10. R. R. TUMMALA *Ceramic and Glass-Ceramic Packaging in the 1990s* J. Am. Ceram. Soc. **74**(5), 895–908 (1991).
11. A. V. VIRKAR *Thermodynamic and Kinetic Effects of Oxygen Removal on the Thermal Conductivity of Aluminum Nitride* J. Am. Ceram. Soc. **72**(11), 2031–2042 (1989).

12. L. M. Sheppard *Aluminum Nitride: A Versatile but Challenging Material* Am. Ceram. Soc. Bull. **69**(11), 1801–1812 (1990).
13. G. W. Prohaska and G. R. Miller *Aluminum Nitride: A Review of the Knowledge Base for Physical Property Development* Mat. Res. Soc. Symp. Proc. **167**, 215–227 (1990).
14. G. A. Slack and S. F. Bartram *Thermal Expansion of Some Diamondlike Crystals* J. Appl. Phys. **1**, 89–98 (1975).
15. K. Watari, T. Tsugoshi, T. Nagaoka, K. Ishizaki, S. Ca, and K. Mori in *Proceedings of the 18th International Japan-Korea Seminar on Ceramics* Edited by A. Kato, H. Tateyama and H. Hasuyama (TIC, Japan 2001), 98–101.
16. K. Watari, M. Kawamoto, and K. Ishizaki *Sintering Chemical Reactions to Increase Thermal Conductivity of Aluminum Nitride* J. Mater. Sc. **26**(17), 4727–4732 (1991).
17. K. Watari, H. Nakano, K. Urabe, K. Ishizaki, S. Cao, and K. Mori *Thermal Conductivity of AlN Ceramic with a Very Low Amount of Grain Boundary Phase at 4 to 1000 K* J. Mater. Res. **17**(11), 2940–2944 (2002).
18. T. B. Jackson, A. V. Virkar, K. L. More, R. B. Dinwiddie, Jr., and R. A. Cutler *High-Thermal-Conductivity Aluminum Nitride Ceramics: the Effect of Thermodynamic, Kinetic, and Microstructural Factors* J. Amr. Ceram. Soc. **80**(6), 1421–1435 (1997).
19. X. Xu, H. Zhuang, W. Li, S. Xu, B. Zhang, and X. Fu *Improving Thermal Conductivity of Sm_2O_3-doped AlN Ceramics by Changing Sintering Conditions* Mat. Sci. Eng. A **342**, 104–108 (2003).
20. J. Jarrige, J. P. Lecompte, J. Mullot, and G. Miller *Effect of Oxygen on the Thermal Conductivity of Aluminium Nitride Ceramics* J. Europ. Ceram. Soc. **17**, 1891–1895 (1997).
21. K. Watari, M. E. Brito, T. Nagaoka, M. Toriyama, and S. Kanzaki *Additives for Low-Temperature Sintering of AlN Ceramics with High Thermal Conductivity and High Strength* Key Engin. Mater. **159-160**, 205–208 (1999).
22. Y. Liu, H. Zhou, Y. Wu, and L. Qiao *Improving Thermal Conductivity of Aluminum Nitride Ceramics by Refining Microstructure* Mater. Lett. **43**(3), 114–117 (2000).
23. G. Pezzotti, A. Nakahira, and M. Tajika *Effect of Extended Annealing Cycles on the Thermal Conductivity of AlN/Y_2O_3 Ceramics* J. Europ. Ceram. Soc. **20**(9), 1319–1325 (2000).
24. L. Qiao, H. Zhou, H. Xue, and S. Wang *Effect of Y_2O_3 on Low Temperature Sintering and Thermal Conductivity of AlN Ceramics* J. Europ. Ceram. Soc. **23**, 61–67 (2003).
25. J. S. Haggerty and A. Lightfoot *Opportunities for Enhancing the Thermal Conductivities of SiC and Si_3N_4 Ceramics Through Improved Processing* Ceram. Eng. Sci. Proc. **16**(4), 475–487 (1995).
26. M. Mitayama, K. Hirao, M. Toriyama, and S. Kanzaki *Thermal Conductivity of β-Si_3N_4: I, Effects of Various Microstructural Factors* J. Am. Ceram. Soc. **82**(11), 3105–3112 (1999).
27. G. Ziegler and D. P. H. Hasselman *Effect of Phase Composition and Microstructure on the Thermal Diffusivity of Silicon Nitride* J. Mater. Sci. **16**, 495–503 (1981).
28. M. Kuriyama, Y. Inomata, T. Kijima, and Y. Hasegawa *Thermal Conductivity of Hot-Pressed Si_3N_4 by Laser-Flash Method*, Amer. Ceram. Soc. Bull. **57**(12), 1119–1122 (1978).
29. K. Tsukuma, M. Shimada and M. Koizumi *Thermal Conductivity and Microhardness of Si_3N_4 with and without Additives* Am. Ceram. Soc. Bull. **60**(9), 910–912 (1981).
30. K. Watari, K. Hirao and M. Toriyama *Effect of Grain Size on the Thermal Conductivity of Si_3N_4* J. Am. Ceram. Soc. **82**(3), 777–779 (1999).
31. Y. Okamoto, N. Hirosaki, M. Ando, F. Munakata, and Y. Akimune *Thermal Conductivity of Self-reinforced Silicon Nitride Containing Large Grains Aligned by Extrusion Pressing* J. Ceram. Soc. Jpn. **105**, 631–633 (1997).
32. S. W. Lee, H. B. Chae, D. S. Park, Y. H. Choa, K. Niihara, and B. J. Hockey *Thermal Conductivity of Unidirectionally Oriented Si_3N_4w/Si_3N_4 Composites* J. Mater. Sci. **35**, 4487–4493 (2000).
33. N. Hirosaki, Y. Okamoto, F. Munakata, and Y. Akimune *Effect of Seeding on the Thermal Conductivity of Self-reinforced Silicon Nitride* J. Europ. Ceram. Soc. **19**, 2183–2187 (1999).
34. Y. Lin, X.-S. Ning, H. Zhou, and W. Xu *Study on the Thermal Conductivity of Silicon Nitride Ceramics with Magnesia and Yttria as Sintering Additives* Mat. Lett. **57**, 15–19 (2002).
35. N. Hirosaki, Y. Okamoto, M. Ando, F. Munakata and Y. Akimune *Effect of Grain Growth on the Thermal Conductivity of Silicon Nitride* J. Ceram. Soc. Jpn. **104**, 49–53 (1996).
36. H. Hubner and E. Dorre *Alumina: Processing, Properties and Applications* (Springer-Verlag, Berlin, 1984), pp. 220–265.
37. W. Nunes Dos Santos, P. I. P. Filho, and R. Taylor *Effect of Addition of Niobium Oxide on the Thermal Conductivity of Alumina,* J. Europ. Ceram. Soc. **18**, 807–811 (1998).
38. R. S. Roth, T. Nagas, and L. P. Cook *Phase Diagrams for Ceramics*, The American Ceramic Society, Columbus, OH (1981), **4**, 117.

Sec. 5 · REFERENCES

39. Y. LIAO, R.C. FANG, Z. Y. YE, N.G. SHANG, S. J. HAN, Q. Y. SHAO, and S.Z. JI *Investigation of the Thermal Conductivity of Diamond Film on Aluminum Nitride Ceramic* App. Phys. A **69**, 101–103 (1999).
40. I. J. DAVIES, T. ISHIKAWA, N. SUZUKI, M. SHIBUYA, and T. HIROKAWA Proc. 5th Japan Int. SAMPE Symp. (Japan Chapter of SAMPE, Yokohama (1997), pp. 1672–1632.
41. M. TAKEDA, Y. IMAI, H. ICHIKAWA, Y. KAGAWA, H. IBA, and H. KAKISAWA *Some Mechanical Properties of SiC (Hi-Nicalon) Fiber-Reinforced SiC Matrix Nicaloceram Composites* Ceram. Eng. Sci. Proc. **18**, 779–786 (1997).
42. T. ISHIKAWA, S. KAJII, K. MATSUNAGA, T. HOGAMI, Y. KOHTOKU, and T. NAGASAWA *A Tough, Thermally Conductive Silicon Carbide Composite with High Strength up to $1600°C$ in Air* Science **282**, 1295–1297 (1998).
43. J. MA and H. H. HNG *High Thermal Conductivity Ceramic Layered System Substrates for Microelectronic Applications* J. Mater. Sci.: Mater. Electron. **13**, 461–464 (2002).
44. A. MAQSOOD, M. ANIS-UR-REHMAN, V. GUMEN, and ANWAR-UR-HAQ *Thermal Conductivity of Ceramic Fibers as a Function of Temperature and Press Load* J. Phys. D: Appl. Phys. **33**, 2057–2063 (2000).
45. M. A. WHITE *Properties of Materials* (Oxford, New York, 1999), pp. 270–271.
46. Y. BALCI, M. E. YAKINCI, M. A. AKSAN, A. ÖZDES, and H. ATES *Thermal Conductivity Properties of Glass-Ceramic $(Bi_{2-\delta-\gamma}Ga_\delta Tl_\gamma)Sr_2Ca_2Cu_3O_{10+x}$ High-T_c Superconductors* J. Low Temp. Phys. **117**, 963–967 (1999).
47. G. SURESH, G. SEENIVASAN, M. V. KRISHNAIAH, and P. S. MURTI *Investigation of the Thermal Conductivity of Selected Compounds of Gadolinium and Lanthanum* J. Nucl. Mater. **249**, 259–261 (1997).
48. G. SURESH, G. SEENIVASAN, M. V. KRISHNAIAH, and P. S. MURTI *Investigation of the Thermal Conductivity of Selected Compounds of Lanthanum, Samarium and Europium* J. Alloys Comp. **269**, L9-L12 (1998).
49. W. A. GROEN, M. J. KRAAN, and G. DE WITH *Preparation, Microstructure and Properties of Magnesium Silicon Nitride ($MgSiN_2$) Ceramics* J. Europ. Ceram. Soc. **12**, 413–420 (1993).
50. W. A. GROEN, M. J. KRAAN, G. DE WITH, and M. P. A. VIEGERS *New Covalent Ceramics: $MgSiN_2$* Mat. Res. Soc. Symp. Proc. **327**, 239–244 (1994).
51. R. J. BRULS, H. T. HINTZEN, and R. METSELAAR *Proceedings of the Twenty Fourth International Thermal Conductivity Conference and Twelfth International Thermal Expansion Symposium* (Pittsburgh, 1997), edited by P. S. Gaal, and D. E. APOSTOLESEU (Technomics, Pennsylvania, 1999), 3.
52. R. J. BRULS, A. A. KUDYBA-JANSEN, P. GERHARTS, H. T. HINTZEN, and R. METSELAAR *Preparation, Characterization and Properties of $MgSiN_2$ Ceramics* J. Mater. Sci.: Mater. Electron. **13**, 63–75 (2002).
53. E. IGUCHI, T. ITOGA, H. NAKATSUGAWA, F. MUNAKATA, and K. FURUYA *Thermoelectric Properties in $Bi_{2-x}Pb_x Sr_{3-y}Y_yCo_2O_{9-\delta}$ Ceramics* J. Phys. D: Appl. Phys. **34**, 1017–1024 (2001).
54. S. KATSUYAMA, Y. TAKAGI, M. ITO, K. MAJIMA, H. NAGAI, H. SAKAI, K. YOSHIMURA, and K. KOSUGE *Thermoelectric Properties of $(Zn_{1-y}Mg_y)_{1-x}Al_xO$ Ceramics Prepared by the Polymerized Complex Method* J. Appl. Phys. **92**, 1391–1398 (2002).
55. N. F. MOTT and E. A. DAVIS *Electronic Processes in Non-Crystalline Materials* (Clarendon, Oxford, 1979).
56. D. ADLER *Amorphous-Semiconductor Devices* Sci. Am. **236**(5), 36–48 (1977).
57. D. G. CAHILL, J. R. OLSON, H. E. FISCHER, S. K. WATSON, R. B. STEPHENS, R. H. TAIT, T. ASHWORTH, and R. O. POHL *Thermal Conductivity and Specific Heat of Glass Ceramics* Phys. Rev. B **44**, 12226–12232 (1991).
58. T. VELINOV and M. GATESHIKI *Thermal Conductivity of Ge-As-Se(S) Glasses* Phys. Rev. B: Condensed Matter Mater. Phys. **55**(17), 11014–11017 (1997).
59. N. A. HEGAB, M. FADEL, M. A. AFIFI, and M. F. SHAWER *Temperature Dependence of Electrical and Thermal Properties of $Te_{82.2}Ge_{13.22}Si_{4.58}$ Glassy Alloy* J. Phys. D: Appl. Phys. **33**, 2223–2229 (2000).
60. J. PHILIP, R. RAJESH and C. P. MENON *Carrier-Type Reversal in Pb-Ge-Se Glasses: Photopyroelectric Measurements of Thermal Conductivity and Heat Capacity* Appl. Phys. Lett. **78**, 745–747 (2001).
61. A. SRINIVASAN, K. N. MADHUSOODANAN, E. S. R. GOPAL, and J. PHILIP *Observation of a Threshold Behavior in the Optical Band Gap and Thermal Diffusivity of Ge-Sb-Se Glasses* Phys. Rev. B **45**, 8112–8115 (1992).
62. J. C. DE LIMA, N. CELLA, L. C. M. MIRANDA, C. CHYING AN, A. H. FRANZAN, and N. F. LEITE *Photoacoustic Characterization of Chalcogenide Glasses: Thermal Diffusivity of Ge_xTe_{1-x}* Phys. Rev. B **46**, 14186–14189 (1992).

63. M. F. Thorpe *Continuous Deformations in Random Networks* J. Non-Cryst. Solids **57**, 350–370 (1983).
64. J. C. Phillips and M.F. Thorpe *Constraint Theory, Vector Percolation and Glass Formation* Solid State Commun. **53**, 699–702 (1985).
65. J. C. Phillips *Vibrational Thresholds Near Critical Average Coordination in Alloy Network Glasses* Phys. Rev. B **31**, 8157–8163 (1985).
66. N. K. Abrikosov, V. F. Bankina, L. V. Poretskaya, L. E. Shelimova, and E. V. Skudnova *Semiconducting II-VI, IV-VI and V-VI Compounds* (Plenum, New York, 1969), pp. 67.
67. M. L. Baesso, A. C. Bento, A. R. Duarte, A. M. Neto, L.C.M. Miranda, J.A. Sampaio, T. Catunda, S. Gama and F. C. G. Gandra Nd_2O_3 *Doped Low Silica Calcium Aluminosilicate Glasses: Thermomechanical Properties* J. Appl. Phys. **85**, 8112–8118 (1999).

Chapter 3.2

THERMAL CONDUCTIVITY OF QUASICRYSTALLINE MATERIALS

A. L. Pope and Terry M. Tritt

Department of Physics and Astronomy, Clemson University, Clemson, SC, USA

1. INTRODUCTION

For many decades x-ray diffraction peaks were understood as reflections from the periodic lattice planes in a solid-state material.[1] These planes could be clearly identified and the crystal structure could be effectively indexed and cataloged. Much of our understanding of structural determination of crystals was seriously challenged in the mid-1980s with the discovery of quasicrystals, or quasicrystalline materials,[2] materials exhibiting fivefold symmetry, which were forbidden to exist in nature.[3] Quasicrystals lack the long-range periodicity of a crystal and yet they exhibit "structural order" which leads to very sharp and distinct x-ray diffraction peaks, in contrast to amorphous materials. These sharp and distinct x-ray diffraction peaks could not be indexed by the then-existing crystallographic techniques which had been developed for periodic crystal structures.

Quasicrystals display long-range positional order without short-range rotational symmetry.[2] Over 100 quasicrystalline systems exist at present and are seen to have 5-, 8-, 10-, or 12-fold symmetries, all of which are classically forbidden. It is striking to note the high structural quality of the quasicrystals when compared with their thermal transport properties, which are more reminiscent of a glass. Quasicrystals typically exhibit thermal conductivity values on the order of $\kappa \approx 1\text{--}3$ W m^{-1} K^{-1}. They are also very hard materials, and this property coupled with their low thermal conductivity has made them attractive for use as thermal barrier coatings. They have been used to coat frying pans to replace the more standard Teflon coatings. We give a brief overview of thermal transport in the two most prominently measured classes of quasicrystals.

Many quasicrystalline systems exist, but the most common quasicrystals are the AlPdMn and AlCuFe quasicrystal systems. AlPdMn quasicrystals can be synthesized in a 5-fold or 10-fold symmetry. Compositional variations exist even within a specific type of quasicrystal, so the following statements are designed to be taken as generalization for the fivefold symmetric crystals aforementioned. The fivefold symmetric quasicrystals are stable and have a wealth of information available on them. At low temperatures ($T < 2$ K) the thermal conductivity is observed to increase as approximately T^2. The thermal conductivity then increases with increasing temperature until a phonon-saturation plateau is observed to occur between 20 K and 100 K.[4] This plateau is observed at much higher temperatures in quasicrystals than in amorphous materials (Fig. 1).[5] Above 100 K the thermal conductivity is observed to begin to increase again.[6,7] It is observed in AlPdMn quasicrystalline systems that the lattice thermal conductivity is nearly constant above 150 K, with the small increase in thermal conduction being due to the electronic contribution. AlPdMn quasicrystals have thermal conductivity values on the order of $\kappa \approx 1$–3 W m^{-1} K^{-1} at room temperature. Once again, the variation in room temperature thermal conductivity is due to different sample composition. It is also noted that the electronic thermal conductivity continues to increase as temperatures are further elevated. In fact, thermal conductivity increases until the material dissociates.

AlCuFe quasicrystals are five-fold symmetric and also have thermal conductivity values about $\kappa \approx 1$ and 3 W m^{-1} K^{-1} at room temperature. Thermal conductivity in these quasicrystals behaves in much the same way as AlPdMn quasicrystals. Thermal conductivity at 1000 K has been observed to be less than 10 W m^{-1} K^{-1} for AlCuFe quasicrystals.[8] Perrot has calculated that the Wiedemann–Franz relation holds at high temperatures for these materials, and is within 15% of the accepted value. The validity of the Wiedemann–Franz relation indicates that the scattering rate of the electrons is proportional to that of the phonons. Perrot demonstrates that the lattice and electronic contributions to thermal conductivity in AlCuFe increase with increasing temperature.

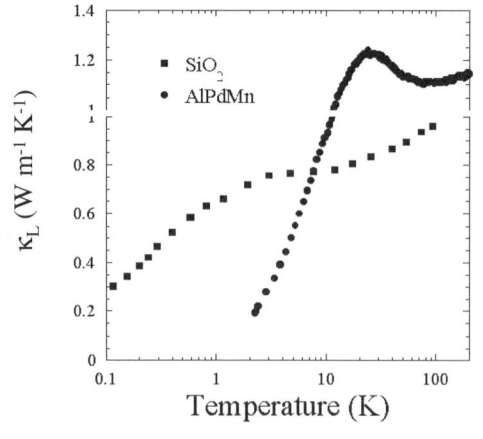

FIGURE 1 Lattice thermal conductivity increases with increasing temperature until a phonon saturation plateau is observed in both AlPdMn quasicrystals and amorphous SiO$_2$.

2. CONTRIBUTIONS TO THERMAL CONDUCTIVITY

The thermal conductivity in these materials is essentially sample independent.[9] Small changes in composition, different annealing practices, and sample composition are observed to have little effect on the overall magnitude of thermal conductivity, changing it by less than a factor of 2. Temperature-dependent thermal conductivity measured from 2 K to 1000 K has been observed to have values below 10 W m^{-1} K^{-1} for the entire temperature range.[10]

In general, the total thermal conductivity of a material can be written as $\kappa_T = \kappa_E + \kappa_L$, where κ_E and κ_L are the electronic and lattice contributions respectively. The Wiedemann–Franz relationship provides a ratio of the electrical resistivity (ρ) to the electronic part of the thermal conductivity (κ_E) at a given temperature, which is the same for most metals ($\kappa_E = L_0 T/\rho$), where L_0 is the Lorentz number ($L_0 = 2.45 \times 10^{-8}$ (V/K)^2). The Wiedemann–Franz relationship, which is well behaved in most metallic systems, has also been shown to hold in many quasicrystalline systems.[11,12]

To investigate the electronic contribution of the thermal conductivity utilizing the Wiedemann–Franz relationship, one analyzes the electrical conductivity of AlPdMn and AlCuFe. As is typical with many quasicrystalline systems, the electrical conductivity increases with increasing temperature. The increase in electrical conduction is contrary to Matthiessen's rule, implying that the weak scattering approximation does not hold for quasicrystals. Electrical conductivity of these quasicrystalline systems (AlPdMn and AlCuFe) is about $\sigma \approx 10^2$ Ω^{-1} cm^{-1}.[9,13,14] Applying the Wiedemann–Franz relationship to these values of electrical conductivity, we observe that the electronic contribution to thermal conduction is negligible below 150 K and begins to influence the temperature dependence of the total thermal conductivity above 150 K. Thus, the thermal conductivity in the aforementioned quasicrystalline systems will be considered to be governed primarily by the lattice vibrations below 150 K. Lattice contributions continue to be significant above 150 K, but due to the continually increasing electronic contribution one must also consider the electronic portion.

While electrical conduction in a quasicrystal is similar in many respects to electronic conduction in crystalline materials, thermal conduction in quasicrystals is most easily compared to that of a glass (a-SiO$_2$), as seen in Fig. 1. In perfect crystals the lattice vibrations are described by phonons. In amorphous materials, where there is no lattice, heat is transported through localized vibrations or excitations, which are often referred to as tunneling states. These localized vibrations lead to the minimum thermal conductivity of a system.[15]

3. LOW-TEMPERATURE THERMAL CONDUCTION IN QUASICRYSTALS

The difference in scattering mechanism between crystalline and amorphous materials can be seen most acutely at low temperatures. At these temperatures ($T < 2$ K), phonon modes are frozen out and boundary scattering and grain size effects limit the lattice thermal conductivity.[16] This causes lattice thermal conductivity in crystals to behave as T^3 and lattice thermal conductivity in some glasses to behave as T^2, due to boundary scattering or scattering of phonons by tunneling states.[17–19] Thompson has shown that both phonon scattering by electrons and tunneling states exist in quasicrystalline materials.[20] Therefore, quasicrystals can be classed as two-level systems. As such, any scattering by electrons will be observed only at higher temperatures.

4. POOR THERMAL CONDUCTION IN QUASICRYSTALS

Thermal conductivity in quasicrystals is two orders of magnitude lower than aluminum, which is somewhat surprising since many quasicrystals are composed primarily (~70) of aluminum. The question arises, as to what mechanisms allow the thermal conduction to be so close to the minimum thermal conductivity. In amorphous materials large amounts of phonon scattering occur due to the irregular structure of these materials. The mean free path is very small and essentially temperature independent. This leads to low values of thermal conductivity, typically about 1 W m^{-1} K^{-1} at 300 K. Thompson has observed in several AlPdMn quasicrystals that above 100 K the thermal conductivities of these materials approach the minimum thermal conductivity.[20]

The thermal conductivity in quasicrystals can be described as glasslike due to the phonon saturation peak seen as well as the small magnitude of thermal conductivity. This glasslike thermal conductivity is attributed to the large structural coherence observed in these materials; in AlPdMn the structural coherence can be up to 8000 Å. This structural coherence gives AlPdMn a large unit cell, which is a component typically observed in low-thermal-conductivity materials.[21]

Janot has explained the poor thermal conductivity observed in quasicrystalline materials as being due to the reduced range of phonons due to variable-range hopping.[19] An alternative explanation is that the quasiperiodic lattice scatters phonons as if a point defect exists at every atomic site.[22] This gives rise to scattering such as is observed in amorphous materials.

5. GLASSLIKE PLATEAU IN QUASICRYSTALLINE MATERIALS

Thermal conductivity increases with increasing temperature until a phonon saturation plateau is observed to occur between 20 K and 100 K.[23] This plateau is observed at much higher temperatures in quasicrystals than in amorphous materials.[24] Kalugin et al. have explained the glasslike plateau utilizing a generalized Umklapp process.[25] Umklapp processes are a consequence of interplay between two scattering processes. The natural length scale for an Umklapp process is the reciprocal lattice spacing where the phonons are scattered outside the first Brillioun zone and in a reduced zone scheme appears as a backward scattering process. Of course, a reciprocal lattice or Brillioun zone does not exist for a quasicrystal. Umklapp processes in crystals will decrease exponentially, but in quasicrystals these Umklapp processes will lead to a power-law behavior of the mean free path. Kalugin et al. compared this theory with quasicrystal data and observed that it provides a reasonable explanation for the plateau-like feature observed in quasicrystalline materials.

6. SUMMARY

Overall, the thermal conduction in quasicrystalline materials appears to behave in a manner much as one would expect a glass to behave. At low temperatures, a T^2 temperature dependence which is sometimes associated with some amorphous materials is evident in quasicrystals. As temperature increases, a phonon saturation plateau is observed; as temperatures increase above 150 K, the electronic contribution of the thermal conductivity begins to become important. Quasicrystals are fascinating materials displaying electronic properties most closely related to crystalline materials and thermal properties most closely associated with amorphous materials.

7. REFERENCES

1. J. D. WINEFORDNER, ed., *Introduction to X-Ray Powder Diffractometry* (Wiley, New York, 1996).
2. D. SHECHTMAN, I. BLECH, D. GRATIAS, and J. W. CAHN Phys. Rev. Lett. **53**, 1951 (1974).
3. C. KITTEL *Solid State Physics* (Wiley, New York)
4. M. CHERNIKOV, A. BIANCHI, and H. OTT. Phys. (Wiley, New York, 7th Edition, 1996) Rev. B. **51**, 153 (1995).
5. W. HÄNSCH and G. D. MAHAN Phys. Rev. B **28**, 1886–1901 (1983).
6. F. CYROT-LACKMANN Metastable in Nanocrystalline Mater. **1**, 43–46 (1999).
7. D. CAHILL *Lattice Vibrations in Glass* Ph.D. Dissertation, Cornell University, 1989.
8. A. PERROT, J. M. DUBOISE, M. CASSART, and J. P. ISSI Quasicrystals in *Proceedings of the 6th International Conference* Tokyo, Japan, 1997, edited by S. Takeuchi and T. Fujiwara (World Scientific, Singapore, 1998).
9. A. L. POPE *Thermal and Electrical Transport in Quasicrystals* Thesis, Department of Physics, Clemson University, Clemson, SC (2002).
10. A. PERROT, J. M. DUBOISE, M. CASSART, and J. ISSI *Quasicrystals in Proceedings of the 6th International Conference* Tokyo, Japan, 1997, edited by S. Takeuchi and T. Fujiwara (World Scientific, Singapore, 1998).
11. A. PERROT, J. M. DUBOIS, M. CASSART. J. P. ISSI Proc. 6th Int. Conf. on Quasicrystals (1997).
12. K. GIANNO, A. V. SOLOGUBENKO, M. A. CHERNIKOV, H. R. OTT, I. R. FISHER, and P. C. CANFIELD *Low-temperature Thermal Conductivity of a Single-Grain Y-Mg-Zn Icosahedral Quasicrystal* Phys. Rev. B **62**, 292 (2000).
13. A. L. POPE, T. M. TRITT, M. A. CHERNIKOV, and M. FEUERBACHER Appl. Phys. Lett. **75**, 1854 (1999).
14. M. A. CHERNIKOV, A, BERNASCONI, C. BEELI, and H. R. OTT Europhys. Lett. **21**, 767 (1993).
15. G. A. SLACK in *Solid State Physics* edited by F. Seitz and D. Turnbull (Academic Press, New York, 1979), Vol. **34**, p. 57.
16. D. K. MACDONALD *Thermoelectricity: An Introduction to the Principles* (Wiley, New York, 1962).
17. R. R. HEIKES and R. URE *Thermoelectricity: Science and Engineering* (Interscience, New York, 1961).
18. N. F. MOTT *Conduction in Non-Crystalline Materials* (Oxford Science, 1987).
19. C. JANOT *Quasicrystals: A Primer* (Clarendon Press, Oxford, 1992).
20. E. THOMPSON, P. VU, and R. POHL Phys. Rev. B. **62**, 11473 (2000).
21. P. A. BANCEL *Quasicrystals: The State of Art* (World Scientific, Singapore 1991).
22. S. LEGAULT *Heat Transport in Quasicrystals*, Thesis, Department of Physics, McGill University, Montreal, Quebec (1999).
23. M. CHERNIKOV, A. BIANCHI, and H. OTT Phys. Rev. B. **51**, 153 (1995).
24. W. HÄNSCH and G. D. MAHAN Phys. Rev. B **28**, 1886–1901 (1983).
25. P. KALUGIN, M. CHERNIKOV, A. BIANCHI, and H. R. OTT Phys. Rev. B **53**, 14145 (1996).

Chapter 3.3

THERMAL PROPERTIES OF NANOMATERIALS AND NANOCOMPOSITES

T. Savage and A. M. Rao

Department of Physics and Astronomy, Clemson University, Clemson, SC USA

Interest in the science and technology of nanomaterials has exploded in the past decade mainly due to their extraordinary physical and chemical properties relative to the corresponding properties present in bulk materials. A nanostructure is characterized by its size, as indicated by the term, having its crucial dimensions on the order of 1–100 nm. There is an assortment of groupings in which nanostructures are often categorized. In this chapter our discussion on nanostructures will primarily be divided into two groups: (a) nanomaterials, consisting of such structures as nanotubes, nanowires or nanorods, and nanoparticles, and (b) nanocomposites, which include a composite material incorporating any of the aforementioned nanostructures in a matrix, in particular, thin films of polymer/nanotube composites.

The extremely small "dimension" is what in many instances gives nanostructures their unique physical and chemical properties. Kittel lists several reasons for these unusual properties:[1] (a) a significant fraction of the atoms in a nanomaterial is composed of surface atoms, as opposed to a large fraction of interior atoms present in bulk materials; (b) the ratio of surface energy to total energy may be of the order of unity; (c) the wavelength of electrons in the conduction or valence band is restricted by geometric size and is shorter than the wavelength in the bulk solid; (d) a wavelength or boundary condition shift will affect optical absorption phenomena; and (e) clearly defined boundaries in magnetic monolayers such as, alternating films of ferromagnetic iron and of paramagnetic chromium, present an opportunity for greater control over magnetic properties by tunneling of the magnetization through the chromium barrier.

262 Chap. 3.3 • THERMAL PROPERTIES OF NANOMATERIALS AND NANOCOMPOSITES

This chapter will focus on heat transfer in nanomaterials, which depend primarily on energy conduction due to electrons as well as phonons (lattice vibrations) and the scattering effects that accompany both (these include impurities, vacancies, and defects). Thermal conductivity, κ, the electrical conductivity, σ, the thermoelectric power (TEP) or Seebeck coefficient, S (as it is often called), and the heat capacity, C, all provide valuable insight into important physical characteristics of the nanomaterial. Values for such quantities in nanomaterials and nanocomposites are often useful for drawing comparisons to measurements in corresponding bulk materials. These types of measurements are also important for when nanostructures are used in various applications, such as nanoelectronic devices and gas sensors.

1. NANOMATERIALS

1.1. Carbon Nanotubes

We begin with the discussion of thermal properties of carbon nanotubes, which are the latest molecular form of carbon, discovered in 1991 by Iijima.[2] A nanotube can be viewed as a graphene sheet rolled into a seamless cylinder with a typical aspect ratio (length/diameter) exceeding 1000. Typically, carbon nanotubes are cast into two groups: (a) single-walled (SWNTs) or (b) multiwalled (MWNTs), which are basically a set of concentric SWNTs. Carbon nanotubes are generally synthesized by one of three commonly used methods. The first is the electric arc discharge in which a catalyst-impregnated graphite electrode is vaporized by an electric arc (under inert atmosphere of ~500 Torr), yielding carbon soot that deposits on the inner surface of the water-cooled arc chamber.[3] In the second method, commonly known as laser vaporization, a pulsed laser beam is focused onto a catalyst-impregnated graphite target (maintained in an inert atmosphere of ~500 Torr and ~1200°C) to generate the soot that collects on a water-cooled finger.[4] Both techniques are widely used for producing SWNTs and typically yield soot with 60–70% of the sample as nanotubes and the remainder as nanoparticles. The third technique, and perhaps the one with the most promise for producing bulk quantities of nanotubes, is chemical vapor deposition (CVD). The nanotubes are formed by the pyrolysis of a hydrocarbon source seeded by catalyst particles (typically ferromagnetic particles such as Fe, Co, Ni, etc., are used) in an inert atmosphere and 700–1200°C. The CVD process is usually the method of choice for preparing MWNTs, although reports have shown that SWNTs can also be grown by this means.[5]

1.1.1. Electrical Conductivity, σ

Since the conduction by electrons is one of the two main ways in which heat is transferred in a solid, it is important to have an understanding of the electrical conductivity, σ. Electron transport in nanotubes can be discussed one-dimensionally (1D) due to their high aspect ratios (i.e., diameters ranging from <1 nm to 4 nm and lengths of up to several microns).[6] This characteristic of nanotubes makes them excellent candidates for use as molecular wires. Another important aspect of nanotubes to electron transport is that they can be metallic or semiconducting, depending on their diameter and chirality (a measure of the amount of "twist" in the lattice).[7] All of the carbon atoms in the nanotube's lattice lie entirely on the surface, leaving the tube hollow as it were. The conduction of electrons in a 1D

Sec. 1 · NANOMATERIALS

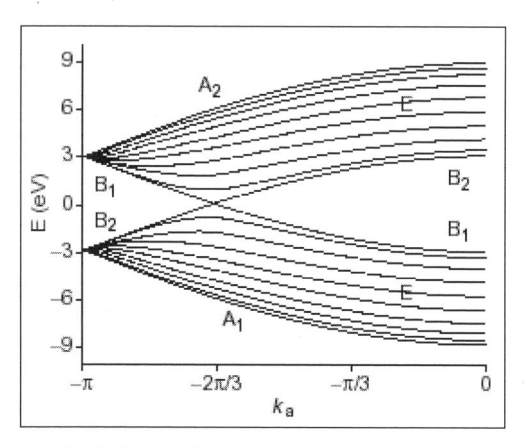

FIGURE 1 Band structure of (10,10) armchair nanotubes. The two central bands (labeled by their different distinct symmetries) intersect at the Fermi energy, giving clear evidence of their metallic nature.[9]

system such as a nanotube occurs ballistically or diffusively.[8] When the electrons travel without being scattered, they are said to conduct ballistically. This type of transport will only occur in a very small nanotube segment such that its length is much shorter than the mean free path of the electron. However, in most instances this is not the case, and the conduction of electrons is highly dependent on scattering effects due to such things as phonons, impurities, and structural defects. The performance of electronic devices based on carbon nanotubes is limited by the Schottky energy barrier which the electrons must cross from the metal onto the semiconducting nanotube.

The electrical conductivity, σ, is a measure of how easily electrons are able to flow through a material. In metals,

$$\sigma = en\mu, \qquad (1)$$

where e is the charge, n is the electron concentration (number of electrons per unit volume), and μ is known as the carrier mobility. The carrier mobility is the factor in σ that takes into account the scattering effects mentioned. The sp^2 carbon bonds in the nanotube lattice give the π electrons (electrons that contribute to conduction) a large mobility, but it is only the "armchair" tubes [designated by chiral vector indices (n,n) or chiral angle = 30°[6]] that have a large carrier concentration, making it a good electrical conductor and thus metallic in nature. This metallic behavior is clearly seen by the band structure diagram in Fig. 1, in which the valence and conduction bands intersect through the Fermi energy approximately one third of the distance to the first Brillouin zone edge.[9] Conversely, "zigzag" [indices $(n,0)$ or angle = 0°[6]] or other chiral nanotubes have medium to small band gaps, making them semiconducting to semimetallic. An interesting aspect of MWNTs is that the individual shells of a given tube can have different chiralities. So, a single MWNT can have both metallic and semiconducting layers. Upon multiple measurements of bundles of MWNTs prepared using the arc discharge method, the average σ has been estimated to be ~1000–2000 S/cm.[10]

Since the electrical conductivity is equal to the reciprocal of the electrical resistivity, ρ (i.e., $\rho = 1/\sigma$), σ is often determined experimentally in nanotubes through measurements of the temperature dependence of ρ by using the standard four-probe geometry. (Note: In many cases, measurements of ρ can be challenging because of

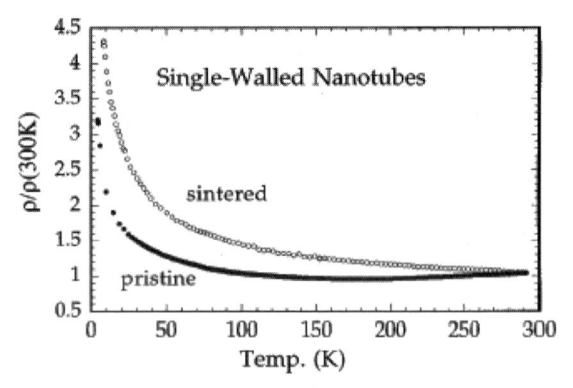

FIGURE 2 Normalized $\rho(T)$ data to room temperature value for pristine and sintered SWNT films.[11]

its dependence on sample dimensions. Because samples are often prepared by pressing nanotube material into mats or random arrays of nanotubes, defining quantities such as cross-sectional area and thickness can be quite difficult. Therefore, electrical resistance, R, is frequently reported rather than ρ, since $R \propto \rho$.) Recalling Matthiessen's rule, we see that the net resistivity of a solid is due primarily to charge carrier scattering and is given by

$$\rho = \rho_L + \rho_R. \qquad (2)$$

The first term is resistivity due to scattering by phonons. The second term accounts for any other effects that contribute to the scattering cross section, such as impurities and defects, and is known as the residual resistivity.

Figure 2 displays normalized resistivity data collected from thin films of SWNTs. The samples were synthesized by the laser ablation method.[11] Both samples were annealed in vacuum to 1000°C. One was then measured, and the data are labeled as pristine. The other sample was sintered by pressing it between electrodes and passing a high current ($\geq$200 A/cm^2) through it while under vacuum. The pristine sample demonstrates metallic behavior ($d\rho/dT > 0$) at higher temperature but crosses over to negative $d\rho/dT$ behavior at $\sim$180 K. This crossover point varies widely over this temperature from sample to sample. The slope of the sintered data, however, remains negative over all temperatures, arguing in favor of semiconducting or activated hopping behavior.

In a separate study similar results were seen for pristine SWNTs in the normalized resistance data with a minimum at $\sim$190 K, as shown by Fig. 3.[12] Introducing impurity K atoms into the lattice results in a significant contribution from the second term in Eq. (2) to the measured $R(T)$ due to scattering effects. The bottom trace in Fig. 3 shows that K-doping has curbed the sharp upturn in the resistance at low T, and, in fact, the slope remains strongly positive.

Another independent set of $\rho(T)$ data is shown in Fig. 4.[13] This set was taken from samples of purified SWNTs that have had their hollow interiors filled with C_{60} molecules. These structures have been nicknamed "peapods" and are often denoted C_{60}@SWNTs. Interactions between adjacent molecules of C_{60} as well as between the C_{60} and the SWNT are believed to play an important role in electron transport in these structures. In Fig. 4 the C_{60}@SWNTs (filled circles) are seen to decrease ρ (i.e. increase σ) at lower temperatures, compared to the "empty" SWNTs (open squares). The argument for this behavior is that the C_{60} molecules form new con-

Sec. 1 · NANOMATERIALS 265

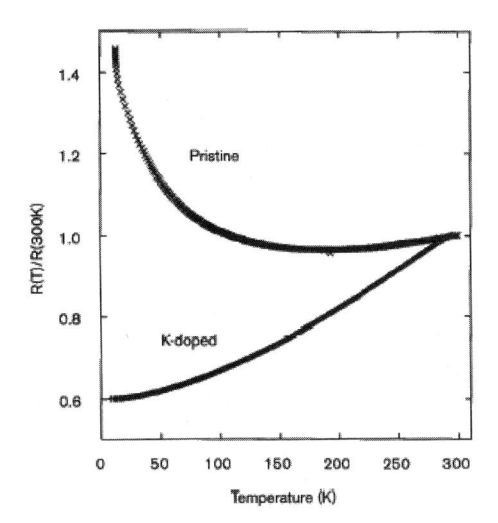

FIGURE 3 Normalized four-probe $R(T)$ data to room temperature value for pristine and K-doped SWNTs.[12] Note the change in sign of the slope of the pristine sample, but the slope of the K-doped sample remains positive.

duction paths that are able to bridge localized defect sites that typically impede electron transport in regular SWNTs at low T.

1.1.2. Thermoelectric Power (TEP)

In nanotubes thermoelectric power (or thermopower) measurements are often used for determining the sign of the dominant charge carrier and for probing their sensitivity to gas adsorption. The TEP is temperature dependent; however, this dependence varies for metals and semiconductors. At any given temperature it is the manner in which the charge carriers are transported and scattered that will determine the TEP of the metal or semiconductor.

Since in most metals the TEP is sensitive to the curvature of the band structure near the Fermi level through the Mott relation

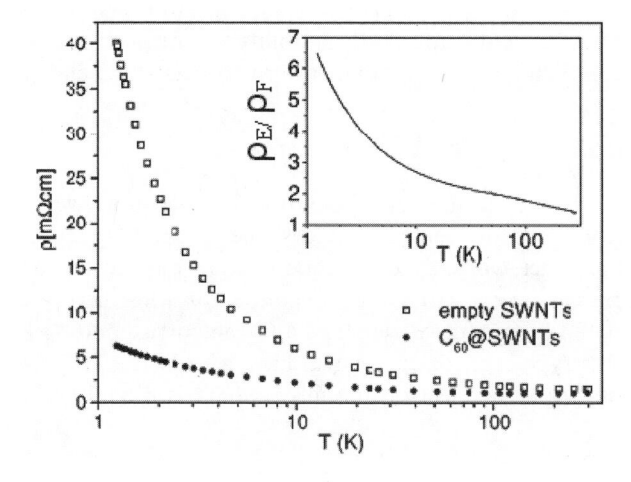

FIGURE 4 Temperature-dependent ρ data for regular SWNTs and "peapods".[13] The ρ ratio for empty (ρ_E) and filled (ρ_F) SWNTs is given in the inset.

$$S_d = \frac{\pi^2 k^2 T}{3e} \left(\frac{\partial \ln \sigma}{\partial \varepsilon}\right)_{\varepsilon=E_F}, \qquad (3)$$

it is a good indicator of dominant carrier type. In this expression k is the Boltzmann constant, e is the electron charge, and E_F is the Fermi energy. The subscript d indicates that this is the contribution of the diffusion thermopower to the total TEP, and clearly we see that the TEP is linearly dependent on T. In many TEP measurements a nonlinear dependence is often also observed in a plot of S versus T. This nonlinear dependency is usually attributed to a combination of effects, which include contributions from semiconducting tubes, abrupt variations in the density of states, or phonon drag.[14] Phonon drag, denoted S_g, is an additional term in the thermopower,[15] and the total TEP in metals can be expressed as

$$S_{\text{tot,met}} = S_d + S_g. \qquad (4)$$

In semiconductors, the mechanism that determines the TEP is somewhat different than that of metals. The general expression for the TEP in semiconductors is

$$S_{\text{sem}} \approx \frac{E_g}{2eT} = \frac{k}{2e}\frac{E_g}{kT}, \qquad (5)$$

where E_g is the gap energy (also known as the band gap).[16] Written in the form of the expression on the far right, we see that $S_{\text{sem}} \propto E_g/kT$. In contrast to S_d in metals, the TEP in semiconductors is inversely proportional to T and increases with decreasing temperature. However, the total TEP deviates from this relationship near absolute zero and goes to zero due to its dependence on σ.[16]

Whereas in metals the only charge carriers that contribute to σ are electrons, semiconductors have two types of free carriers–electrons and holes. A hole is simply the absence of an electron in the valence band that behaves like a positively charged electron. If electrons are the dominant charge carriers in a material, then the material is said to be n-type. Conversely, if holes are the dominant carriers, the material is said to be p-type. The total electrical conductivity, σ_{tot}, of a semiconductor is the sum of the electrical conductivity due to electrons and holes (i.e., $\sigma_{\text{tot}} = \sigma_e + \sigma_h$, where σ_e and σ_h are given by Eq. (1) and n and μ account for the corresponding hole concentration p and mobility μ_h, respectively, in σ_h). The total TEP of a semiconductor can be written in terms of its σ dependences as

$$S_{\text{tot,sem}} = \frac{S_e \sigma_e + S_h \sigma_h}{\sigma_{\text{tot}}}, \qquad (6)$$

where S_e and S_h are the electron and hole diffusion thermopowers, respectively, of the semiconductor. In many semiconductors one of the two terms in Eq. (6) will dominate and the other will only contribute a small part to the overall TEP. The true control variables in the TEP of a semiconductor will, therefore, be the carrier concentration of the dominant carrier type and the corresponding carrier mobility. For example, an n-type semiconductor will have a negative TEP with a magnitude depending primarily on the electron mobility, which is directly related to the electron energy.

The electronic properties of nanotubes are affected greatly by their gas exposure history, chirality, and diameter. The TEP has been observed to be particularly sensitive to such changes. Exposure to oxygen (or air) has been shown to have a dramatic effect on the TEP in nanotubes.[17–19] In Fig. 5a the TEP is plotted as a

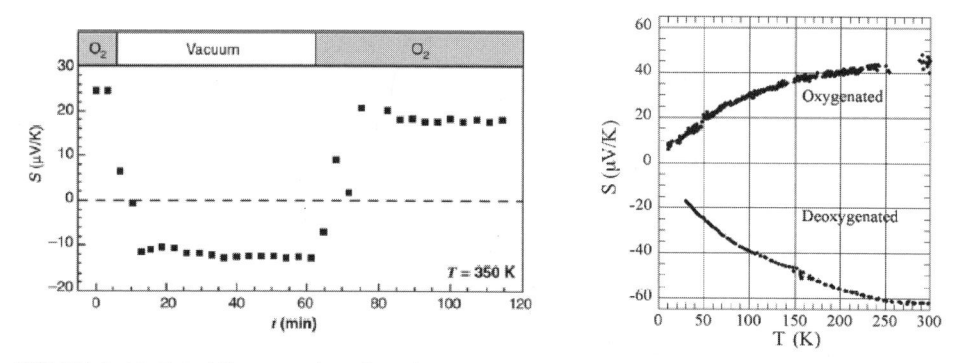

FIGURE 5 (a) TEP (S) versus time data for a SWNT film cycled between vacuum ($S < 0$) and O_2 saturation ($S > 0$).[17] (b) The T-dependent TEP data for a SWNT film that has been O_2 saturated and then deoxygenated.[18]

function of time as samples of purified SWNTs were cycled between vacuum and O_2.[17] The temperature was held constant at 350 K during this experiment. The sign of the TEP goes reversibly from positive to negative in a time frame of ~15–20 min. as O_2 is removed and then reintroduced into the system. The magnitude of the TEP swings from ~+20 μV/K in the presence of O_2 to ~ -12 μV/K in vacuum. The sign reversal in the TEP is indicative of a change in the dominant charge carrier switching from n-type (in vacuum) to p-type (in O_2). A separate study showed that as T decreases the saturation time due to O_2 adsorption increases, going from an order of minutes at $T = 350$ K to several days at $T = 300$ K and even longer for $T < 300$ K.[18] Their TEP versus T data also clearly demonstrate the result of oxygen exposure as shown in Fig. 5b. Here TEP data for a film of purified SWNTs in its O_2-saturated state as well as after being completely deoxygenated is given. Both curves approach zero as $T \to 0$ and have roughly similar values of magnitude at room temperature but are opposite in sign.

Another independent study also displays the role of O_2 adsorption on nanotubes as well as the effect of exposure to various other gases.[19] Interestingly, they reported that collisions between gas particles and the nanotube wall can have a significant

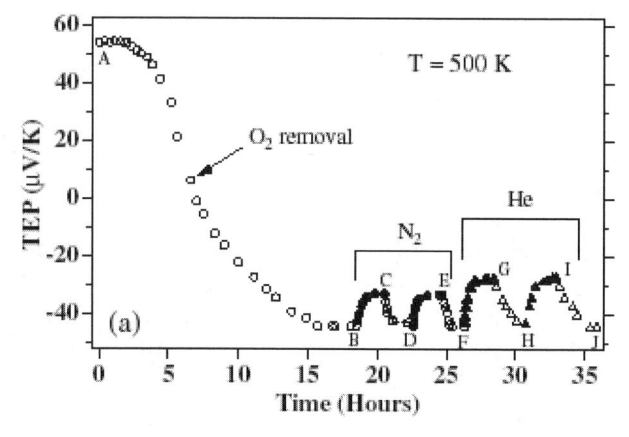

FIGURE 6 TEP versus time data for a SWNT mat at $T = 500$ K. The sample was initially air saturated. It was then degassed and cycled between N_2 and He (dark symbols) and vacuum (open symbols).[19]

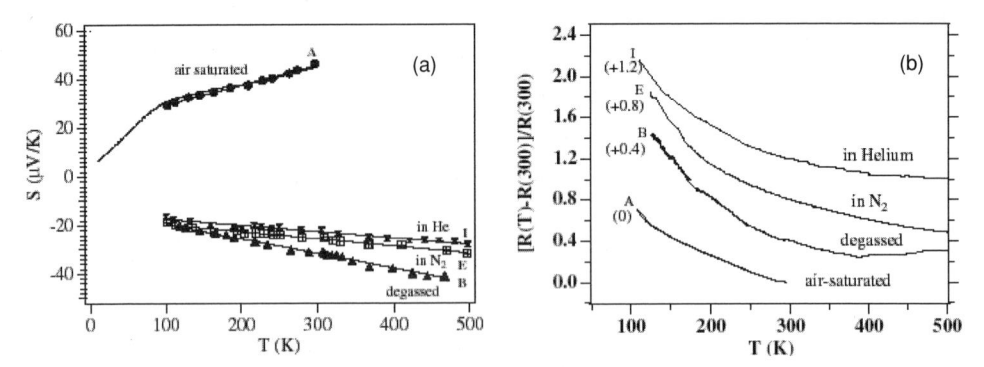

FIGURE 7 The T dependence of (a) TEP and (b) normalized R for SWNT mats. The letters marking each trace in both figures (A, B, E, I) correspond with the letters in Fig. 6 that signify the condition of the samples when they were measured. Values in parentheses indicate the amount of vertical shift in each trace.[19]

impact on the TEP. Figure 6 shows the time evolution of the TEP for a mat of as-prepared SWNTs as it was exposed to a number of different environments at $T = 500$ K after it had first been air saturated. The air saturated SWNTs show an initial TEP value of $\sim$+54 μV/K. As the sample chamber was evacuated to $\sim 10^{-6}$ Torr, the TEP is seen to switch signs and then eventually flatten out over a period of $\sim$15 to a degassed (or deoxygenated) value of $\sim$-44 μV/K. The behavior of the TEP upon repeated cycling of the sample between N_2/vacuum and He/vacuum is shown in the figure. A peak corresponding with the presence of each gas can be observed in the TEP data. The time needed to remove the O_2 was much longer than the time it took to rid the system of N_2 or He. This fact is attributed to a stronger binding between O_2 and the walls of the SWNTs, suggesting an exchange of electrons between the two (i.e., chemisorption).

The TEP and corresponding normalized R data of the SWNT sample used in Fig. 6 are given in Fig. 7 as a function of T. The letters correspond with those in Fig. 6, signifying the particular environment of the sample chamber during the measurement. Note that each of the TEP plots in Fig. 7a demonstrates a linear T dependence, with the exception of the air-saturated trace, which appears to have a peak at $\sim$100 K, giving support for the dominance of the contribution to the thermopower from metallic tubes. Recognizing that each of the normalized R traces have been upshifted by the number indicated in parentheses and, in fact, nearly lie on top of each other, a comparison of Figs. a and b clearly shows that the TEP is a much more sensitive and reliable measurement than R for probing the effects of gas adsorption in nanotubes. The n-type nature of the TEP in pristine or degassed SWNTs is believed to arise from the asymmetry in the band structure in metallic tubes brought about by tube–tube interactions[20] or lattice defects.[21]

There has also been evidence for the existence of the Kondo effect through TEP measurements of SWNTs.[22] The TEP (T) data for as-prepared SWNT mats that were synthesized with a variety of catalyst particles are given in Fig. 8. The broad peak in the temperature range of 70–100 K has been attributed to the occurrence of magnetic impurities in the SWNTs. This peak is often reported in SWNTs and is typically associated with the Kondo effect. The position and size of the peak depend on the type of catalyst particle used. Each of the traces from SWNTs obtained from Fe–Y, Co–Y, Ni–Y, or Ni–Co catalysts exhibit noticeable Kondo peaks (the Fe–Y being the most significant at $\sim$80 μV/K at 80 K). On the other

Sec. 1 · NANOMATERIALS

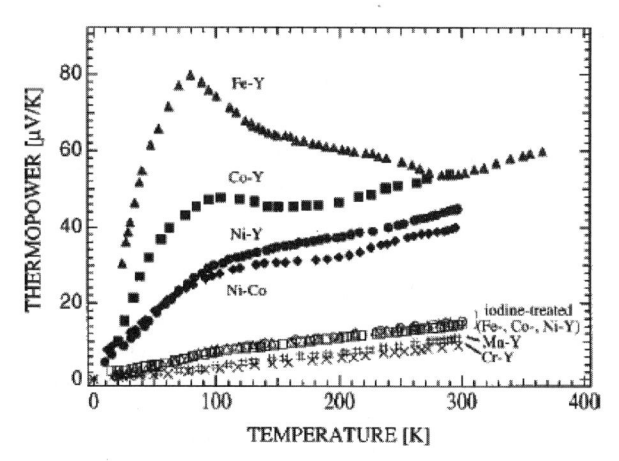

FIGURE 8 TEP versus T data for SWNTs synthesized using assorted catalyst particles.[22] The broad Kondo peak is evidenced in the upper four traces.

hand, the ones from Mn–Y and Cr–Y do not. Also, by treating the samples that do display a Kondo peak with iodine, we can virtually eliminate this behavior, making the TEP completely linear in T.

The effect of filling a SWNT with C_{60} molecules on the TEP is also worthy of discussion. One study comparing purified regular SWNTs with C_{60}@SWNTs showed a marked decrease in the TEP as a function of T for C_{60}@SWNTs.[13] The regular SWNTs show a room temperature TEP value of ~60 μV/K after air saturation, whereas the C_{60}@SWNTs saturate at ~40 μV/K at 300 K. This is argued as evidence for a significant phonon drag contribution to the thermopower in pristine SWNTs, since the presence of C_{60} inside C_{60}@SWNTs increases the probability for phonon scattering and thus decreases the phonon relaxation time. This effectively reduces the response of the TEP because S_g is directly proportional to the phonon relaxation time. Also, the interior of an empty SWNT is available for oxygen adsorption, while C_{60} blocks the inner adsorption sites in a filled C_{60}@SWNT. Both of these factors are thought to play a key role in determining the TEP in C_{60}@SWNTs.

Since much of the previous discussion in this section has been on SWNTs, a general comparison between the TEP(T) of SWNTs and MWNTs should prove quite useful (cf. Fig. 9).[23] Plots of the T dependence of the TEP for four different carbon samples are given in Fig. 9, which include data from as-prepared SWNTs produced by the arc discharge (solid circles) and the laser ablation (open boxes) techniques, as well as an as-prepared film of MWNTs grown by CVD (open diamonds) and a sample of highly oriented pyrolytic graphite (HOPG). All four samples were exposed to room air and room light for an extended period of time and are, therefore, considered to be sufficiently O_2-doped. Clearly, SWNTs have a much higher room temperature TEP value (~+40 and +45 μV/K) than the MWNTs (~+17 μV/K). Notice that the sign of the TEP of both the SWNTs and the MWNTs is positive, indicating p-type behavior in oxygen–doped nanotubes. The bulk graphite displays a room temperature value of ~ −4 μV/K. This figure serves as a quick reference guide for TEP(T) values in all three air-exposed carbon forms.

Finally, recent advances in mesoscopic thermoelectric measuring devices have made measuring single nanotubes possible.[24] Most TEP measurements to date have been made on a collection of SWNTs or MWNTs in the form of mats or

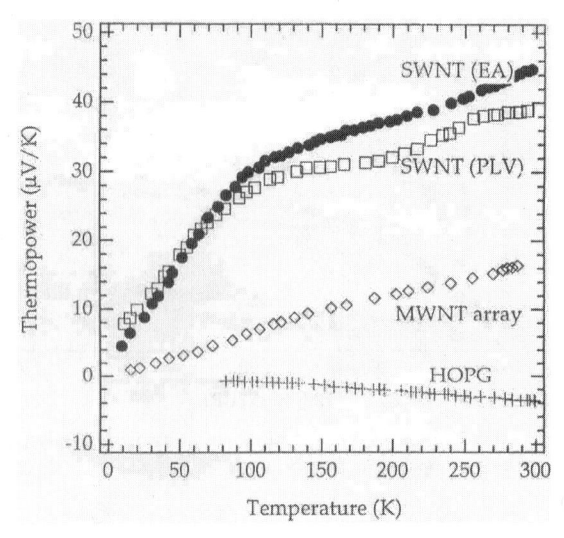

270 Chap. 3.3 • THERMAL PROPERTIES OF NANOMATERIALS AND NANOCOMPOSITES

FIGURE 9 TEP versus T data for air-saturated SWNTs, MWNTs, and highly oriented pyrolytic graphite (HOPG).[23]

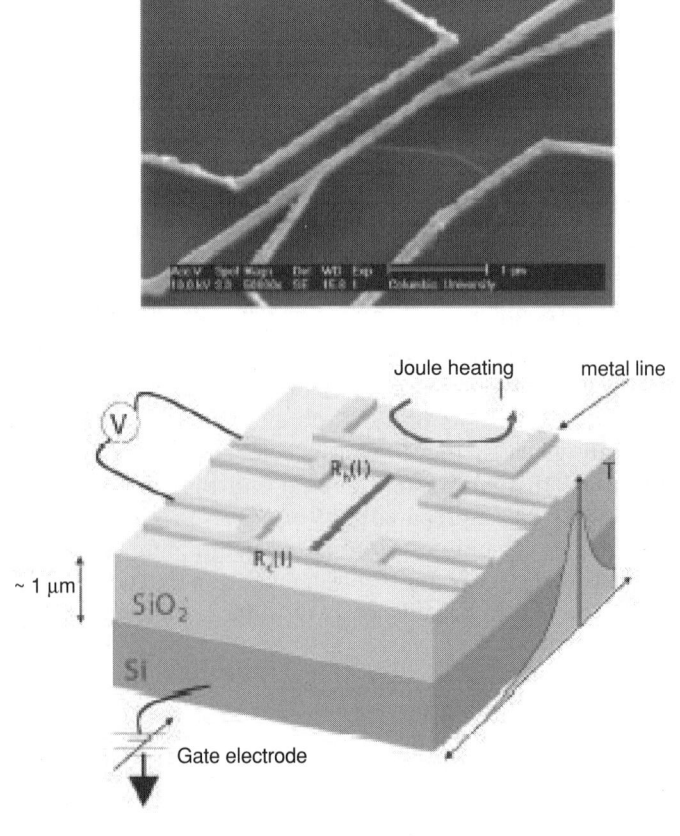

FIGURE 10 SEM image and schematic of novel mesoscopic device used for measuring TEP in individual nanotubes.[24]

thin films and are thus not necessarily intrinsic to an individual tube. Figure 10 displays a SEM image and the schematic of a device currently being used to measure the TEP of isolated nanotubes. The CVD process is used to grow an isolated nanotube on a silicon oxide/silicon substrate. Contact electrodes are then formed on the nanotube by electron-beam lithography. A microheater near one end of the nanotubes is used to establish a temperature gradient of 0.1–0.5 K/μm. The thermoelectric voltage is then measured from the electrode contacts using a high-input-impedance voltage preamplifier. Measurements of this kind have indicated that individual nanotubes are capable of TEP values >200 μV/K at $T < 30$ K. This type of technology should be able to provide valuable insight into the inherent thermoelectric properties of individual nanotubes and opens the door for a variety of different studies.

1.1.3. Thermal Conductivity, κ

Thermal conductivity, κ, is simply a measure of the ease with which heat energy can be transferred through a material. It is a material-dependent property and is defined as the constant of proportionality relating the rate of heat flux, j_Q (i.e., heat energy per unit area per unit time), through a solid to the temperature gradient across it:

$$j_Q = -\kappa \frac{\delta T}{\delta x}. \tag{7}$$

An expression that relates κ to the phonon mean free path, λ, and the heat capacity per unit volume, C, can be derived through the standard kinetic theory of gases and is

$$\kappa = \frac{1}{3} C v \ell, \tag{8}$$

where v is the phonon velocity. Early work done in CNTs predicted thermal conductivities that exceeded those seen in either diamond or graphite (two materials with largest known κ).[25] Some of the first measurements in low-density mats of as-prepared SWNT bundles produced a room temperature κ of 0.7 W/m-K.[26] The $\kappa(T)$ data for this sample are shown in Fig. 11 over a temperature range of 8–350 K. The data exhibit very close to linear behavior over all T with a slight upturn in $d\kappa/dT$ around 25 to 40 K. The low-temperature (<25 K) data shown in the inset are quite clearly linear in T. Calculations of the Lorenz ratio ($\kappa/\sigma T$) indicated that the electron contribution to κ is very small, and therefore only phonon contributions were used to fit the data. Since their model agrees well with general behavior of the measured data, the authors argue that phonons dominate κ over all T due to the one-dimensionality of the SWNTs [i.e., phonon–phonon scattering (umklapp processes) is reduced at high T in a 1D system].

Later measurements performed on high-density, thick films ($\sim$5 μm) of annealed SWNT bundles after being aligned in a high magnetic field (H) showed a significant increase in κ.[27] Also, the anisotropic nature of κ in SWNTs is clearly evidenced by the data in Fig. 12, where measurements have been made parallel to the direction of H-alignment. Here κ increases smoothly with T in both traces, but the room temperature value is an order of magnitude greater in aligned SWNTs ($\sim$220 W/m-K), compared to unaligned ($\sim$30 W/m-K). This experimental value for κ in aligned tubes is within an order of magnitude of what is observed in graphite or even diamond.

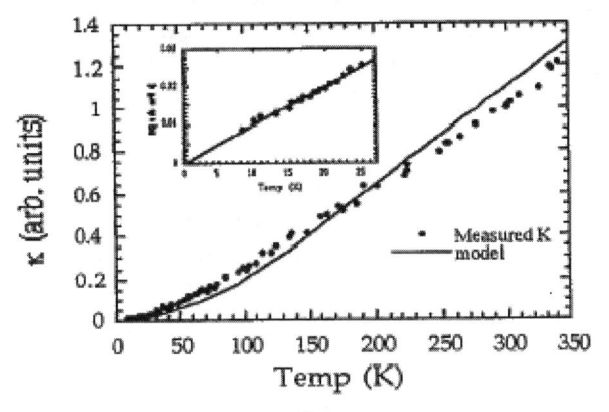

272 Chap. 3.3 • THERMAL PROPERTIES OF NANOMATERIALS AND NANOCOMPOSITES

FIGURE 11 $\kappa(T)$ data for a mat of SWNT bundles.[26] The low-temperature behavior is given in the inset.

Theory predicts extraordinarily large values of κ for isolated nanotubes. Room temperature values of ~6600 W/m-K have been calculated for an individual metallic (10,10) armchair nanotubes[28] which is nearly twice as much as the 3320 W/m-K that has been reported in diamond.[29] But much like the TEP, most κ measurements have been made in bulk samples of nanotubes in the form of mats or films, making it difficult to be certain of the intrinsic value of the individual tubes. However, recent developments in microscale devices for the purpose of thermal measurements have made it possible to measure the κ of a single MWNT.[30] Figure 13 gives the T dependence of κ for a MWNT with diameter ~14 nm. The behavior is very similar to that in Fig. 12 for a SWNT mat, with the exception of the downturn near $T = 325$ K. The most notable feature though, is the large room temperature value of ~3000 W/m-K which is greater than that of the SWNTs in Fig. 12 by an order of magnitude but akin to the predicted value for an individual SWNT mentioned earlier.

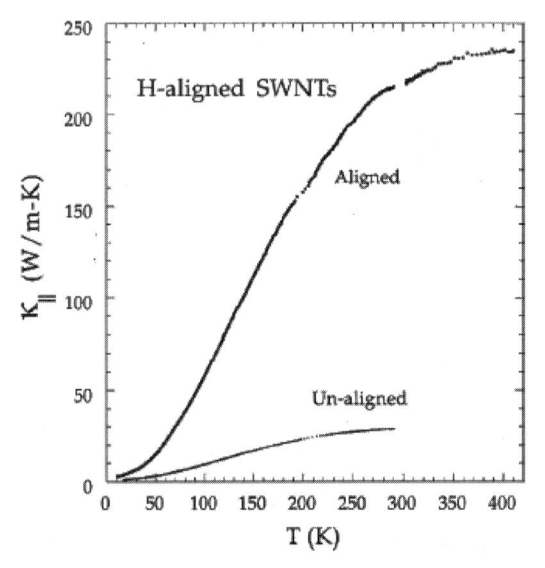

FIGURE 12 Anisotropic nature $\kappa(T)$ of dense SWNT mats.[27] The "aligned" sample was done with a high magnetic field.

Sec. 1 • NANOMATERIALS

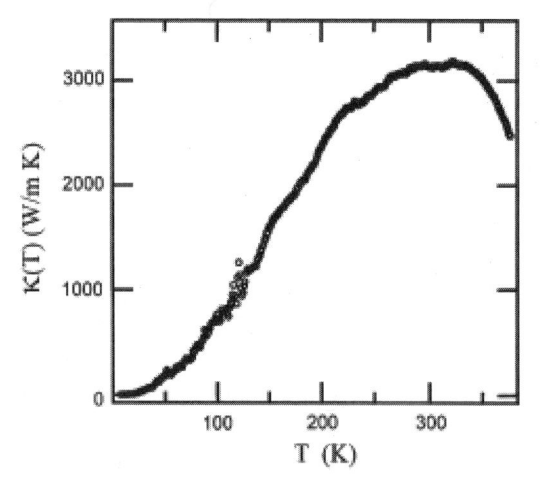

FIGURE 13 The T dependence of κ for an individual MWNT of diameter 14 nm.[30]

The reported κ of C$_{60}$@SWNT "peapod" structures is also of interest.[13] One might expect that C$_{60}$ molecules inside the hollow lattice of a SWNT host would act like "rattlers." In the presence of bulk materials, rattlers can have the effect of reducing the thermal conductivity and thereby enhancing the figure of merit, $ZT = S^2\sigma T/\kappa$ (ZT is a dimensionless value that essentially measures a material's effectiveness as a thermoelectric). However, this may not necessarily be the case in a 1D system such as a nanotube. The $\kappa(T)$ data for air-exposed empty and C$_{60}$-filled SWNT bundles are given in Fig. 14. The general behavior as a function of T is again very similar to that observed in Fig. 12, but interestingly, the C$_{60}$@SWNTs show very little variation in $\kappa(T)$, and the change that is observed is actually a slight increase in $\kappa(T)$. The authors' argument for this result is threefold: (a) the one vibrational mode that could contribute to κ produces a negligible sound velocity compared to that which originates from the LA mode of the tube; (b) the interaction forces between the C$_{60}$ and the tube are not sufficient to affect the tube stiffness by a sizable amount; and (c) κ is reduced by localized effects generated by

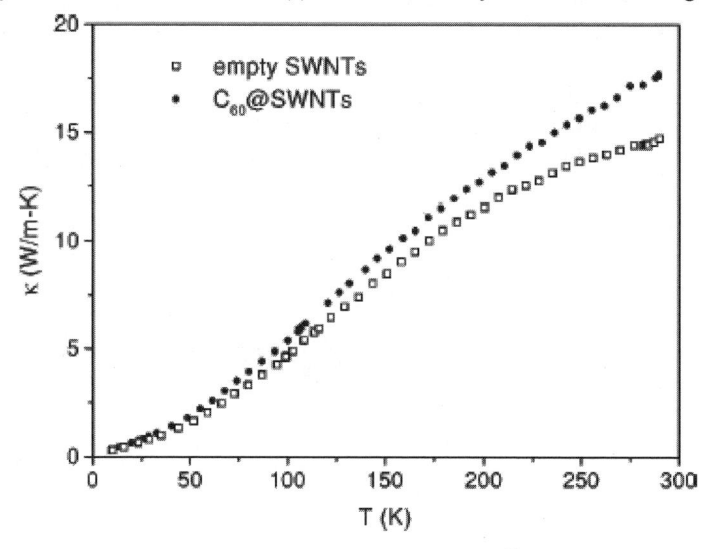

FIGURE 14 $\kappa(T)$ data comparing empty SWNTs to C$_{60}$@SWNTs.[13]

shifts in the C$_{60}$ molecules. Consequently, the presence of C$_{60}$ inside a nanotube does not substantially alter κ.

1.1.4. Heat Capacity, C

The heat capacity, C, of a solid is defined by the temperature derivative of the energy, U:

$$C_\alpha = \left(\frac{\partial U}{\partial T}\right)_\alpha, \tag{9}$$

where the subscript α indicates the parameter being held constant (i.e., volume or pressure). Usually, volume is held constant for theoretical discussions, but most measurements are made at constant pressure. At low temperatures it can yield important information on phonon and electron behavior and system dimensionality. In many solids the total heat capacity at constant volume (C_V) is made up of a phonon term and an electron term:

$$C_V = C_{\text{ph}} + C_{\text{el}}. \tag{10}$$

Calculations indicate that contributions due to phonons should dominate C_V over all T in nanotubes.[31] From the Debye approximation at low T, C_{ph} should scale as T raised to the dimensionality of the system (i.e., $C_{ph} \propto T^n$, where n is the dimension of the system).

Theory has shown that a 2D sheet of graphene indeed scales as T^2; however, when this sheet is rolled into a seamless cylinder (i.e., a SWNT), $C_{ph} \propto T$ at sufficiently small radius.[31] The result is suggestive of the quasi-1D nature of nanotubes since the C_{ph} behavior will cross over to 2D if the radius is larger than allowable for a given value of T. This inverse relationship between the radius and temperature on C_{ph} can be seen in Fig. 15.[31] The temperature at which C_{ph} (and C_V by virtue of the dominance of the contribution due to phonons) scales as T will increase with decreasing radius in SWNTs. In MWNTs it is the number of shells as well as the radius that determines this behavior. As the number of walls and radii increase, C_V is expected to scale between T^2 and T^3, much closer to that of three-dimensional graphite.

Measurements of the heat capacity at constant pressure (C_P) for bundles of purified SWNTs plotted as the specific heat (=C_P/mass) are shown in Fig. 16.[32] The $C_P(T)$ data in Fig. 16a show the behavior of C_P from 2–300 K. Figure 16b

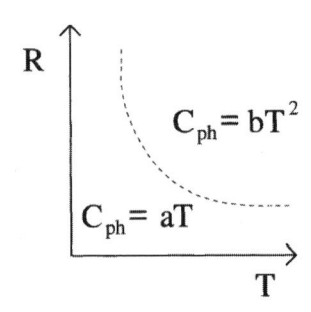

FIGURE 15 Schematic demonstrating the relationship between the radius (R), T, and C_{ph} in SWNTs.[31] For sufficiently small R and T; C_{ph} scales as T; otherwise it scales as T^2.

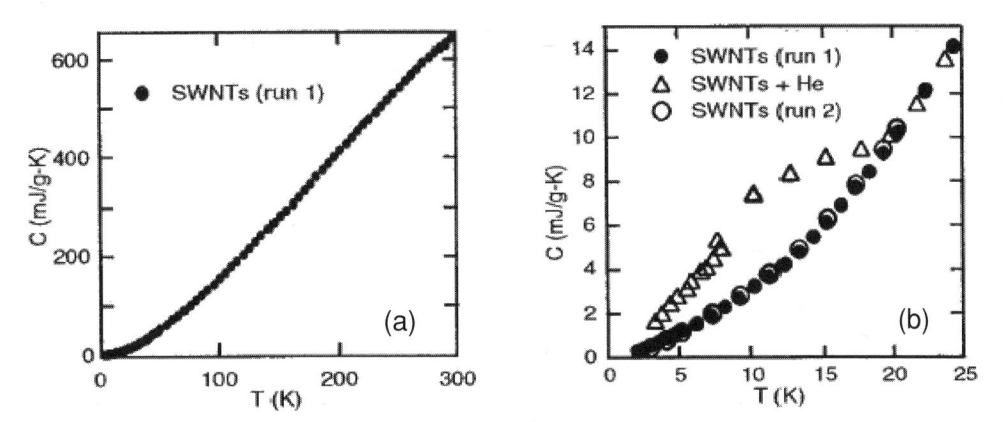

FIGURE 16 The T dependence of C_P per unit mass for SWNT bundles.[32] (a) The cooling data from room temperature to 2 K and (b) the low-T data showing the effects of He exposure on C_P.

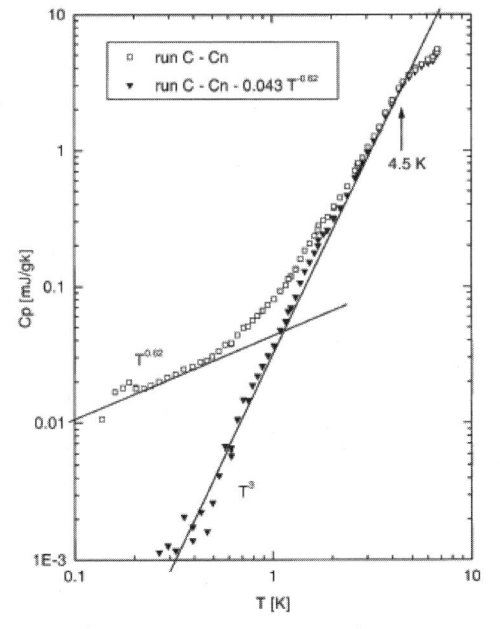

FIG. 17 Upon subtracting the T^{-2} term, C_P in bundles of SWNTs is seen to be the addition of a $T^{0.62}$ term and a T^3 term from 0.3–4 K.[34]

displays the low-temperature (<25 K) data. The presence of He is seen here to have the effect of increasing C_P at low T (triangles). This result is in good agreement with the theory, which predicts this rise in C_P upon adsorption of He into the interstitial channels of the SWNT bundles.[33] A log–log plot of the data in Fig. 16a (not shown here) was in very good agreement with that modeled from an individual SWNT down to 4 K, demonstrating the 1D nature of the samples.[32]

Finally, more recent measurements of C_P down to 0.1 K show that the heat capacity is determined by three terms with different powers of T in samples of purified SWNT bundles.[34] Their data fit very well with the relation $C_P = \alpha T^{-2} + \beta T^{0.62} + \gamma T^3$ over 0.3–4 K. The T^{-2} term is attributed to ferromagnetic catalyst particles in the sample. Upon accounting for this contribution, the data are seen to be in reasonable agreement with the remaining two terms, as shown by Fig. 17. The source of the $T^{0.62}$ term is not explicitly understood by the authors, however, the T^3 term is said to originate from the tube–tube coupling within the SWNT bundles, thereby giving rise to the 3D behavior at low T.

1.2. Nanowires

Whereas carbon nanotubes are hollow graphene cylinders with large aspect ratios, nanowires are solid cylinders with dimensions of the same order as nanotubes (nanowires of shorter lengths having aspect ratios of ∼10 are typically referred to as nanorods) and come in a wide variety of different materials. Several techniques have been developed for nanowire synthesis. These include the template method in which tiny pores in a chemically stable material are filled by one of an assortment of processes (e.g., vapor deposition, electrochemical deposition, pressure injection, etc.) with the desired material of the nanowires, the vapor–liquid–solid (VLS) method that uses the supersaturation of a liquid catalyst particle by gaseous material to precipitate a solid in one direction, and the solution phase growth of nanowires through the use of surfactants. Some types of nanowires that have been synthesized with these methods include Bi, Bi_2Te_3, BN, CdS, Cu, Fe, GaAs, GaN, Ga_2O_3, GaP, GaS, Ge, In, InP, InAs, MgO, Si, SiGe, SiO_2, Sn, SnO_2, and ZnO. For a review on these and other techniques as well as the many more types of nanowires that have been studied to date, see Ref. 8. Because, like nanotubes, they can often be thought of as 1D systems, nanowires demonstrate very unique properties compared to their bulk counterparts and show great potential for nanoscale devices.

1.2.1. Electrical Conductivity

The diameter size in nanowires is perhaps the most significant parameter in electrical conduction. Electrical transport is expected to be comparable to that of bulk material except when the diameter becomes small enough (on the order of the electron wavelength).[8] In this case 1D quantum size effects can be expected. As in nanotubes, electron transport in nanowires is expected to be primarily diffusive, except for extremely short wire segments.

Measurements in some semiconducting nanowires have exhibited particularly unique properties. Silicon nanowires, one of the most extensively studied types of nanowires, have shown an increase from 45 to 800 nS in the average room temperature transconductance, with a peak value of 2000 nS across a single nanowire with a diameter of 10–20 nm and a length of 800–2000 nm after treatment to reduce the effects of oxidation.[35] In accordance with Eq. (1), they also showed an increase in the average μ from 30 to 560 cm^2/V-s, with a peak value of 1350 cm^2/V-s. Both results are improvements on what is seen in planar silicon. Another study found that GaN nanowires (diameter ∼30 nm; length ∼330 nm) had an average value of μ = 2.15 cm^2/V-s at room temperature, compared to 380 cm^2/V-s found in bulk GaN.[36]

Measurements of the T dependence of R can also show the dramatic contrast between nanowires and their bulk parents. For example, Fig. 18 clearly shows the difference in the $R(T)$ data for bulk Bi and Bi nanowires of various diameters.[37] Bismuth nanowires were calculated to crossover from semimetallic to semiconducting behavior at a diameter of ∼50 nm due to the effects of quantum confinement.[38]

Sec. 1 • NANOMATERIALS 277

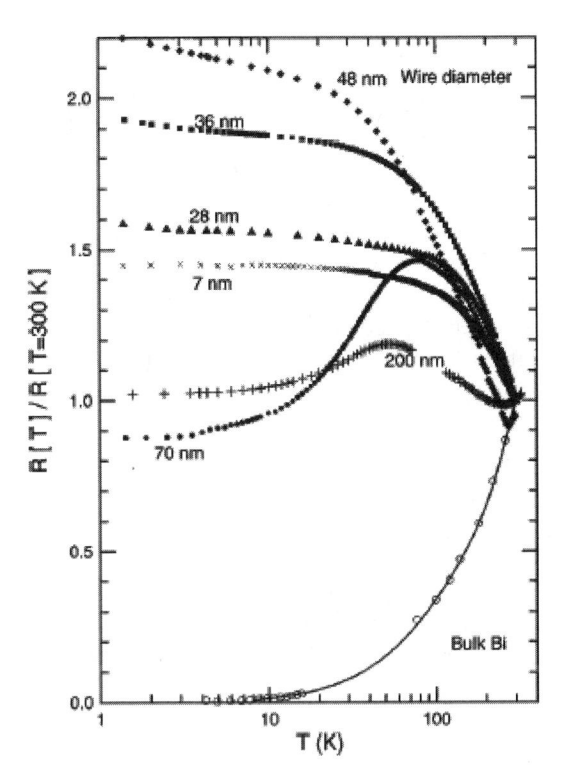

FIGURE 18 Normalized resistance $R(T)$ data for bulk Bi and several Bi nanowires of assorted diameters.[37]

The experiments in Fig. 18 confirmed this prediction by showing a transition at 48 nm. This crossover effect is purely a result of the extremely small size of the Bi nanowires and cannot be observed in its bulk form.

1.2.2. Thermoelectric Power

Thermoelectric measurements in certain types of nanowires demonstrate the distinct

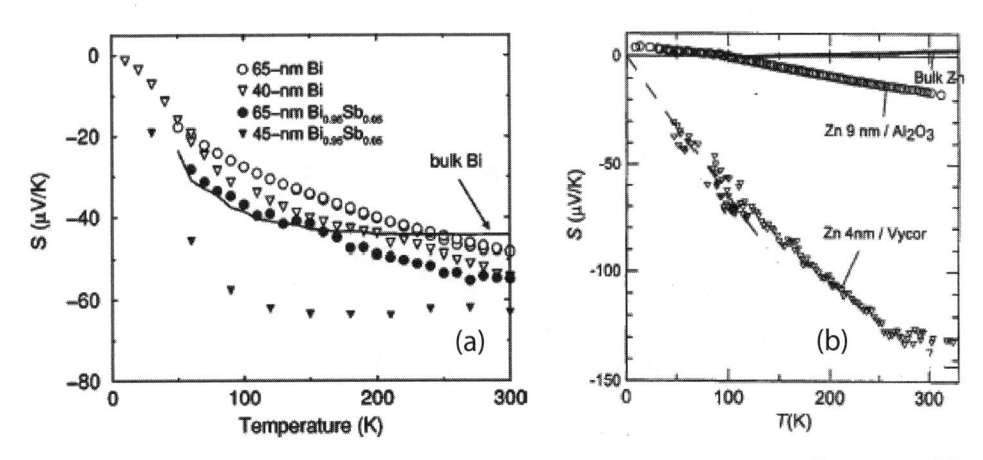

FIGURE 19 T dependence of the TEP comparing bulk and nanowire samples of (a) Bi[39] and (b) Zn.[8,40]

advantages of being extremely small as well. The TEP data in Fig. 19 present clear evidence of this. Figure 19a shows the $S(T)$ data for bulk Bi and Bi nanowires as well as Bi nanowires alloyed with Sb (5 at.%).[39] The magnitude of the TEP over all T is increased by Sb alloying. Also, in both the Bi and $Bi_{0.95}Sb_{0.05}$ nanowires a decrease in diameter corresponds to an increase in TEP magnitude. Each of the four nanowire samples shows an enhanced room temperature value, but only the $Bi_{0.95}Sb_{0.05}$ nanowire with a diameter of 45 nm is greater over the entire range of T. A comparison of the measured TEP in bulk Zn and samples of Al_2O_3 and Vycor glass with Zn nanowires embedded is given in Fig. 19b.[8,40] Both samples containing Zn nanowires exhibit enhancements in the TEP with respect to the bulk sample. However, the smaller-diameter Zn nanowire (4 nm) samples show a huge increase in TEP magnitude, in particular at $T = 300$ K.

1.2.3. Thermal Conductivity and Heat Capacity

Very little has been reported to date on the intrinsic thermal conductivity and heat capacity in nanowires, due to the difficulty of these types of measurements. Measurements show a strong dependence on diameter and surface oxidation.[8] Some $\kappa(T)$ data have shown that Si nanowires (diameter ~22 nm) scale as T. Indicative of their 1D nature, the $C(T)$ data were linear in T as well.

1.3. Nanoparticles

Nanoparticles can be thought of as zero-dimensional (i.e., constrained in all three directions) fine particles usually containing 10 to 1000 atoms.[1] They are sometimes referred to as nanoclusters. They can be made of virtually any kind of material. An understanding of their thermal properties is important for their use in the development of nanoscale devices. For example, the use of thin films and coatings of nanodiamond clusters in order to take advantage of the large thermal conductivity of diamond shows promise in a variety of thermal management applications.

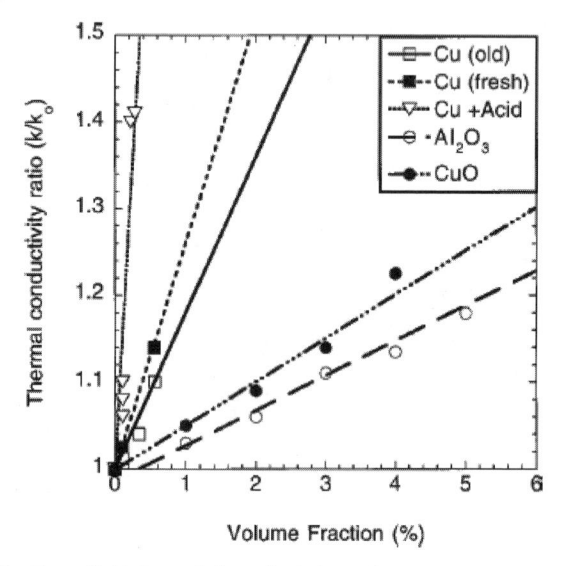

FIGURE 20 κ data for "nanofluids" consisting of ethylene glycol and Cu nanoparticles normalized to the κ of regular ethylene glycol.[41]

Nanoparticles have been reported to enhance the thermal conductivity in certain fluids. Figure 20 shows the κ versus nanoparticle concentration data for samples of ethylene glycol containing Cu nanoparticles normalized to regular ethylene glycol.[41] Each of these "nanofluids" exhibits improvements in κ data. The most significant increase in κ (~40%) can be seen in the sample containing ~0.3 vol.% Cu nanoparticles and thioglycolic acid (triangles), which was used to stabilize the nanoparticles. The average diameter of the nanoparticles was less than 10 nm. Note that the enhancement in κ appears to be time dependent, as evidenced by the decrease in κ for the "old" samples (~2 months) as compared to the "fresh" ones (~2 days). The authors do not specify the cause of this behavior, and it is believed that further investigations are needed for a more thorough understanding of this phenomenon.

Most of the work reported on the thermal properties of nanoparticles has been in the area of heat capacity. Calculations predict that C in nanoparticles, due to their tiny size, will not follow Debye's T^3 law seen in most solids at low temperatures but an exponential behavior in T and particle size.[42] However, measurements in certain ferromagnetic nanoparticles like $MnFe_2O_4$ do show a T^3 dependence with the contribution of an additional $T^{3/2}$ magnetic term.[43] The $C_P(T)$ data from 2 to 300 K are given for a sample of $MnFe_2O_4$ nanoparticles in Fig. 21. The data increase with temperature over all T but show a definite change at ~ T = 19 K. For $T < 19$ K, the data (not show here) do indeed fit well an expression for C consisting of T^3 and $T^{3/2}$ terms.

2. NANOCOMPOSITES

In this chapter we refer to nanocomposites as materials that have any of the previously mentioned nanostructures embedded in them. These types of materials have generated a lot of interest, since it is believed that the incorporation of certain nanosized structures can enhance the host material's various physical and chemical properties and, of particular interest, their thermal properties.

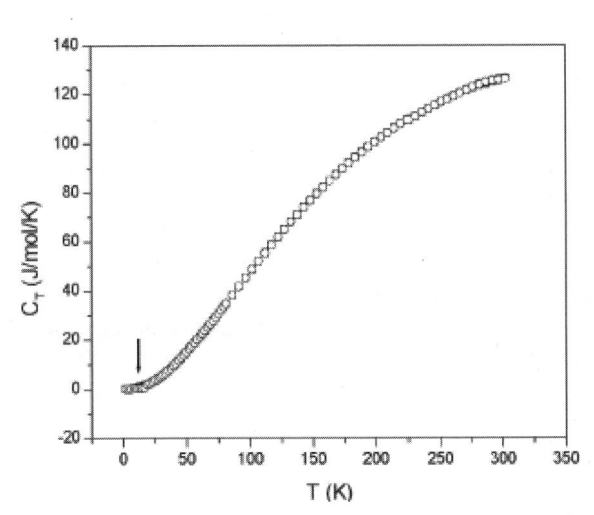

FIGURE 21 $C(T)$ data for a sample of $MnFe_2O_4$ nanoparticles.[43] The arrow indicates a sharp change in the data at ~ T = 19 K.

280 Chap. 3.3 • THERMAL PROPERTIES OF NANOMATERIALS AND NANOCOMPOSITES

a.

b.

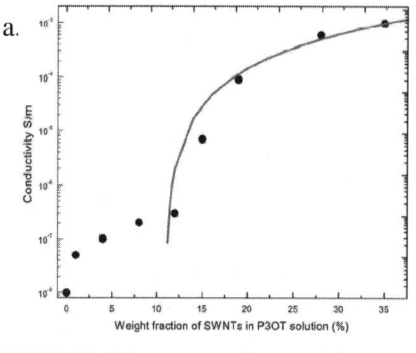

	Conductivity (S/cm)	Solvent Content (%)	Degradation Temp (°C)	RMS Roughness (nm)
P-1	2.35	16	509	0.209
P-2	0.79	12	511	0.658
P-3	4.92	5.4	518	4.159
P-4	6.02	6.4	529	34.04

FIGURE 22 Enhancements in σ seen in many nanotube/polymer composites is evidenced by σ data in films of both (a) P3OT and SWNTs[45] and (b) PANI and MWNTs.[46]

2.1. Electrical Conductivity

Enhancements in σ have been reported in a number of materials that include nanostructures in their composition, specifically nanotube/polymer composites.[44-46] The σ data as a function of SWNT content (wt.%) in films (thickness ~100 nm) of poly(3-octylthiophene) (P3OT) and SWNTs is shown in Fig. 22a.[45] An increase in σ at room temperature by nearly five orders of magnitude can be seen as the SWNT concentration is increased from 0 to 35 wt.%. A particularly sharp change occurs as the amount of SWNTs increased from 12 to 20 wt.%. A significant change in σ was also seen in films (thickness ~20 μm) of polyaniline (PANI) and MWNTs.[46] The table in Fig. 22b shows several measured quantities of the PANI/MWNT nanocomposites, including the room temperature σ. The first column in this table corresponds to the MWNT concentration in the film: P-1 = 0 wt.%, P-2 = 0.5 wt.%, P-3 = 1.0 wt.%, and P-4 = 5 wt.%. Both the 1 and 5 wt.% samples show an increase in σ, compared to films without MWNTs. It has also been reported that assorted metal-oxide composites change from insulating to conducting upon evenly dispersing carbon nanotubes throughout the sample.[47]

2.2. Thermal Conductivity

The incorporation of carbon nanotubes into various materials can also have the effect of improving thermal conductivity. The $\kappa(T)$ data in Fig. 23 show a comparison between samples of industrial epoxy with and without bundles of SWNTs mixed into the composite.[48] The epoxy with SWNTs (~1 wt.%) displays an increase in κ over the entire range of T. Most notably, the room temperature κ value of the SWNT epoxy exhibits an enhancement of 125%. This gives clear evidence of the advantages in thermal management that can be attained by simply adding carbon nanotubes to the given material.

3. APPLICATIONS

We close this chapter with a brief discussion of burgeoning applications of nanomaterials and nanocomposites which have demonstrated extraordinary potential due to their exceptional properties. However, due to the nature of this chapter,

Sec. 3 · APPLICATIONS

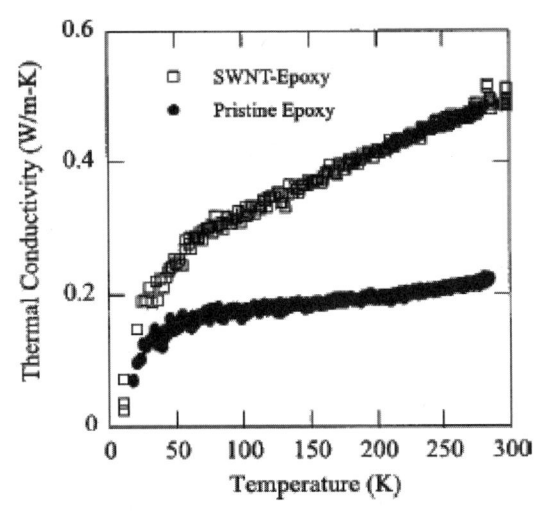

FIGURE 23 $\kappa(T)$ data for pristine epoxy and SWNT-epoxy nanocomposite.[48]

this discussion will be restricted to only a couple of applications involving *thermal* properties. One of the most recently reported applications for nanotubes is their use as a "nanothermometer."[49] Liquid Ga was seen to expand linearly with T over a range of 50–500°C when placed inside a nanotube (diameter ~75 nm and length ~10 μm). The electron microscope images in Fig. 24a–c display the Ga menisci at 58°C, 490°C, and 45°C, respectively. In Fig. 24d the reproducibility and precision of the measurements are demonstrated by the height of the meniscus versus T data.

A second application that has shown a lot of promise is the use of nanotubes as gas sensors. As discussed earlier, the TEP in nanotubes is extremely sensitive to gas exposure. Based upon this fact the thermoelectric "nano-nose" was developed.[50] By using isothermal Nordheim–Gorter (S vs. ρ) plots like those in Fig. 25a, the adsorption of various gases as well as the particular adsorption mechanism (i.e., physisorption or chemisorption) can be detected. In Fig. 25a the linear behavior of H_2, He, and N_2 is indicative of physisorption of these gases onto the various nanotube surfaces. On the other hand, the nonlinear response of NH_3 and O_2 shown in the inset signifies electron transfer between the gas molecules and the nanotube (i.e., chemisorption). Figure 25b displays the T dependence of the TEP for a purified SWNT mat after being degassed (S_0) and then after being exposed to

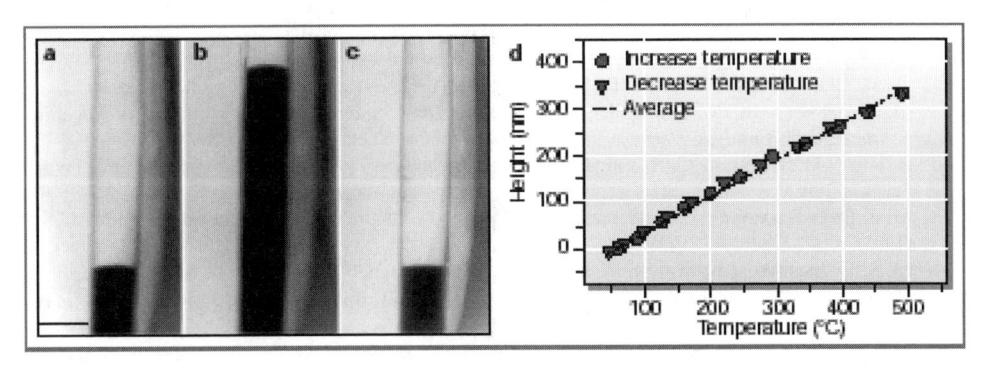

FIGURE 24 Images of the linearly varying Ga meniscus at (a) 58°C, (b) 490°C, and (c) 45°C inside a carbon nanotube.[49] The scale bar in (a) corresponds to 75 nm. (d) Reproducibility established by the height versus T plot of the meniscus for both warming and cooling cycles.

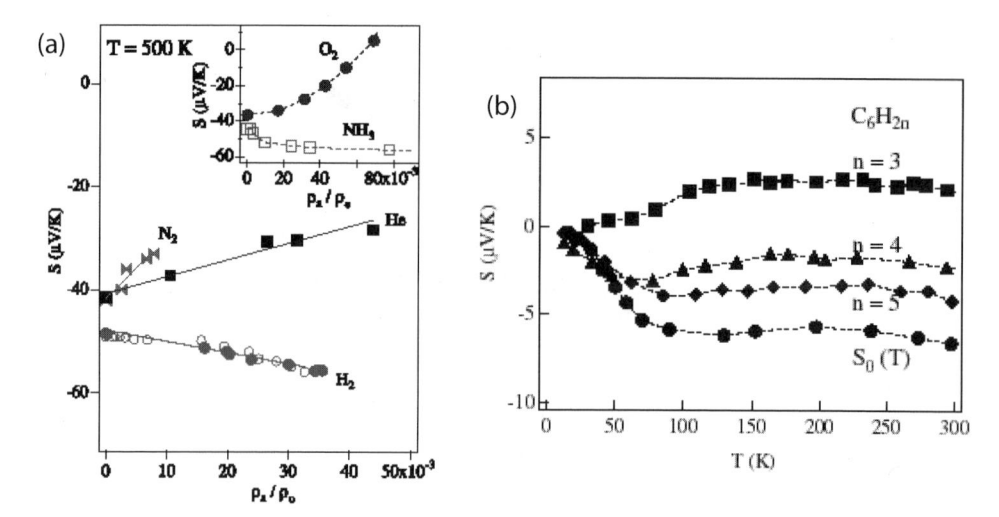

FIGURE 25 (a) Nordheim–Gorter (S vs. ρ) plots demonstrating the senstitivity of nanotubes to exposure to various gases as well as indicating the type of adsorption process involved.[50] (b) $S(T)$ data for a degassed SWNT mat and as it was exposed to different hydrocarbon vapors.[51]

different hydrocarbon vapors (n = 3–5).[51] After each measurement the sample was degassed until it reached its initial value. As clearly seen in the figure, exposure to each hydrocarbon vapor renders a different $S(T)$ signature. The results from both of these studies demonstrate the potential for using nanotubes as highly sensitive gas sensors. As these types of techniques and different technologies continue to progress, the pace to exploit these nanomaterials appears to quicken, and the list of their possible uses seems to grow exponentially.

4. REFERENCES

1. C. Kittel *Introduction to Solid State Physics,* 7th ed., (Wiley, New York 1996), p. 168–169.
2. S. Iijima *Helical Microtubules of Graphitic Carbon* Nature **354**, 56 (1991).
3. C. Journet, L. Alvarez, V. Micholet, T. Guillard, M. Lamy De La Chapelle, E. Anglaret, J. L. Sauvajol, S. Lefrant, P. Bernier, D. Laplaze, G. Flamant, and A. Loiseau *Single Wall Carbon Nanotubes: Two Ways of Production* Syn. Met. **103**, 2488 (1999).
4. A. Thess, R. Lee, P. Nicolaev, H. Dai, P. Petit, J. Robert, C. Xu, Y. H. Lee, S. G. Kim, A. G. Rinzler, D. T. Colbert, G. E. Scuseria, D. Tomanek, J. E. Fischer, and R. E. Smalley *Crystalline Ropes of Metallic Carbon Nanotubes* Science **273**, 483 (1996).
5. H. M. Cheng, F. Li, X. Sun, S. D. M. Brown, M. A. Pimenta, A. Marucci, G. Dresselhaus, and M. S. Dresselhaus *Bulk Morphology and Diameter Distribution of Single-Walled Carbon Nanotubes Synthesized by Catalytic Decomposition of Hydrocarbons* Chem. Phys. Lett. **289**, 602 (1998).
6. R. Saito, M. S. Dresselhaus, and G. Dresselhaus *Physical Properties of Carbon Nanotubes* (World Science, 1998).
7. M. S. Dresselhaus, G. Dresselhaus, and R. Saito C_{60}-*Related Tubules* Solid State Comm. **84**, 201 (1992).
8. M. S. Dresselhaus, Y.-M. Lin, O. Rabin, M. R. Black, and G. Dresselhaus *Nanowires*, Springer Handboek of Nanotechnology, edited by Bharat Bhushan, Springer-Verlag Heidelberg, Germany, 99-145 (2004).
9. D. T. Colbert and R. E. Smalley *Fullerene Nanotubes for Molecular Electronics* Trends In Biotech. **17**, 46 (1999).
10. K. Kaneto, M. Tsuruta, G. Sakai, W. Y. Cho, and Y. Ando *Electrical Conductivities of Multi-Wall Carbon Nanotubes,* Synth. Met. **103**, 2543 (1999).

Sec. 4 • REFERENCES

11. J. HONE, I. ELLWOOD, M. MUNO, A. MIZEL, M. L. COHEN, and A. ZETTL *Thermoelectric Power of Single-Walled Carbon Nanotubes* Phys. Rev. Lett. **80**(5), 1042 (1998).
12. J. E. FISCHER and A. T. JOHNSON *Electronic Properties of Carbon Nanotubes,* Curr. Opinion Solid State Mater. Sci. **4**, 28 (1999).
13. J. VAVRO, M. C. LLAGUNO, B. C. SATISHKUMAR, D. E. LUZZI, and J. E. FISCHER *Electrical and Thermal Properties of C_{60}-Filled Single-Wall Carbon Nanotubes,* Appl. Phys. Lett. **80**(8), 1450 (2002).
14. A. B. KAISER, Y. W. PARK, G. T. KIM, E. S. CHOI, G. DÜSBERG, and S. ROTH *Electronic Transport in Carbon Nanotube Ropes and Mats,* Synth. Met. **103**, 2547 (1999).
15. F. J. BLATT, P. A. SCHROEDER, C. L. FOILES, and D. GREIG *Thermoelectric Power of Metals* (Plenum Press, 1996).
16. P. M. CHAIKIN *Organic Superconductivity* (Plenum Press, New York, 1990).
17. P. G. COLLINS, K. BRADLEY, M. ISHIGAMI, and A. ZETTL *Extreme Oxygen Sensitivity of Electronic Properties of Carbon Nanotubes,* Science **287**, 1801 (2000).
18. K. BRADLEY, S.-H. JHI, P.G. COLLINS, J. HONE, M. L. COHEN, S. G. LOUIE, and A. ZETTL *Is the Intrinsic Thermoelectric Power of Carbon Nanotubes Positive?* Phys. Rev. Lett. **85**(20), 4361 (2000).
19. G. U. SUMANASEKERA, C. K. W. ADU, S. FANG, and P. C. EKLUND *Effects of Gas Adsorption and Collisions on Electrical Transport in Single-Walled Carbon Nanotubes* Phys. Rev. Lett. **85**(5), 1096 (2000).
20. D. TOMANEK and M. A. SCHLUTER *Growth Regimes of Carbon Clusters* Phys. Rev. Lett. **67**(17), 2331 (1991).
21. T. KOSTYRKO, M. BARTKOWIAK, and G. D. MAHAN *Reflection by Defects in a Tight-Binding Model of Nanotubes* Phys. Rev. B **59**(4), 3241 (1999).
22. L. GRIGORIAN, G. U. SUMANASEKERA, A. L. LOPER, S. L. FANG, J. L. ALLEN, and P. C. EKLUND *Giant Thermopower in Carbon Nanotubes: A One-Dimensional Kondo System* Phys. Rev. B **60**(16), R11309 (1999).
23. A. M. RAO (unpublished).
24. J. P. SMALL, L. SHI, and P. KIM *Mesoscopic Thermal and Thermoelectric Measurements of Individual Carbon Nanotubes* Solid State Comm. **127**(2), 181–186 (2003).
25. R. S. RUOFF and D. C. LORENTS *Mechanical and Thermal Properties of Carbon Nanotubes* Carbon **33**(7), 925 (1995).
26. J. HONE, M. WHITNEY, and A. ZETTL *Thermal Conductivity of Single-Walled Carbon Nanotubes* Synth. Met. **103**, 2498 (1999).
27. J. HONE, M. C. LLAGUNO, N. M. NEMES, A. T. JOHNSON, J. E. FISCHER, D. A. WALTERS, M. J. CASAVANT, J. SCHMIDT, and R. E. SMALLEY *Electrical and Thermal Transport Properties of Magnetically Aligned Single Wall Carbon Nanotube Films* Appl. Phys. Lett. **77**(5), 666 (2000).
28. S. BERBER, Y.-K. KWON, and D. TOMANEK *Unusually High Thermal Conductivity of Carbon Nanotubes* Phys. Rev. Lett. **84**(20), 4613 (2000).
29. T. R. ANTHONY, W. F. BANHOLZER, J. F. FLEISCHER, L. WEI, P. K. KUO, R. L. THOMAS, and R. W. PRYOR *Thermal Diffusivity of Isotropically Enriched ^{12}C Diamond* Phys. Rev. B **42**, 1104 (1990).
30. P. KIM, L. SHI, A. MAJUMDAR, and P. L. MCEUEN *Mesoscopic Thermal Transport and Energy Dissipation in Carbon Nanotubes* Physica B **323**, 67 (2002).
31. L. X. BENEDICT, S. G. LOUIE, and M. L. COHEN *Heat Capacity of Carbon Nanotubes* Solid State Commun. **100**(3), 177 (1996).
32. J. HONE, B. BATLOGG, Z. BENES, A. T. JOHNSON, and J. E. FISCHER *Quantized Phonon Spectrum of Single-Wall Carbon Nanotubes* Science **289**, 1730 (2000).
33. M. W. COLE, V. H. CRESPI, G. STAN, C. EBNER, J. M. HARTMAN, S. MORONI, and M. BONINSEGNI *Condensation of Helium in Nanotube Bundles* Phys. Rev. Lett. **84**(17), 3883 (2000).
34. J. C. LASJAUNIAS, K. BILJAKOVIC, Z. BENES, and J. E. FISCHER *Low-Temperature Specific Heat of Single-Wall Carbon Nanotubes* Physica B **316-317**, 468 (2002).
35. Y. CUI, Z. ZHONG, D. WANG, W. U. WANG, and C. M. LIEBER *High Performance Silicon Nanowire Field Effect Transistors* Nano Lett. **3**(2), 149 (2003).
36. J.-R. KIM, H. M. SO, J. W. PARK, J.-J. KIM, J. KIM, C. J. LEE, and S. C. LYU *Electrical Transport Properties of Individual Gallium Nitride Nanowires Synthesized by Chemical-Vapor-Deposition* Appl. Phys. Lett. **80**(19), 3548 (2002).
37. J. HEREMANS, C. M. THRUSH, Y.-M. LIN, S. CRONIN, Z. ZHANG, M. S. DRESSELHAUS, and J. F. MANSFIELD *Bismuth Nanowire Arrays: Synthesis and Galvanometric Properties* Phys. Rev. B **61**, 2921 (2000).
38. Y.-M. LIN, X. SUN, and M. S. DRESSELHAUS *Theoretical Investigation of Thermoelectric Transport Properties of Cylindrical Bi Nanowires* Phys. Rev. B **62**, 4610 (2000).

39. Y.-M. Lin, O. Rabin, S. B. Cronin, J. Y. Ying, and M. S. Dresselhaus *Semimetal-Semiconductor Transition in $Bi_{1-x}Sb_x$ Alloy Nanowires and Their Thermoelectric Properties* Appl. Phys. Lett. **81**(13), 2403 (2002).
40. J. P. Heremans, C. M. Thrush, D. T. Morelli, and M.-C. Wu *Thermoelectric Power of Bismuth Nanocomposites* Phys. Rev. Lett. **88**, 216801-1 (2002).
41. J. A. Eastman, S. U. S. Choi, S. Li, W. Yu, and L. J. Thompson *Anomalously Increased Effective Thermal Conductivities of Ethylene Glycol-Based Nanofluids Containing Copper Nanoparticles* Appl. Phys. Lett. **78**(6), 718 (2001).
42. K. N. Shrivastava *Specific Heat of Nanocrystals* Nano Lett. **2**(1), 21 (2002).
43. G. Balaji, N. S. Gajbhiye, G. Wilde, and J. Weissmüller *Magnetic Properties of $MnFe_2O_4$ Nanoparticles* J. Magnetism and Magnetic Mat. **242-245**, 617 (2002).
44. J. M. Benoit, B. Corraze, S. Lefrant, W. J. Blau, P. Bernier, and O. Chauvet *Transport Properties of PMMA-Carbon Nanotubes Composites* Syn. Met. **121**, 1215 (2001).
45. E. Kymakis, I. Alexandou, and G. A. J. Amaratunga *Single-Walled Carbon Nanotube-Polymer Composites: Electrical, Optical and Structural Investigation* Syn. Met. **127**, 59 (2002).
46. P. C. Ramamurthy, W. R. Harrell, R. V. Gregory, B. Sadanadan, and A. M. Rao *Electronic Properties of Polyaniline/Carbon Nanotube Composites* Syn. Met. **137**, 1497 (2003).
47. E. Flahaut, A. Peigney, Ch. Laurent, Ch. Marlière, F. Chastel, and A. Rousset *Carbon Nanotube-Metal-Oxide Nanocomposites: Microstructure, Electrical Conductivity and Mechanical Properties* Acta Mater. **48**, 3803 (2000).
48. M. J. Biercuk, M. C. Llagune, M. Radosavljevic, J. K. Hyun, A. T. Johnson, and J. E. Fischer *Carbon Nanotube Composites for Thermal Management* Appl. Phys. Lett. **80**(15), 2767 (2002).
49. Y. Gao and Y. Bando *Carbon Nanothermometer Containing Gallium* Nature **415**, 599 (2002).
50. C. K. W. Adu, G. U. Sumanasekera, B. K. Pradhan, H. E. Romero, and P. C. Eklund *Carbon Nanotubes: A Thermoelectric Nano-Nose* Chem. Phys. Lett. **337**, 31 (2001).
51. G. U. Sumanasekera, B. K. Pradhan, H. E. Romero, K. W. Adu, and P. C. Eklund *Giant Thermopower Effects from Molecular Physisorption on Carbon Nanotubes* Phys. Rev. Lett. **89**(16), 166801-1 (2002).

INDEX

Absolute technique: *see* Steady-state thermal conductivity technique
AlCuFe quasicrystals, 256, 257
n-Alkanes, 98–99
Alloys, 86–87
AlPdMn quasicrystals, 256–258
Alumina (Al$_2$O$_3$), 244, 245
Aluminum nitride (AlN), 240–243
 diamond film on, 244
Amorphous silicon dioxide (a-SiO$_2$), 171–172
Annealing, 142
Antimony telluride (Sb$_2$Te$_3$), 124–12
Atomic displacement parameters (ADPs), 134, 139

Bare substrate specimens, 218–219
Bipolar conduction, 110–112
Bipolar diffusion, 8
Bismuth and bismuth-antimony alloys, 126–127
Bismuth selenide (Bi$_2$Se$_3$), 124–125
Bismuth telluride (Bi$_2$Te$_3$), 112–114
 and its alloys, 124–126
Bi$_2$Te$_3$/Sb$_2$Te$_3$ superlattices (SLs), 178–181
Bloch-Grüneisen formula, 58
Bloch-Wilson theory, 83–85
Boltzmann equation (BE), 10, 12, 32–35, 43, 95, 162
 linearization, 35, 47
Born-Oppenheimer approximation, 52
Boundary scattering, 78
Bridge method, 215, 222–224

Bulk materials, determining thermal conductivity of, 187–188, 202; *see also specific topics*

Callaway model, 12, 16
Carbon-doped silicon dioxide (CDO), 172
Carbon fiber-incorporated alumina ceramics, 245
Ceramic composites, 244–245
Ceramic fibers, 245
Ceramics, 239, 251
 novel materials with various applications, 244–248
 rare-earth based, 246–247
 thermoelectric, 247–248
 traditional materials with high thermal conductivity, 240–244
Chalcogenide glasses, 248–250
Chalcogenides and oxides, novel, 145–148
Chemical vapor deposition (CVD), 262
Clathrates, 137–141
Collisions, 24
Comparative technique, 193–195
Copper (CU) thin films, 169, 170
Crystalline insulators: *see under* Insulators
Crystals, 100; *see also* Quasicrystals
 boundary scattering, 117–118
 phonon scattering by impurities, 115–117
 phonon scattering in pure, 114–115

CsBi$_4$Te$_6$, 147
Current density: *see* Electric current density

Debye approximation, 11–12, 94, 135–137
Debye equation of state, 119
Debye frequency, 51
Debye model
  for phonon distribution, 118
  of thermal conductivity of solids, 100
Debye temperature, 81, 119
Debye theory, 11
Debye's equation for heat transfer in gases, 94
Diamond, impurities in, 97
Diamond films, 174
  on aluminum nitride, 244
Dielectric films, 171, 174
  amorphous SiO$_2$ films, 171–172
  thin film coatings, 173–174
Diffusion, bipolar, 8
Diffusivity method, laser-flash thermal, 197
Dislocations, 78
Drude formula, 25, 40
Drude model, 24–29

Elastic electron scattering, 42
Electric arc discharge, 262
Electric current density (J$_e$), 24, 35–36
Electrical conductivity (σ), 40–44
  of carbon nanotubes, 262–265
  of nanocomposites, 279–280
  of nanowires, 276–277
Electrical heating and sensing, 208
  cross-plane thermal conductivity measurements of thin films, 208
    steady-state method, 212–214
    3ω method, 208–213
  in-plane thermal conductivity measurements, 214–216
    bridge method, 215, 222–224
    membrane method, 215–222
    without substrate removal, 225
Electrical resistivity, 170; *see also* Electronic thermal resistivity; Scattering processes
Electron-hole scattering, 68

Electron scattering; *see also* Scattering processes
  elastic, 42
Electronic thermal conduction, 3–9; *see also* Lattice thermal conductivity
Electronic thermal conductivity, 23, 44–46, 80
Electronic thermal resistivity, 69; *see also* Electrical resistivity
Electrons, 24

Fermi-Dirac distribution, 29–30
Fermi energy, 40
Fermi liquid theory, 65, 66
Fermi surfaces, 40–42, 68, 84–85
Film-on-substrate system, 218–219
Fourier law, 26
Friedel sum rule, 50
Fuchs theory, 170

GaAs/AlAs superlattices (SLs), 178–181
Gas sensors, use of nanotubes as, 281
Germanium, 129, 130
Glass-ceramic superconductor, 245–246
Glasses, 93–94, 239, 248, 250–251
  chalcogenide, 248–250
  comparison with crystals, 100
  detailed models of thermal conductivity, 100–101
  recent amorphous ice results, 101

Half-Heusler compounds, 141–142
  effect of annealing, 142
  effect of grain size reduction, 144–145
  isoelectronic alloying on M and Ni sites, 142, 144
Harman technique, 201–202
Heat current density: *see* Thermal current density
Heat flux, 10
Heat pulse method, 220, 224
Heat spreading effect, methods of measuring, 215, 216
Heat transfer method, 218–219
Heusler alloys, 142; *see also* Half-Heusler compounds
HfNiSn, 142–144
Highly oriented pyrolytic graphite (HOPG), 269, 270

INDEX

Horizontal processes, 46

Ice, 101
  crystalline phases of, 98–99
Ideal resistivity, 58
Ideal thermal resistivity, 60, 84
Impurity scattering, 78, 80
Inclusion compounds, 99
Insulators, 93–94
  minimal thermal conductivity, 101
  phononic thermal conductivity in simple crystalline, 94–97
    acoustic phonons carry heat, 94–96
    impurities, 97
    role of optic modes in more complex, 97–99
    molecular, 97–99
    optic-acoustic coupling, 99–100

Kapaitza resistance, 154–156, 161
Kinetic theory, simple, 2
Klemens coefficient and numerical solution, 83, 84

Laser-flash thermal diffusivity method, 197
Laser vaporization, 262
Lattice thermal conductivity, 9–17, 73
  phonon thermal resistivity limited by electrons, 73–77
  prediction of, 118–120
  in semiconductors, 112–114
Lattice vibrations: see Phonons
Lead selenide (PbSe), 128
Lead telluride (PbTe), 127–128
Liouville theorem, 34
Lorenz number (L), 110, 111, 170–171
Lorenz ratio (L), 45, 61, 85–86

Maggi-Righi-Leduc effect, 77
Magnesium silicon nitride ($MgSiN_2$), 247
'Maldonado' technique, 197–199
Mass-defect scattering, 117
Matthiessen's rule, 5, 43, 44, 46
Membrane method, 215–222
Metallic thin films: see under Thin films
Metals, 21, 87–88; see also specific topics
  carriers of heat in, 22–23

Metals (cont.)
  processes limiting phonon thermal conductivity in, 73–79
  pure, 28–29, 79–86
  specific heat, 29–32
  thermal conductivity of real, 79–87
  transport parameters for monovalent, 82–83

Microfabrication techniques, 219
Minimal thermal conductivity, 101
Mirage method, 231
MNiSn, 141–142
Modulation heating techniques, 220–221, 224
Molecular dynamics (MD), 163
Mott relation, 265–266
Multiwalled nanotubes (MWNTs), 262, 263, 269, 270, 272, 280
MX_3, 129

$NaCo_2O_4$, 147–148
Nanocomposites, 261, 279
  applications, 280–282
  electrical conductivity, 279–280
  thermal conductivity, 280
'Nanofluids,' 278–279
Nanomaterials, 261
  applications, 280–282
Nanoparticles, 278–279
'Nanothermometer,' 280–281
Nanotubes, carbon, 262
  electrical conductivity, 262–265
  heat capacity (C), 274–275
  thermal conductivity (κ), 271–273
  thermoelectric power (TEP), 265–271
Nanowires, 276
  electrical conductivity, 276–277
  thermal conductivity and heat capacity, 278
  thermoelectric power, 277–278
Non-Kapitzic heat flow, 161–162
  analytic theory, 162–163
Normal processes (N-processes), 12, 115–116

Optic-acoustic coupling, 99–100
Optical-electrical hybrid methods, 232–233

Optical heating methods, 225–226
  frequency-domain photothermal and photoacoustic methods, 230–232
  time-domain pump-and-probe methods, 226–229
Parallel thermal conductance (PTC) technique, 200–201
Partially stabilized zirconia (PSZ), 174
Pauli exclusion principle, 29, 30, 38, 62
Peltier coefficient, 39
Peltier effect, 39
Phonon-boundary scattering rate, 14–16
Phonon-dislocation scattering, 15
Phonon dispersion curves, 9–10
Phonon distribution, Debye model for, 118
Phonon distribution function, 10, 12
Phonon method, dominant, 17
Phonon-phonon normal scattering, 14
Phonon-point-defect scattering, 14
Phonon scattering(s), 14, 16; *see also under* Crystals; Scattering processes
  processes, 11–12, 17
  relaxation rate, 15–16
Phonon spectrum, contribution to thermal conductivity of different parts of, 117, 118
Phonons, 9, 156–157; *see also specific topics*
  acoustic, 94–96
  mean free path, 11
Photothermal deflection method, 231
Photothermal displacement method, 231
Photothermal emission method, 230–231
Photothermal reflectance method, 230
Photothermal reflectance signal, 228–229
Photothermoelectric methods, 232–233
Point-defect scattering, 115
Polyaniline (PANI), 280
Polysilicon films, 175–176
Pulse heating, 220, 224
Pulse-power method, 197–199
Pump-and-probe methods, time-domain, 226–229
Pyrolytic graphite, highly oriented, 269, 270

Quasicrystalline materials, 255–256, 259
  contributions to thermal conductivity, 257
  glasslike plateau in, 258
Quasicrystals
  low-temperature thermal conduction, 257–258
  poor thermal conduction, 258

Radial flow method, 194–197
Radial methods, classes of apparatus in, 195
Radiation, 102
Radiation loss, 191, 192
Relaxation time, 74, 116; *see also under* Phonon scattering(s)
Relaxation time approximation, 34–35, 42–44, 68

Scattering processes, 46; *see also specific topics*
  electron-electron scattering, 61–64
  e-e processes and electrical resistivity, 64–69
  e-e processes and thermal resistivity, 69–73
  electron-phonon scattering, 50–61
  impurity scattering, 46–50
Screened Coulomb interaction, 47
Seebeck coefficient (S), 39, 45, 106, 107, 109, 111, 112, 142
Semiconductor lasers, 167–168
Semiconductor superlattices, 178–182
Semiconductors, 105–106, 149; *see also* Clathrates; Half-Heusler compounds
  electronic thermal conductivity in bipolar conduction, 110–112
  nondegenerate and degenerate, 109–110
  separation of lattice thermal conductivity and, 112–114
  transport coefficients for a single band, 106–109
Semimetal thin films, 174, 176–177
Silicon, 128–129
Silicon carbide fiber-reinforced ceramic matrix composite (SiC-CMC), 244–245

INDEX

Silicon dioxide (SiO$_2$), 171–172, 256
Silicon-germanium (Si-Ge) alloys, 129
Silicon nitride (Si$_3$N$_4$), 243–244
Silicon thin films, 175–177
Silver, 6
Single-walled nanotubes (SWNTs), 262, 264, 267–275, 279–282
Size effects, 168; *see also* Thin films
Skutterudite thin films and superlattices, cross-plane thermal conductivity of, 177
Skutterudites, 129–130
  binary (unfilled), 130–131
  effect of doping on Co site, 132–133
  filled, 133–137
Smith-Palmer equation, 87
Steady-state methods, 212–214, 216–220, 223
Steady-state thermal conductivity technique, 188–189
  heat loss and thermal contact issues, 189–191
  heat loss terms, 191–193
Superlattices (SLs), 153–154, 163–164, 168, 182; *see also* Semiconductor superlattices
  parallel to layers, 154
  perpendicular to layers, 154
    multilayer interference, 156–157
    temperature, 157–159
    thermal boundary resistance, 154–156
    with thick layers, 159–160
  semiconductor, 178–182

TAGS-85, 128
Temperature, defining, 157–159
Temperature sensors: *see* Electrical heating and sensing
Thermal barrier coatings (TBCs), 173–174
Thermal conductivity, 1–2, 17
  defined, 2
  standard formula for, 163
Thermal conductivity coefficient (κ), 95–97
  temperature dependence, 96–97
Thermal conductivity tensor, 94
Thermal current density (J$_Q$), 25–26, 36
Thermal parameters, 134

Thermal resistivity; *see also* Ideal thermal resistivity; Lattice thermal conductivity; Scattering processes
  electron-electron (scattering) processes and, 69–73
  electronic, 69; *see also* Electrical resistivity
Thermoelectric materials, 123–124, 149; *see also specific materials*
  Groups IV and Group IV elements, 127–128
Thermoelectric power (TEP), 39, 265–271, 277–278
Thin film coatings on dielectric films, 173–174
Thin-film thermal conductivity characterization, techniques for, 205–208, 233–234; *see also specific techniques*
Thin films, 167–168, 182; *see also* Dielectric films; Electrical heating and sensing
  metallic, 169–171
  semiconductor, 174–177
  semimetal, 174, 176–177
Thomas-Fermi screening parameter, 48, 53
Time-domain pump-and-probe methods, 226–229
TiNiSn$_{1-x}$Sb$_x$, 145
TiNiSn$_{0.95}$Sb$_{0.05}$, 145
Tl$_2$GeTe$_5$, 146–147
Tl$_9$GeTe$_6$, 146
Tl$_2$SnTe$_5$, 146–147
Transient heating methods, 220–222, 224
Transport coefficients, 35–40

Umklapp processes (U-processes), 12, 77–80, 115–116

Vapor deposition, chemical, 262
Vaporization, laser, 262
Vertical processes, 46

Wiedemann-Franz law, 4–7, 23, 27, 39, 46, 86, 88, 106, 257

Yukawa potential, 47–48

Z-meters, 201–202
$Zr_{0.5}FH^{0.5}NiSn$, 142–144
ZrNiSn, 142–144